# AUTOCAD® 2025

# BEGINNING

# AND

# INTERMEDIATE

## LICENSE, DISCLAIMER OF LIABILITY, AND LIMITED WARRANTY

# AutoCAD® 2025
# Beginning
# and
# Intermediate

**Munir M. Hamad**
*Autodesk™ Approved Instructor*

**Mercury Learning and Information**
Boston, Massachusetts

PPublisher: David Pallai
Mercury Learning and Information
121 High Street, 3rd Floor
Boston, MA 02110
info@merclearning.com
www.merclearning.com
800-232-0223

Munir M. Hamad, *AutoCAD® 2025 Beginning and Intermediate.*
ISBN: 978-1-50152-316-8

AutoCAD is a trademark of Autodesk, Inc. Version 1.0, 2025

The publisher recognizes and respects all marks used by companies, manufacturers, and developers
as a means to distinguish their products. All brand names and product names mentioned in this book
are trademarks or service marks of their respective companies. Any omission or misuse (of any kind)
of service marks or trademarks, etc. is not an attempt to infringe on the property of others.

Library of Congress Control Number: 2024938281

242526321    This book is printed on acid-free paper in the United States of America.

Our titles are available for adoption, license, or bulk purchase by institutions, corporations, etc.
For additional information, please contact the Customer Service Dept. at 800-232-0223(toll free).

All of our titles are available in digital format at academiccourseware.com and other digital vendors.
*Companion files (figures and code listings) for this title are available with proof of purchase by
contacting info@merclearning.com.* The sole obligation of Mercury Learning and Information
to the purchaser is to replace the files, based on defective materials or faulty workmanship, but not
based on the operation or functionality of the product.

# CONTENTS

# PREFACE

- Since its inception, AutoCAD has enjoyed a large user base, which has made it the most popular CAD software in the world. AutoCAD is widespread due to its logic and simplicity, which makes it easy to learn. AutoCAD evolved through the years to become a comprehensive software application addressing all aspects of engineering drafting and designing.
- This book addresses all levels of AutoCAD 2D drafting, Essentials, Intermediate, and Advanced, which makes it ideal for novice users, and for those seeking to be professional power users ready for industry.
- This book is not a replacement of AutoCAD manuals, but is considered to be complementary source that includes hundreds of hands-on practices to strengthen the knowledge learned and solidify the techniques discussed.
- At the end of each chapter, the reader will find "Chapter Review Questions" which are the same sort of questions you may see in the Autodesk certification exam; so, the reader is invited to solve them all. Correct answers are at the end of each chapter.
- Chapter 10 contains three projects: one is an architectural plan, presented in metric and imperial units. The other two projects are for mechanical engineering. One of them is thoroughly explained and the other is not! These two projects are also presented in both metric and imperial units.

# ABOUT THE COMPANION FILES

Included with the book is a disc which contains the following (*all of the companion files are also available for downloading from the publisher with proof of purchase by writing to info@merclearning.com*):

- A link to retrieve the AutoCAD 2025 trial version, which will last for 30 days starting from the day of installation. This version will help you solve all practices and projects (if you are a college/university student, you can download a student version of the software from www.autodesk.com)
- Practice files which will be your starting point to solve all practices in the book
- Copy the folder named "Practices" into one of the hard drives of your computer
- As for projects, you will find two folders inside, "Metric" for metric units projects, and "Imperial" for projects using imperial units

# About the Book

This is the most comprehensive book about AutoCAD 2025 – 2D drafting on the market. It is divided into three major parts:

- *Essentials:* from Chapter 1 to Chapter 10. It assumes that the reader has no previous experience in AutoCAD; hence it starts from scratch. Chapter 10 contains three projects – one architectural, and two mechanical using *both Imperial and metric units*.
- *Intermediate:* from Chapter 11 to Chapter 18. It contains a deeper discussion on a subject we touched on in the Essentials part, or a new advance feature.
- *Advanced*: from Chapter 19 to Chapter 26. It discusses the most advanced features of AutoCAD 2025.

If you don't have any prior experience in AutoCAD this book is a perfect start, and you can stop at the end of any part. But if you want to be a real power user of AutoCAD, you should go through all 26 chapters, solving all projects and practices.

This book is also a good source to prepare for the *AutoCAD Certified Professional* exam.

The chapters are divided as follows:
- Chapter (1) covers AutoCAD basics along with the interface
- Chapter (2) covers AutoCAD techniques to draw with accuracy
- Chapters (3 & 4) cover all modifying commands
- Chapter (5) covers the AutoCAD method of organizing the drawing using layers and inquiry commands
- Chapter (6) covers the methods of creating and editing blocks, and inserting and editing hatches
- Chapter (7) covers AutoCAD methods of writing text
- Chapter (8) covers how to create and edit dimensions in AutoCAD
- Chapter (9) covers how to plot your drawing
- Chapter (10) includes three projects, one architectural and two mechanical, covering both metric and imperial units
- Chapter (11) covers the creation of more 2D objects
- Chapters (12 & 13) cover advanced practices and techniques
- Chapter (14) covers Block tools and Block Editing
- Chapter (15) covers the creation of Text Style and Table Styles along with Formulas in tables
- Chapter (16) covers the creation of Dimension Style & Multileader style plus adding multileaders

- Chapter (17) covers the creation of Plot styles, the meaning of Annotative, and DWF creation
- Chapter (18) covers how to create a template file and customize AutoCAD interface
- Chapter (19) covers Parametric Constraints
- Chapter (20) covers Dynamic Blocks
- Chapter (21) covers Block Attributes
- Chapter (22) covers External Reference
- Chapter (23) covers Sheets Sets
- Chapter (24) covers CAD Standards and Advanced Layer Commands
- Chapter (25) covers Importing PDF files, Design Views, AutoCAD Web, Mobile Apps
- Chapter (26) covers the Drawing Compare Function

## CHAPTER 1

# AUTOCAD 2025 BASICS

**In This Chapter**

- Starting AutoCAD
- Dealing with the AutoCAD interface
- AutoCAD defaults and drawing units
- Dealing with file-oriented commands
- Undo and Redo commands

## 1.1  STARTING AUTOCAD

AutoCAD was released in 1982 by Autodesk, Inc., which was a small company at that time. It was designed to be used for PCs only. Since then, AutoCAD has enjoyed the largest user base in the world in the CAD business. Users can use AutoCAD for both 2D and 3D drafting and designing. AutoCAD can be used for architectural, structural, mechanical, electrical, environmental, and manufacturing drawings and for road and highway designs.

Though the focus these days is BIM (Building Information Modeling), AutoCAD is still the most profitable software for Autodesk, Inc. due to the ease of use and comprehensiveness, which addresses all of the user needs. Another version of AutoCAD, called AutoCAD LT, is used for 2D drafting only.

To start AutoCAD 2025, double-click the shortcut on your desktop that was created in the installation process. AutoCAD will present the **Welcome** window, as shown in the following:

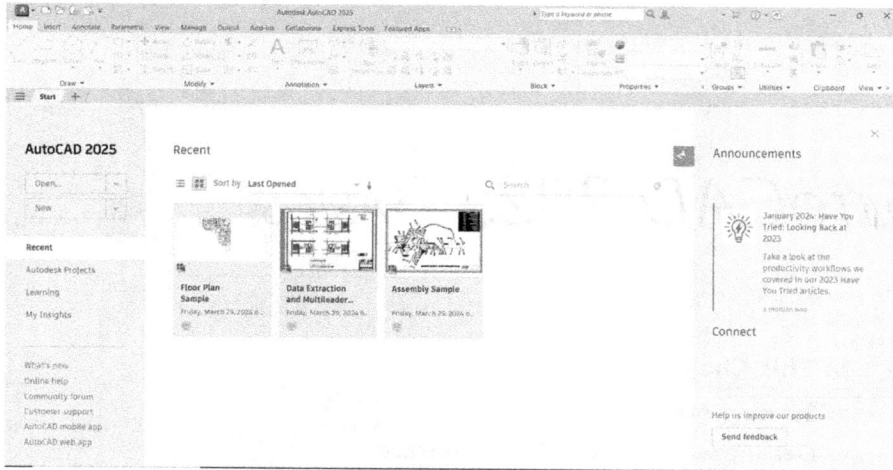

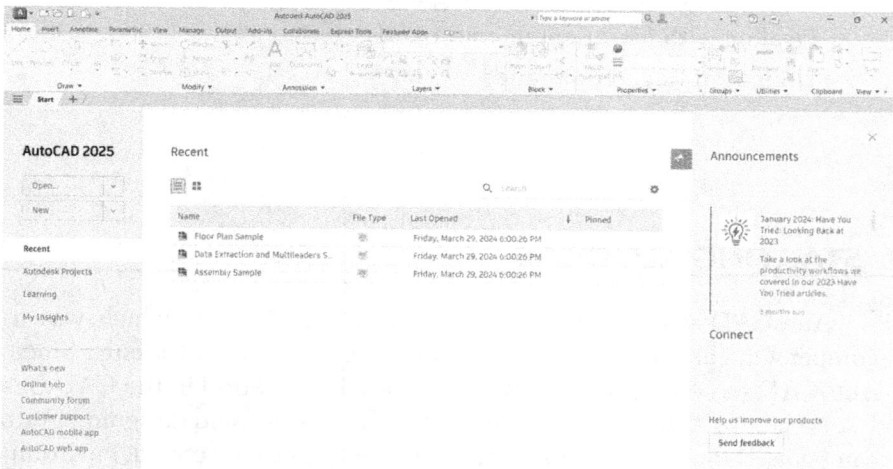

While in the welcome window, you will be able to:

- Open an existing file, Open a Sheet Set, Explore the sample files that come with the software
- Start a new drawing, or Download more templates accessible online
- See the recent files you opened (either as a list, or as icons)

- Check if AutoCAD has an Announcement for you regarding your software/ hardware
- Send your feedback to Autodesk

Starting a new file or opening an existing file will show you the interface of AutoCAD 2025, as shown in the following:

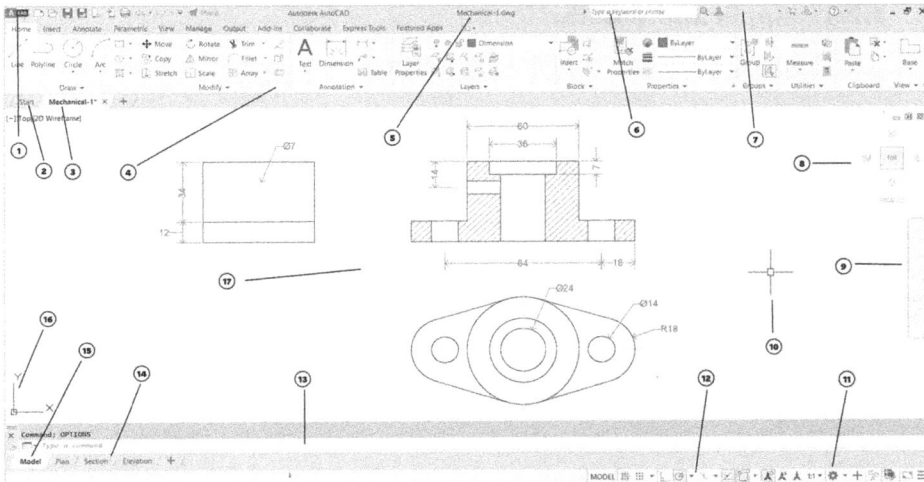

| 1. Application Menu | 2. Start tab | 3. File tab |
|---|---|---|
| 4. Ribbon | 5. Current File Name | 6. Info Center |
| 7. Sign In | 8. ViewCube | 9. Navigation Bar |
| 10. Cross Hairs | 11. Workspace | 12. Status Bar |
| 13. Command Window | 14. Layout tab | 15. Model tab |
| 16. UCS Icon | 17. Graphical Area | |

## 1.2  AUTOCAD 2025 INTERFACE

The AutoCAD interface is based on Ribbons and Application Menu. The most important feature of this interface is that the size of the **Graphical Area** will be larger.

### 1.2.1  Application Menu

The Application menu contains file-related commands:

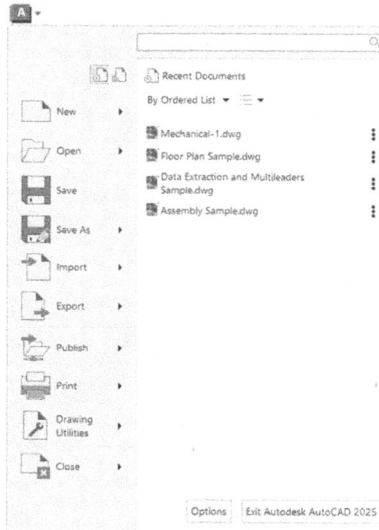

Included are commands such as creating a new file, opening an existing file, saving the current file, saving the current file under a new name and in a different folder, import from different file formats, exporting the current file to a different file format, printing and publishing the current file, etc. We will discuss almost all of these commands in different places in this book. By default, you will see the recent files. You can choose how to display the recent files in the Application Menu using this control:

Also, you can choose how to sort the recent files using this control:

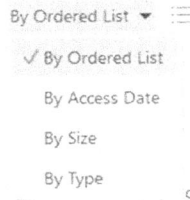

## 1.2.2   Quick Access Toolbar

This toolbar contains all File commands mentioned in the Application Menu along with Workspace and Undo/Redo.

The user can customize this toolbar by clicking the arrow at the end, which will produce the following:

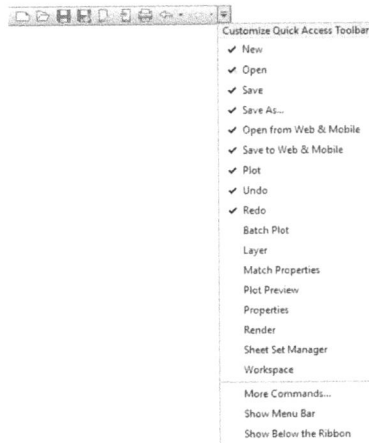

As you can see, the user can add or remove commands; clicking **Show Menu Bar** may also be useful because the ribbon does not include all AutoCAD commands.

### Share Drawing

This command will share a link of the current version of the drawing including its external references in AutoCAD Web, you will see the following message:

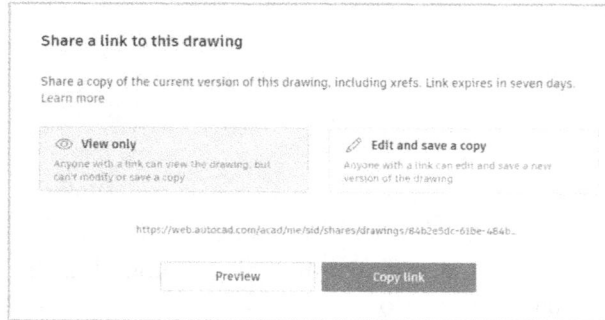

You can create a link which people can view it only, or, with the ability to edit and save copy of it.

### 1.2.3   Ribbons

Ribbons consist of two parts: tabs and panels, as shown below:

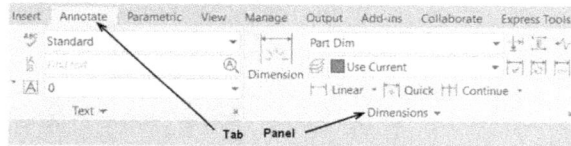

Some panels have more buttons than shown: the following is the **Modify** panel:

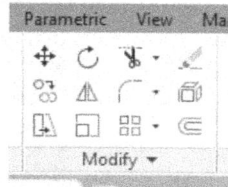

Click the small triangle near the title and you will receive the following:

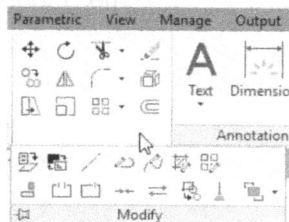

If you move away from the panel, buttons will disappear. To make them visible always, click the push pin. The new view are as follows:

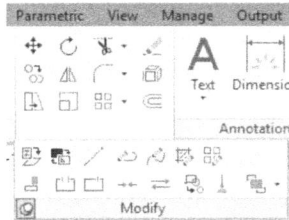

For some commands, there are many options. To make the user's life easier, AutoCAD put all the options in the same button. Refer to the following illustration:

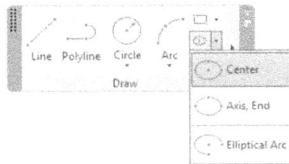

The ribbons have a very simple help feature. If you hover the mouse over a button, a small help screen appears:

If the user leaves the mouse over the button for a longer period of time, AutoCAD will show more detailed help:

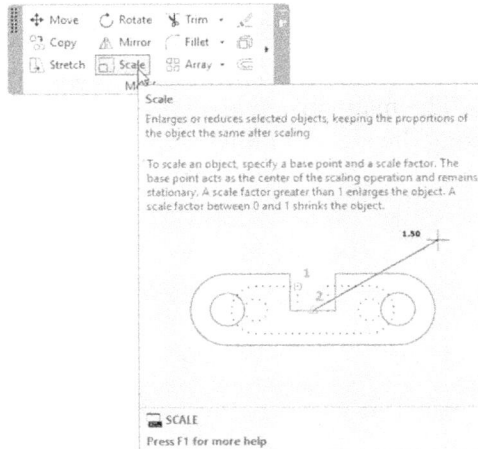

Panels have two states—either *docked* or *floating*. By default, all panels are docked in their respective tab. Drag and drop the panel in the graphical area to make it floating. One important feature of making the panel floating is you will be able to see it while other tabs are active.

You can send the panel back to its respective tab by clicking the small button at the top right side:

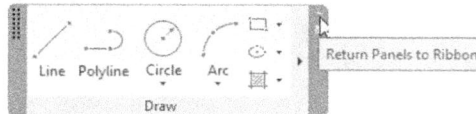

While the panel is floating you can toggle the orientation:

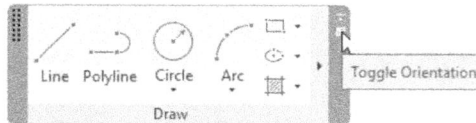

It will either extend to the right:

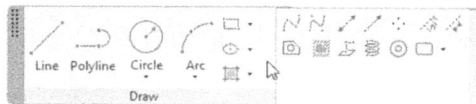

Or down:

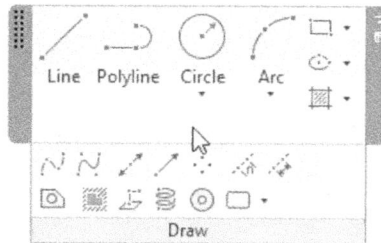

The button at the end of the tab names allows you to cycle through the different states of the ribbons. The main objective of this new feature is to give you yet more graphical area. Clicking the small button will produce the following:

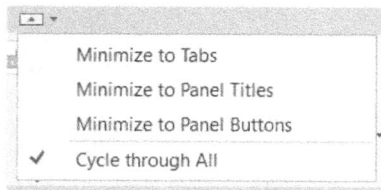

### 1.2.4 InfoCenter

InfoCenter is the place to access help topics online and offline, along with other tools:

For instance, if you type a word or phrase in the field shown below, AutoCAD opens the Autodesk Exchange window and finds all the related topics online and offline. Online means all Autodesk websites along with some popular blogs.

Use Sign In to access Autodesk Account.

### 1.2.5 Command Window

By default, the command window is floating, but you can dock it at the bottom or the top of the screen. Reading the command window all the time will save a significant amount of wasted time trying to determine what AutoCAD wants. AutoCAD will show two items at the command window: your commands and AutoCAD prompts asking you to perform a particular action (e.g., specify a point, input an angle, etc.). Refer to the following illustration:

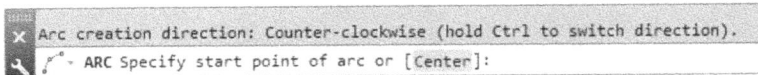

```
Arc creation direction: Counter-clockwise (hold Ctrl to switch direction).
ARC Specify start point of arc or [Center]:
```

### 1.2.6 Graphical Area

The graphical area is your drafting area. This is where you will draw all your lines, arcs, and circles. It is a precise environment with an XYZ space for 3D and an XY plane for 2D. You can monitor coordinates in the left part of the status bar.

### 1.2.7 Status Bar

The status bar in AutoCAD contains coordinates along with some important functions; some of them are for precise drafting in 2D and some of them are for 3D.

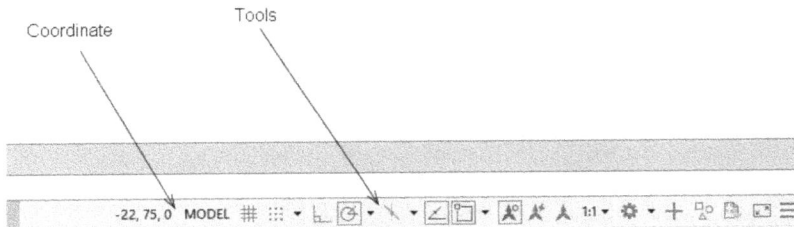

Coordinate          Tools

-22, 75, 0  MODEL

The buttons you will see on the status bar are not necessarily all that are available. To customize this, click on the last button at the right (the one with three horizontal lines) and you will see the following list (showing some of the options):

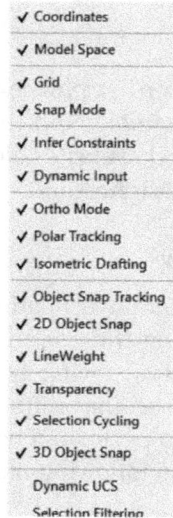

✓ Coordinates

✓ Model Space

✓ Grid

✓ Snap Mode

✓ Infer Constraints

✓ Dynamic Input

✓ Ortho Mode

✓ Polar Tracking

✓ Isometric Drafting

✓ Object Snap Tracking

✓ 2D Object Snap

✓ LineWeight

✓ Transparency

✓ Selection Cycling

✓ 3D Object Snap

Dynamic UCS

Selection Filtering

### 1.2.8   Start Tab

When you click the Start tab at any moment, you will receive the welcome screen again. Hence, you can start a new drawing, open an existing drawing, etc.

At the left of the Start tab, you will see three horizontal lines if you click it, you will see the following menu:

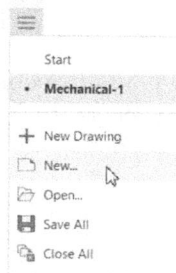

Start

• Mechanical-1

+ New Drawing

New...

Open...

Save All

Close All

This menu will allow you to do the same file functions of the main menu.

## 1.3   AUTOCAD DEFAULTS

Before using the AutoCAD environment, you should be aware of the following AutoCAD settings:

- AutoCAD saves points as Cartesian coordinates (X,Y) whether metric or imperial numbers were used. This is the first method of precise input in AutoCAD to type the coordinates using the keyboard

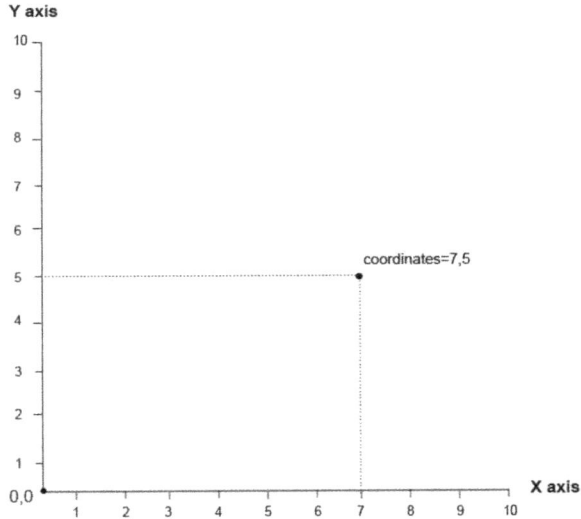

- To specify angles in AutoCAD, assume east (to your right) is your 0° and then go counterclockwise. (This is applicable only for the northern hemisphere; we will learn in Chapter 10 how we can change this setting for the southern hemisphere)

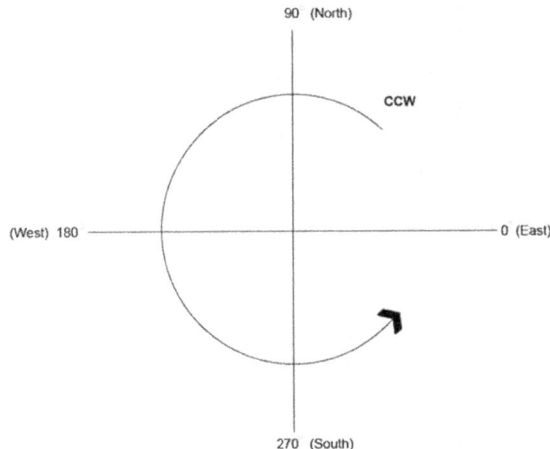

- The wheel in the mouse has four zooming functions: Zoom In (move wheel forward), Zoom Out (move wheel backward), Panning (press and hold the wheel), and Zooming Extents (double-click the wheel)
- Pressing [Enter] or [Spacebar] is equal in AutoCAD
- When pressing [Enter] without typing any command, AutoCAD will repeat the last command. If it is the first thing you do in the current session, it will start Help
- Pressing [Esc] will cancel the current command
- Pressing [F2] will show the Text Window—check the following:

## 1.4   WHAT IS MY DRAWING UNIT?

If you draw a 6-units line in AutoCAD, what does AutoCAD really mean by 6 units? Is it 6 m, 6 ft, or neither? AutoCAD deals with whatever units you choose; if you mean 6 m, so let it be. If you mean 6 ft, AutoCAD will proceed with this assumption as well. What is important is that you be consistent throughout your file. This is true for the Model Space, where you will prepare your drafting, but when it relates to printing, you have to recall your assumption and set your drawing scale accordingly. In Chapter 9, printing will be discussed. At the bottom left of the screen, you can see the model tab and the layouts, as shown in the following:

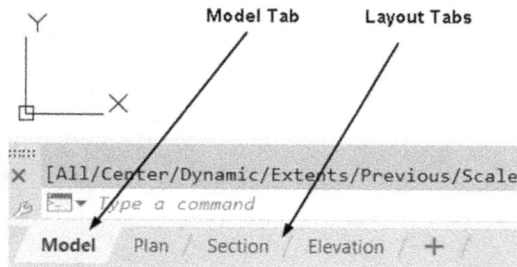

## 1.5    CREATE A NEW AUTOCAD DRAWING

This command will create a new drawing based on a premade template. Use the **Quick Access Toolbar** and click the **New** button:

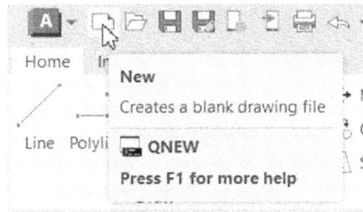

Or you can click the (+) sign in the File tab:

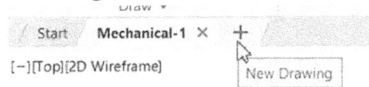

You will see the following dialog box:

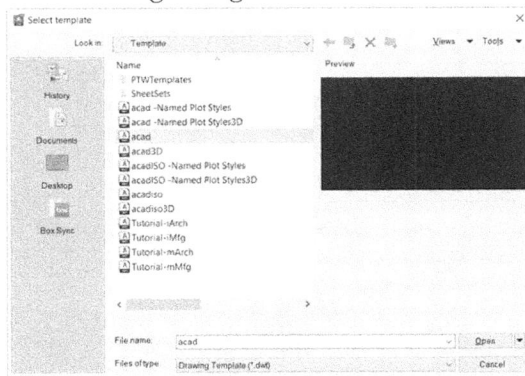

Do the following steps:
- Select the desired template file (AutoCAD template files have the extension *.dwt*). AutoCAD 2025 includes many premade templates that you can use (it is preferable for the companies to create their own template files).
- Once you are done, click the **Open** button
- AutoCAD AutoCAD will start with a new file (which has a temporary name like *Drawing1.dwg*; you should, however, rename it to something meaningful)

*In the Start tab under **New** drop-down, you can do the same, as shown below:*

**NOTE**

## 1.6 OPEN AN EXISTING AUTOCAD DRAWING

This command will open an existing drawing file for additional modifications. From the **Quick Access** Toolbar, click the **Open** button:

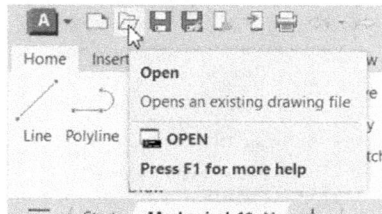

You will see the following dialog box:

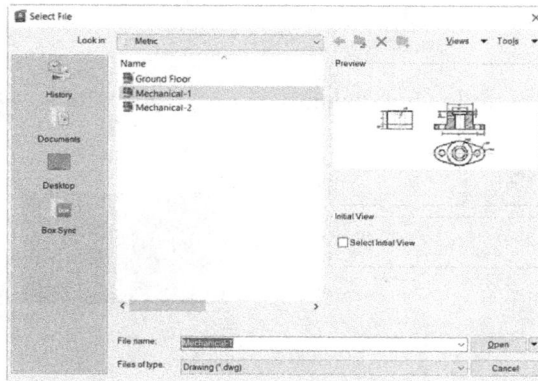

Complete the following steps:
- Specify your desired drive and folder
- You can open a single file by selecting its name from the list and clicking the **Open** button, or you can double-click on the file's name. Or, you can open more than one file by selecting the first file name, holding the [Ctrl] key, clicking the other file names in the list (which is a common MS Windows skill), and then clicking the **Open** button

### 1.6.1 File Tab

Using the File tab beneath the ribbon, you will see a tab for each opened file, as in the following:

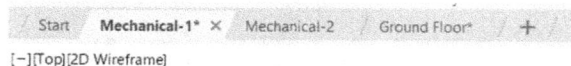

[−][Top][2D Wireframe]

The color used is gray. The current file (tab) will appear in a lighter gray, and the other tabs will appear in a darker gray. If you hover over one of the tabs, two actions will happen:

- The path of this file will appear
- The model space and the layouts of this file will appear beneath

See the following:

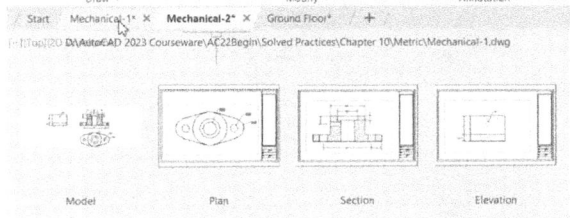

There will be a blue frame around the model space view. Moving your mouse to the right will show the layouts in the graphical area; if you find what you are looking for, click the layout view to move to it.

- *A star beside the name of the file in the file tab means this file was modified and you need to save changes*
- *Click (×) beside the name to close the file, hence closing the tab*
- *The user can customize the File tabs and the Layout tabs by switching them off; all you have to do is to go to the* **View** *tab and locate the* **Interface** *panel; the two buttons will be blue if the tabs are on, but if you want to turn them off, click once on each button:*

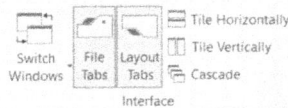

**NOTE**

- *You can now drag a file tab off of the AutoCAD application window, making it a separate window. This is good for multi-display setup. At any moment, you can right-click the title bar of a floating drawing window you will see the following options:*

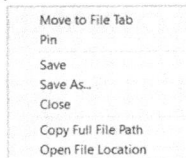

- *These options will help the user to Move the file back to File Tab, pin the floating window, Save and Save As the floating window, close it, and Copy File Path, and File Location*

## 1.7    CLOSING DRAWING FILES

This command will close the current opened files or all opened files, depending on the command you choose. Use the **Application Menu**, move your mouse to the **Close** button, then select either **Current Drawing** to close the current file, or **All Drawings** to close all the opened files in single command:

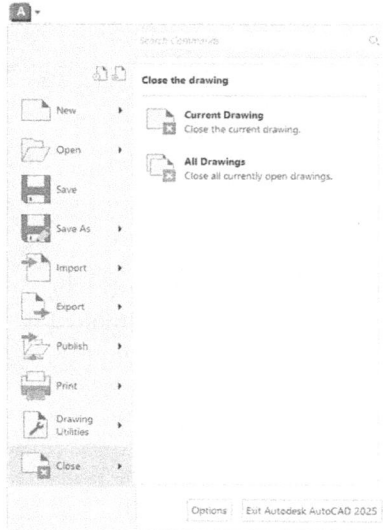

If any of the open files were modified, AutoCAD will ask if you want to save or close without saving, as in the following dialog box:

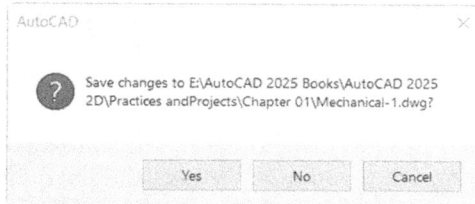

## 1.8    UNDO AND REDO COMMANDS

Undo and Redo are to help you to correct mistakes. They can be used in the current session only.

### 1.8.1   Undo Command

This command will undo the effects of the last command. You can reach this command by going to the **Quick Access toolbar** and clicking the **Undo** button. If you want to undo several commands, click the small arrow at the right. You will see a list of the commands; select the group and undo them:

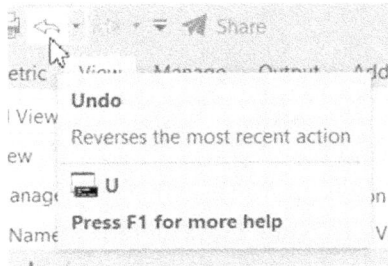

Also, you can type **u** at the command window (do not type **undo**, because it has a different meaning), or press Ctrl + Z at the keyboard.

### 1.8.2   Redo Command

This command will undo the undo command. You can reach this command by going to the **Quick Access toolbar** and clicking the **Redo** button. If you want to redo several commands, click the small arrow at the right. You will see a list of the commands; select the group and redo them:

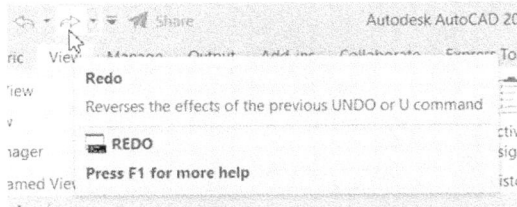

Also, you can type **redo** at the command window or press **Ctrl + Y**.

## PRACTICE 1-1    AUTOCAD BASICS

**1.** Start AutoCAD 2025

**2.** Open the following files:

    **a.** Ground Floor

    **b.** Mechanical-1

    **c.** Mechanical-2

**3.** Using the File tab, check the three files and their layouts

**4.** Use the different zoom techniques with the mouse wheel

**5.** Using the Application menu, close all files without saving

## NOTES

## CHAPTER REVIEW

**1.** The File tab will show the Model space and layouts of the opened files

    **a.** True

    **b.** False

**2.** The AutoCAD template file extension is _____

    **a.** *.dwt

    **b.** *.dwg

    **c.** *.tmp

    **d.** *.temp

**3.** AutoCAD units can be a meter or foot, per user desire

    **a.** True

    **b.** False

**4.** Moving the wheel forward will mean _____

**5.** To undo any command in AutoCAD you can:

    **a.** Click the Undo icon from the Quick Access Toolbar

    **b.** Type **u** at the command window

    **c.** Type [Ctrl] + Z

    **d.** All of the above

**6.** Ribbons consist of _____ and _____

**7.** The Menu bar by default is not shown, but you can show it if you want to:

    **a.** True

    **b.** False

**8.** The AutoCAD drawing file extension is _____

**9.** Positive angles in AutoCAD are created by moving in this direction: _____

## CHAPTER REVIEW ANSWERS

    **1.** a

    **3.** a

    **5.** d

    **7.** a

    **9.** CCW (counterclockwise)

# PRECISE DRAFTING IN AUTOCAD 2025

## In This Chapter

- What are the drafting priorities?
- How to draw lines, circles, and arcs using precise methods
- How to draw polylines using precise methods
- How to convert lines and arcs to polylines and vice versa
- What are Object Snap and Object Track?

## 2.1 DRAFTING PRIORITIES

While you are drafting, there are two main concerns in your mind: the accuracy of your drawing and how fast you can finish it. People crave to finish their drawings fast but without weakening the drawing's accuracy. Experts tend to put accuracy ahead of fast production.

Accuracy is important in the context of the "life cycle" of the drawing. If your drawing is accurate, all the other people who will modify it in the future will be able to do their mission without a hassle. On the other hand, their lives will be extremely difficult if you finish your drawing fast but none of the objects are accurate.

In this chapter, you will learn how to use the four most essential drafting commands, which are:

- Line command, to draw line segments
- Arc command, to draw circular arcs

- Circle command, to draw circles
- Polyline command, to draw lines and arcs jointly

While we are discussing the four drafting commands, we will introduce accuracy tools, which will help you accelerate the drafting process.

## 2.2   DRAWING LINES USING LINE COMMAND

This command will enable you to draw straight lines; each line segment presents a single object. To issue this command go to the **Home** tab, locate the **Draw** panel, and then select the **Line** button:

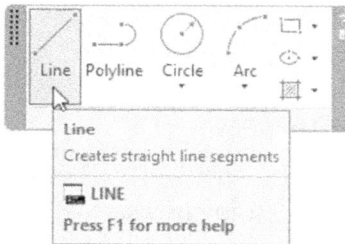

You will see the following AutoCAD prompts:

```
Specify first point:
Specify next point or [eXit/Undo]:
Specify next point or [Close/eXit/Undo]:
Specify next point or [Close/eXit/Undo]:
```

Using the first prompt, specify the coordinates of your first point. Keep specifying points, and when you are done, do one of the following:

- If you want to stop without closing the shape, use either X to exit the command, or press [Enter] ([Esc] will do the job as well, but don't make it a habit, as [Esc] generally means abort)
- If you want to close the shape and finish the command, press C on the keyboard or right-click and select the **Close** option
- If you made any mistakes, you can undo the last point by typing U on the keyboard or right-clicking and selecting the **Undo** option

This is what the right-click menu looks like:

## 2.3 WHAT IS DYNAMIC INPUT IN AUTOCAD?

Dynamic Input has multiple functions:

- It will show prompts at the command window in the graphical area
- It will show the lengths and angles of the lines before drafting, which will specify them accurately

In order to turn on/off the **Dynamic Input**, click the following button on the Status bar:

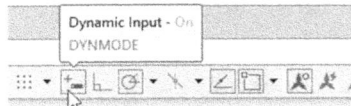

### 2.3.1 Example for Showing Prompts

By default, if you type any command using the command window, AutoCAD will help you by showing all the commands starting with the same letters, as in the following example:

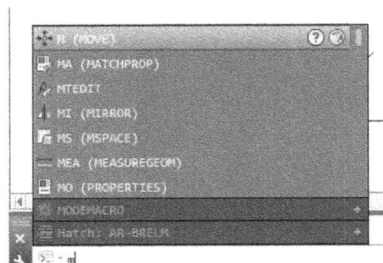

We typed the letter **m**, and accordingly AutoCAD gave us all the commands starting with this letter. While Dynamic Input is on, this is applicable to the crosshairs as well, as in the following:

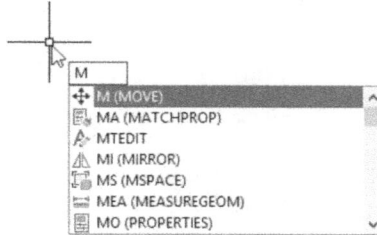

Let's assume we started the command window, using either way. Once you press [Enter], the following prompt will appear:

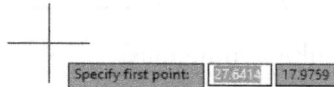

Type in the x and y coordinates, using the [Tab] key to move between the two fields:

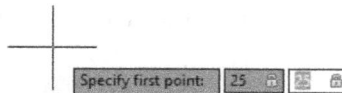

### 2.3.2 Example for Specifying Lengths and Angles

Once you specify the starting point, AutoCAD will use Dynamic Input to show the length and the angle of the line utilizing the rubber band mode:

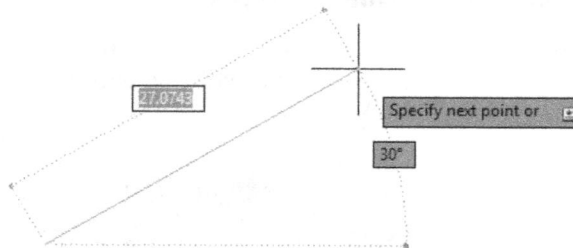

**NOTE** *Angles are measured CCW starting from the east, but only for 180°, unlike the angle system in AutoCAD, which will be for the whole 360°.*

Type the length of the line, then press [Tab] to input the angle (it will give 1o increments); once you are done, press [Enter] to specify the first line. Continue doing the same for the other segments:

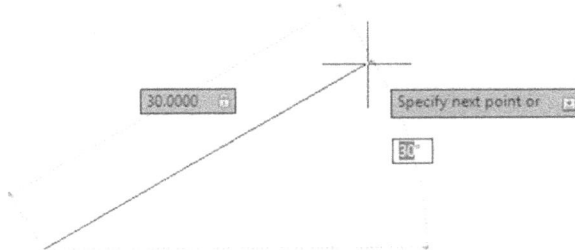

If Dynamic Input button is not shown, click the three lines at the bottom right of the screen and turn it on.

## PRACTICE 2-1    DRAWING LINES USING DYNAMIC INPUT

1. Start AutoCAD 2025

2. Open file **Practice 2-1.dwg**

3. Using status bar, click off, Polar Tracking, Ortho, and Object Snap, and make sure Dynamic Input is on

4. Draw the following shape, using 0,0 as your start point, bearing in mind all sides = 4, and all angles are multiples of 45°:

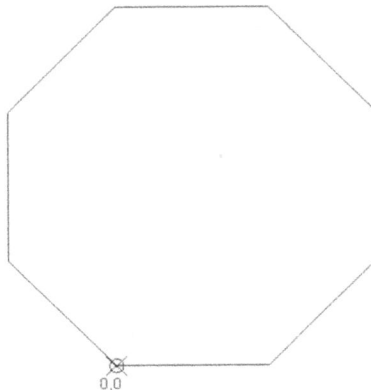

5. Save the file and close it

## 2.4   EXACT ANGLES (ORTHO VS. POLAR TRACKING)

When using Dynamic Input angles are incremented by 1°, but still the user cannot depend on it to specify angles precisely in AutoCAD.

The **Ortho** function will force the lines to be at right angles (orthogonal) using the following angles: 0, 90, 180, and 270.

In order to turn on/off the **Ortho,** use the following button in the status bar:

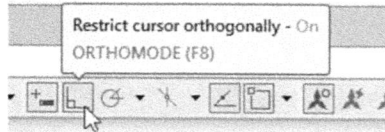

However, what if we want to use other angles such as 30, 45, 60, etc.? Ortho will not help in this case; for this reason, AutoCAD introduced another function called Polar Tracking, which will show in the graphical area rays starting from the current point heading toward angles like 30, 45, etc. Since Ortho and Polar Tracking contradict each other, when you switch one on, the other will be turned off automatically. In order to turn on/off **Polar Tracking**, use the following button in the status bar:

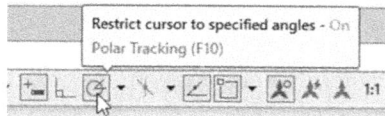

If you click the small triangle at the right of the button, you will see the following menu:

You can select the desired angle or select **Tracking Settings** to change some of the default settings of Polar Tracking. You will see the following dialog box:

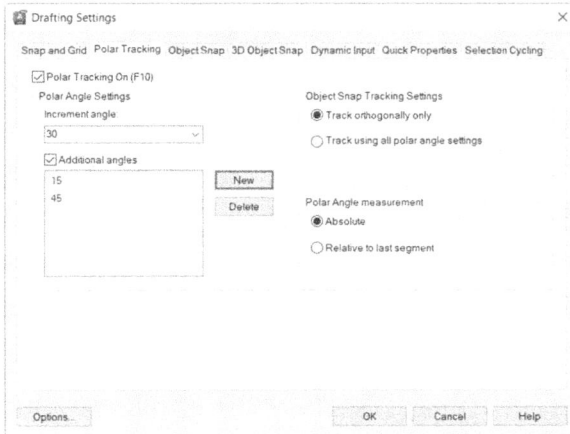

### 2.4.1 Increment Angle

Increment Angle is the angle to be used along with its multiples. Select one from the list or type your own.

### 2.4.2 Additional Angles

If you are using 30 as your increment angle, then 45 will not be among the angles that Polar Tracking will allow you to use. Hence, you will need to specify it as an additional angle. However, be aware you will not use its multiples.

### 2.4.3 Polar Angle Measurement

When you are using Polar Tracking, you have the ability to specify angles as an absolute angle (based on 0° at the east) or use the last line segment to be your 0 angle. Refer to the following illustration:

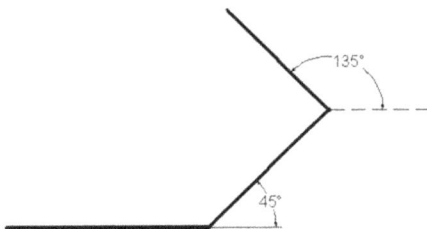

Measured using 0.0 angle          Measured using the last segment

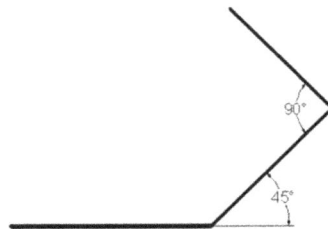

**NOTE** *While you control the angle using either Ortho or Polar Tracking, you can type in the distance desired and then press [Enter] to draw accurate distances. This method is called Direct Distance Entry.*

## PRACTICE 2-2   EXACT ANGLES

**1.** Start AutoCAD 2025

**2.** Open **Practice 2-2.dwg**

**3.** Draw the following shape using Line command, starting from 0,0 as your starting point (or any point you like), bearing in mind you have to use Polar Tracking. All angles are 30 and its multiplications and all lines are 2 units in length, using the Direct Distance Entry method to input the exact distances:

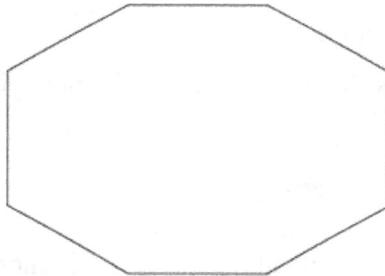

**4.** Save and close the file

## 2.5   PRECISE DRAFTING USING OBJECT SNAP

Object Snap, or OSNAP, is the most important accuracy tool to be used in AutoCAD for 2D and 3D as well. It is a way to specify points on objects precisely using the AutoCAD database stored in the drawing file.

Some of the Object Snaps are:

- **Endpoint**: To catch the Endpoint of a line, or arc
- **Midpoint**: To catch the Midpoint of a line, or arc
- **Intersection**: To catch the Intersection of two objects (any two objects)
- **Perpendicular**: To catch the Perpendicular point on an object
- **Nearest**: To catch a point on an object Nearest to your click point

We will discuss more object snaps when we discuss more drawing objects. Here are graphical presentations of each one of these OSNAPs:

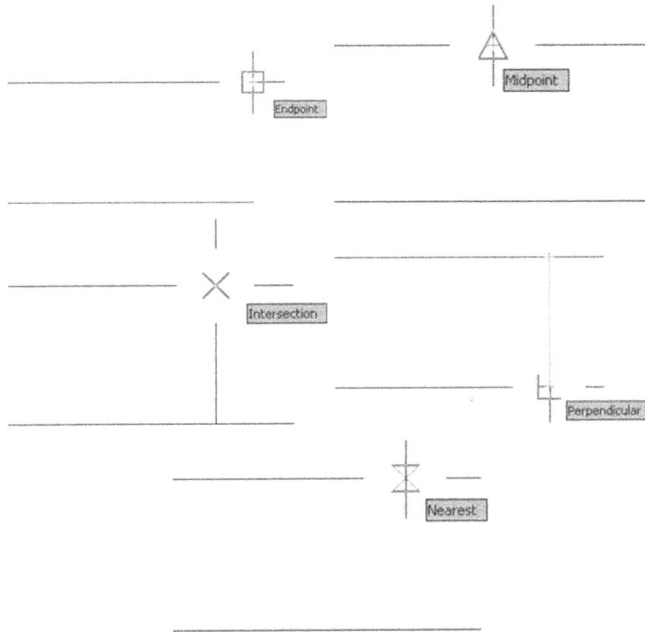

### 2.5.1 Activating Running OSNAPs

To activate running OSNAPs in the drawing, click on the Object Snap button in the status bar:

If you click the small triangle at the right of the button, you will see the following menu:

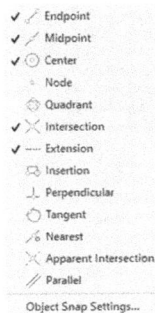

You can switch on the desired OSNAPs one by one. If you want to make a more convenient change, select the **Object Snap Settings** option, and the following dialog box will appear:

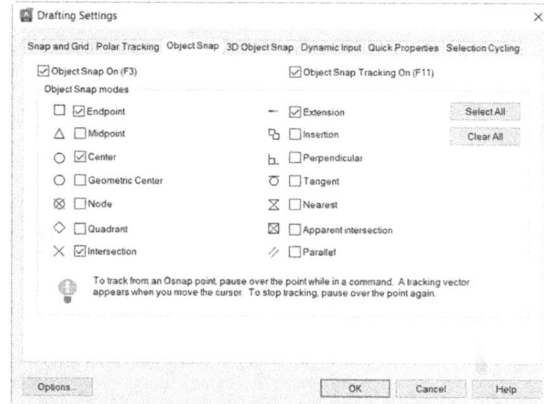

There are two buttons at the right: **Select All**, **Clear All**. We recommend using Clear All first and then selecting the desired OSNAPs. When done, click **OK**.

### 2.5.2    OSNAP Override

While the Object Snap button is on, several OSNAPs are working, and others are not! Suppose you want to temporarily switch all of them off, use a single OSNAP, and then after finishing set everything back to normal. This is called OSNAP Override:

There are two ways to activate an override:

- Using the keyboard, type the first three letters of the desired OSNAP
- Using the keyboard, hold the [Shift] key, and then right-click. You will see the following pop-up menu:

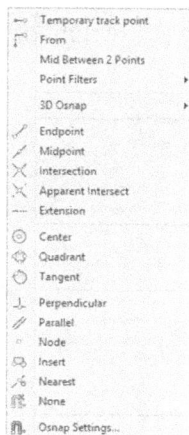

## PRACTICE 2-3   OBJECT SNAP (OSNAP)

**1.** Start AutoCAD 2025

**2.** Open **Practice 2-3.dwg**

**3.** Check the Object Snap at the status bar and make sure that Endpoint, Midpoint, Intersection, and Perpendicular are the only OSNAPs switched on

**4.** Using Object Snap and Line command, draw lines in the drawing to create the following:

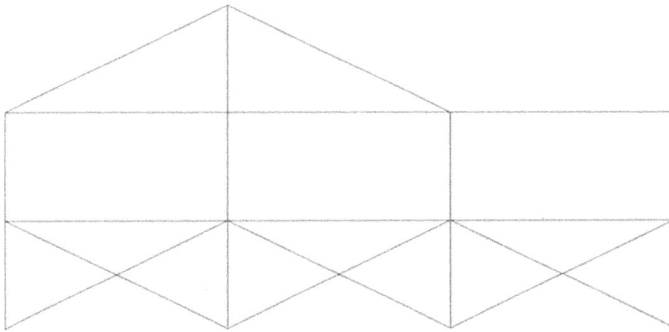

**5.** Save and close the file

## 2.6   DRAWING CIRCLES USING CIRCLE COMMAND

This command will draw a circle using different methods based on the available data. If you know the coordinates of the center, there are two possible methods. If the user knows the coordinates of points at the diameter of the circle, there are another two methods. Finally, if there are drawn objects like lines, arcs, or other circles which can be used as tangents for the to-be-created circles, there are two more methods. The following are the six methods to draw a circle in AutoCAD:

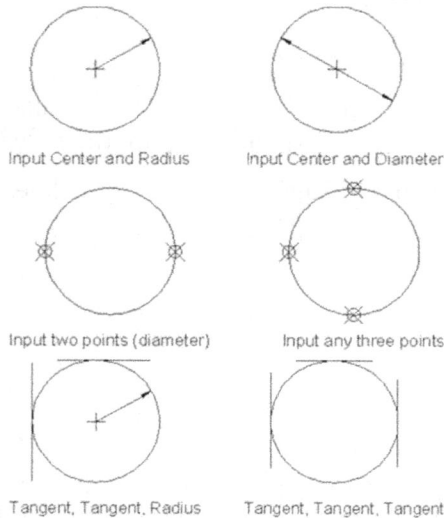

Input Center and Radius       Input Center and Diameter

Input two points (diameter)      Input any three points

Tangent, Tangent, Radius      Tangent, Tangent, Tangent

To issue this command, go to the **Home** tab, locate the **Draw** panel, then select the arrow near the **Circle** button to see all the available methods:

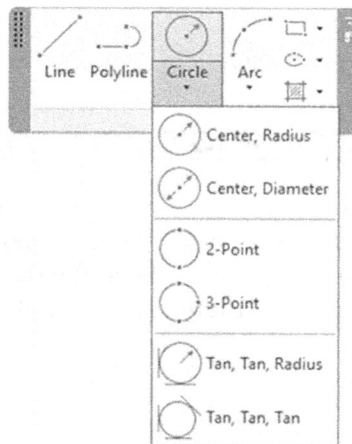

Line   Polyline   Circle   Arc

Center, Radius

Center, Diameter

2-Point

3-Point

Tan, Tan, Radius

Tan, Tan, Tan

## 2.7   DRAWING CIRCULAR ARCS USING ARC COMMAND

This command will draw an arc part of a circle. To make things easier, AutoCAD uses eight pieces of information related to a circular arc. These are:

- The starting point of the arc
- Any point as a second point on the parameter of the arc

- The ending point of the arc
- The Direction of the arc, which is the tangent that passes through the Start point. User should input the angle of the tangent
- The distance between the starting point and the ending point which is called Length of Chord
- The Center point of the arc
- The Radius
- The angle between Start-Center-End which is called Included Angle

Refer to the following illustration:

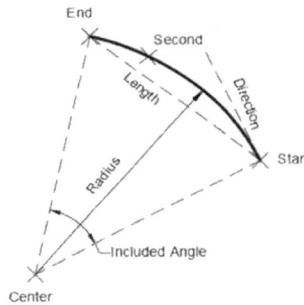

If you provide three of these eight pieces of information, AutoCAD will be able to draw an arc, but not with any three. The combination of the information needed can be found in the **Home** tab, using the **Draw** panel, while clicking the arrow near the **Arc** button to see all the available methods:

As you can see, the Start point is always required information. Normally the user should think counterclockwise when specifying points, but if on the other hand you want to work clockwise, simply hold the [Ctrl] key and it will change.

## 2.8   OBJECT SNAPS RELATED TO CIRCLES AND ARCS

Some of the Object Snaps related to circles and arcs are:

- **Center**: To catch the Center of an arc or circle
- **Quadrant**: To catch the Quadrant of an arc or circle
- **Tangent**: To catch the Tangent of an arc or circle

Here are graphical representations of each one of these OSNAPs:

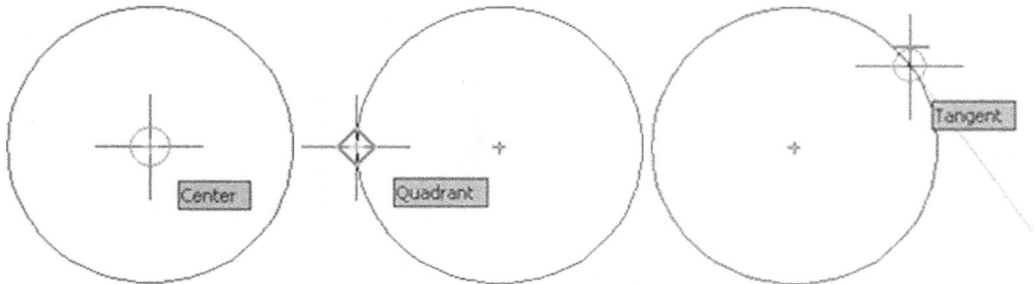

## 2.9   USING OBJECT SNAP TRACKING WITH OSNAP

Sometimes OSNAP alone is not enough to specify desired points, especially if we need complex points. To solve this problem in the past we used to draw dummy objects to help us specify complex points, like in the case when you want to specify the center of the circle at the center of a rectangle. We used to draw a line from the midpoints of the two vertical lines, and the same for the horizontal lines. However, since the introduction of Object Snap Tracking, or OTRACK, in AutoCAD 2000, the drawing of mock objects diminished. OTRACK depends on active OSNAP modes, which means if you want to use the midpoint, OTRACK requires that you switch the midpoint on first. To activate OTRACK go to the status bar and click the **Object Snap Tracking** button on:

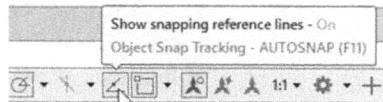

The procedure is very simple:

- Using OSNAP go to the desired point and stand still for a couple of seconds (don't click), then move to the right or left (also up and down depending on the next point), and you will see an infinite line extending in both directions (this line will be horizontal or vertical depending on your movement).

- If you want to use a single point to specify your desired point, move to the needed direction, type in the desired distance, and press [Enter].
- If you need two points, go to the next point, and stay for a couple of seconds, then make the movement toward the desired direction. Another infinite line will appear. Go to the intersection point of the two infinite lines and that will be your point.

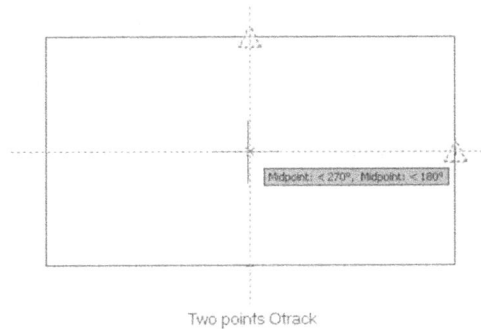

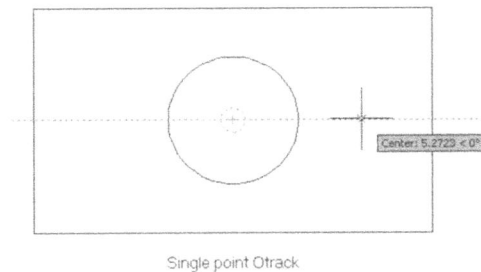

Two points Otrack

Single point Otrack

Polar command will make a major contribution here as well. Let's return to the same dialog box:

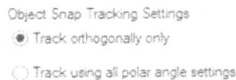

Object Snap Tracking Settings
⦿ Track orthogonally only
◯ Track using all polar angle settings

Under **Object Snap Tracking Settings**, there are two choices:

- Track orthogonally only (default option)
- Track using all polar tracking settings

This means you can use the current Polar angles (increment and additional angles) to specify points using OTRACK.

**NOTE** *To deactivate an OTRACK point, stay at the same point again for a couple of seconds, and it will be deactivated.*

## PRACTICE 2-4A   DRAWING USING OSNAP AND OTRACK

1. Start AutoCAD 2025

2. Open **Practice 2-4a.dwg**

3. Using the proper OSNAP and OTRACK settings, create the four arcs as shown below:

4. Create the two circles as shown below (Radius = 1)

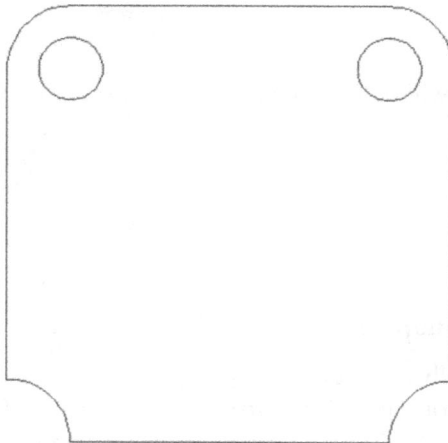

**5.** Using OSNAP and OTRACK (using two points), draw the circle at the center of the shape (Radius = 3.0)

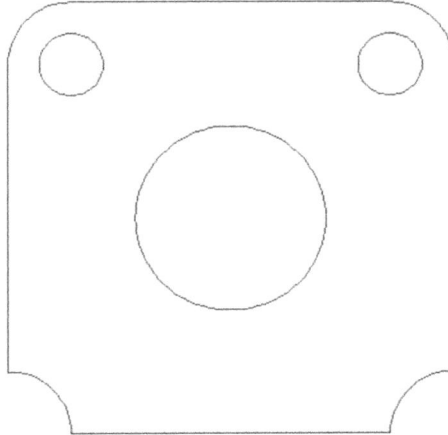

**6.** Using OSNAP and OTRACK (one point) draw the two circles at the right and left (distance center-to-center = 5.0, and Radius = 0.5)

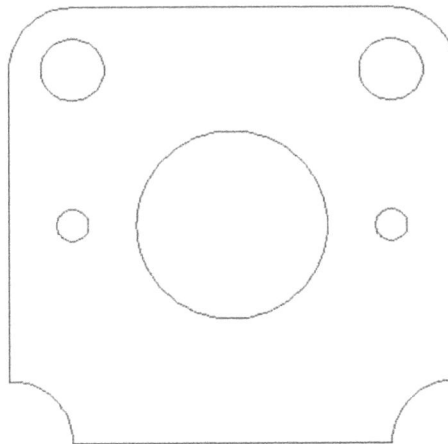

**7.** Change the increment angle in the Polar Tracking dialog box to 45. Make sure that **Track using all polar angle settings** is on, draw a circle (Radius = 0.5), its center specified using OSNAP, OTRACK, and polar tracking as shown below:

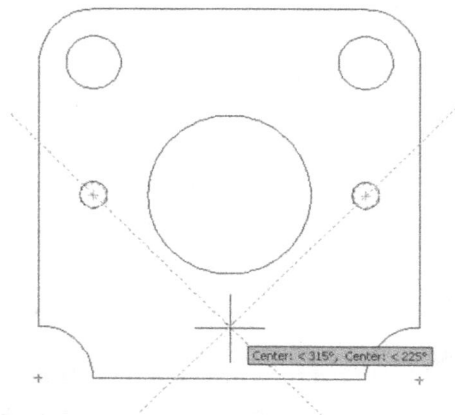

Center: < 315°, Center: < 225°

**8.** Do the same procedure to draw a circle at the top, to acquire the final shape as follows:

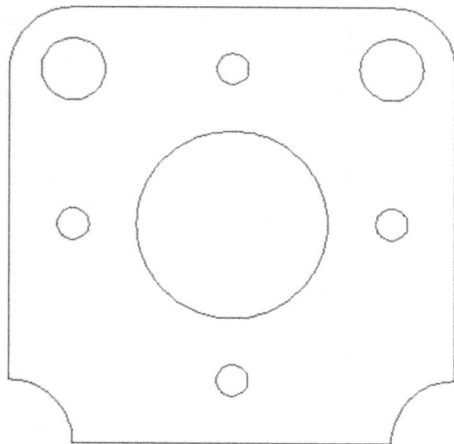

**9.** Save and close the file

## PRACTICE 2-4B  DRAWING USING OSNAP AND OTRACK

**1.** Start AutoCAD 2025

**2.** Open **Practice 2-4b.dwg**

**3.** Using the proper OSNAP and OTRACK settings, add lines and circles to make the shape look similar to the following:

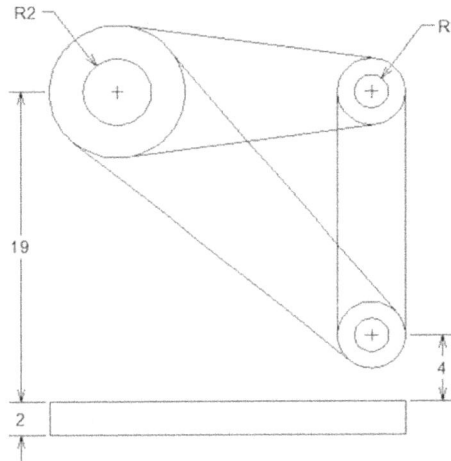

**4.** Save and close the file

## 2.10   DRAWING LINES AND ARCS USING POLYLINE COMMAND

Polyline command will do all or any of the following:

- Draw both line segments and arc segments
- Draw a single object in the same command rather than drawing segments of lines and arcs, like in Line and Arc commands
- Draw lines and arcs with starting and ending widths

To use the command, go to the **Home** tab, locate the **Draw** panel, then select the **Polyline** button:

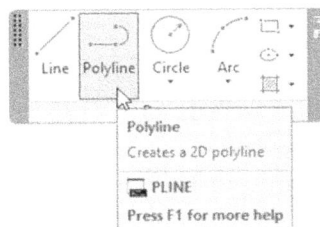

The following prompt will appear:

```
Specify start point:
Current line-width is 1.0000
Specify next point or [Arc/Halfwidth/Length/Undo/ Width]:
```

AutoCAD will ask you to specify the first point, and when you do, AutoCAD will report to you the current line-width; if you like it, continue specifying points using the same method we learned in the Line command. If not, change the width as a first step by typing the letter **W**, or right-clicking and selecting the **Width** option, which will bring up the following prompt:

```
Specify starting width <1.0000>:
Specify ending width <1.0000>:
```

Specify the starting width, press [Enter], and then specify the ending width. The next time you use the same file, AutoCAD will report these values for you when you issue the Polyline command. The process for **Half-width** is the same, but instead of specifying the full width, you specify half-width.

The Undo and Close options are identical to the ones at the Line command.

Length will specify the length of the line using the angle of the last segment.

Arc will draw an arc attached to the line segment; you will see the following prompt:

```
Specify endpoint of arc or [Angle/CEnter/CLose/
Direction/Halfwidth/Line/Radius/Second pt/Undo/Width]:
```

The arc will be attached to the last segment of the line or will be the first object in a Polyline command; using either method, the first point of the arc is already known, so we need two more pieces. AutoCAD will make an assumption (which you have the right to reject): AutoCAD will assume that the angle of the last line segment will be considered the direction (tangent) of the arc. If you accept this assumption, you should specify the endpoint. If not, choose from the following to specify the second piece of information:

- The Angle of the arc
- The Center point of the arc
- Another Direction to the arc
- The Radius of the arc
- The Second point which can be any point on the parameter of the arc

Based on the information selected as the second point, AutoCAD will ask you to supply the third piece of information, but by all means it will not be out of the eight we discussed before.

*Normally the user should think counterclockwise when specifying points; on the other hand, if you want to work clockwise, simply hold the [Ctrl] key and it will change.*

## 2.11 CONVERTING POLYLINES TO LINES AND ARCS AND VICE-VERSA

This is an essential technique which will convert any polyline to lines and arcs and convert lines and arcs to polylines.

### 2.11.1 Converting Polylines to Lines and Arcs

The **Explode** command, will explode a polyline to lines and arcs. To issue this command, go to the **Home** tab, locate the **Modify** panel, then select the **Explode** button:

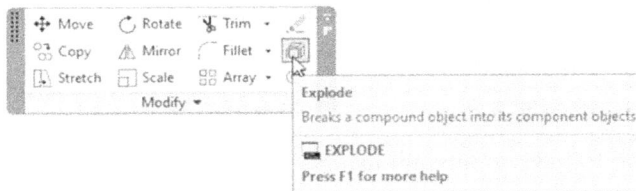

AutoCAD will show the following prompt:

```
Select objects:
```

Select the desired polylines and press [Enter] when done. Check the new shape; you will discover it has been changed to lines and arcs.

### 2.11.2 Joining Lines and Arcs to Form a Polyline

What we will discuss here is an option called **Join**, within a command called **Edit Polyline**. To issue this command, go to the **Home** tab, locate the **Modify** panel, then select the **Edit Polyline** button:

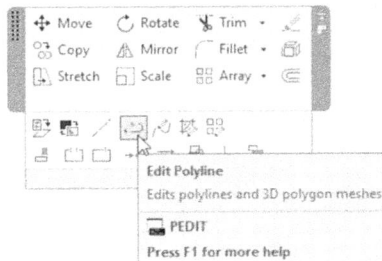

You will see the following prompts:

```
Select polyline or [Multiple]:
Object selected is not a polyline
Do you want to turn it into one? <Y>
Enter an option [Close/Join/Width/Edit vertex/Fit/
Spline/Decurve/Ltype gen/Reverse/Undo]: J
```

Start first by selecting one of the lines or arcs you want to convert. AutoCAD will respond by telling you that the selected object is not a polyline, providing you the option to convert this specific line or arc to a polyline. If you accept this, options will appear, and one of these options is **Join**. Select the **Join** option, then select the rest of the lines and arcs. At the end press [Enter] twice. Check the objects after finishing; you will find that they were converted to polylines.

## PRACTICE 2-5    DRAWING POLYLINES AND CONVERTING

**1.** Start AutoCAD 2025

**2.** Open **Practice 2-5.dwg**

**3.** Draw the following polyline using a start point of 18,5, and width = 0.1

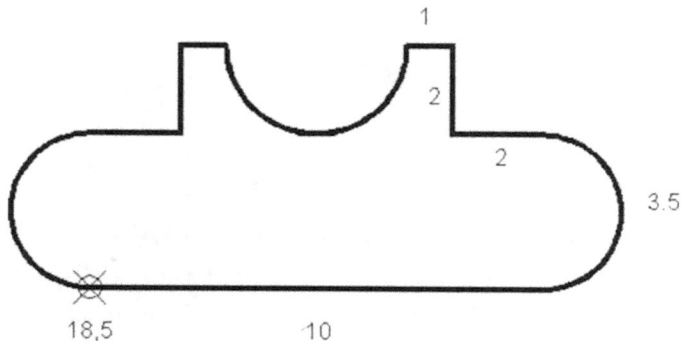

**4.** Then explode the polyline. The width will disappear

**5.** Check the objects after exploding; they are now lines and arcs

**6.** Save and close

## 2.12    USING SNAP AND GRID TO SPECIFY POINTS ACCURATELY

Snap and Grid will be the third method to assist us specify points accurately in the XY plane.

By default, the mouse is not accurate, so we cannot depend on it to identify points. We need to control its movement, which is the sole function of Snap. Snap can control the mouse to jump in the X and Y plane with exact distances.

The grid by itself is not an accurate tool, but it will complement the Snap function. It will show horizontal and vertical lines replicating the drawing sheets. In order to turn **Snap** on/off, use the following button in the status bar:

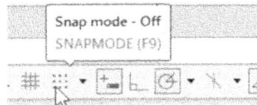

In order to turn the **Grid** on/off, use the following button in the status bar:

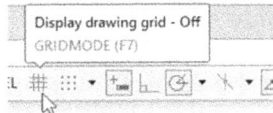

Most likely, switching both on will not help; you will need to modify the settings to set your own requirements. Using the small triangle at the right of the Snap button, you will see the following:

Select the **Snap Settings** option; the Drafting Settings dialog box will pop up::

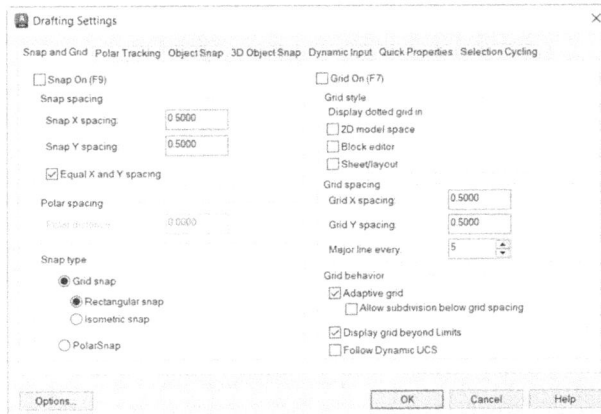

Input the Snap X Spacing and Snap Y Spacing (by default they are equal). Switch off the checkbox to make them unequal. Do the same for the Grid Spacing in X and Y. If you want the Grid to follow Snap, set the Grid spacing to zeros. In the Grid, there are major and minor lines; set the major line frequency.

Set if you want to see the Grid in dots (as it was before AutoCAD 2011) and where (2D Model space, Block editor, or Sheet layout).

Grid behavior is for 3D only.

Specify the type of Snap: Grid Snap or Polar Snap.

The user can use function keys to turn on/off both Snap and Grid:

- F9 = Snap on/off
- F7 = Grid on/off

## 2.13   USING POLAR SNAP

While using the Snap command, the mouse will use the increment distance to specify the exact distance only in horizontal and vertical directions. While going diagonally, Snap will not help you. To solve this problem AutoCAD provided the concept of Polar Snap, which will specify exact increments using all the angles specified in the Polar Tracking dialog box. To activate the Polar Snap, AutoCAD will switch off the normal Snap (which is called Grid Snap). To do that go to the Snap button, click the triangle at the right of the button, and select the **Snap Settings** option, which will bring up the following dialog box:

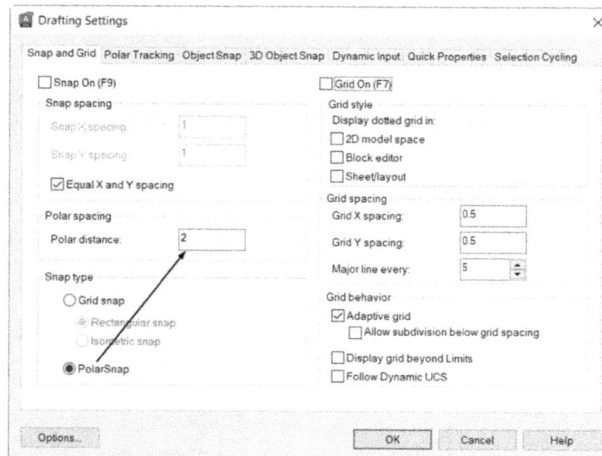

Select the Polar Snap button, then specify the Polar distance as shown in the above illustration.

## PRACTICE 2-6   SNAP AND GRID

**1.** Start AutoCAD 2025

**2.** Open **Practice 2-6.dwg**

**3.** Change Polar Tracking to use angle = 45

**4.** Change the Snap to Polar Snap and set the distance to 0.5

**5.** Draw the following shape, bearing in mind to start from point 22,5, make all segment lengths 6.5, and use an angle of 45:

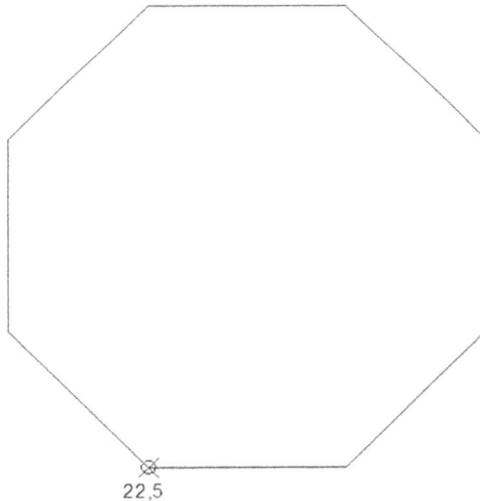

22,5

**6.** Save and close the file

# PRACTICE 2-7    SMALL PROJECT

**1.** Start AutoCAD 2025

**2.** Start a new drawing using acad.dwt

**3.** Using all the commands and techniques you learned in Chapter 2, draw the following drawing (without dimensions) starting from any point you wish:

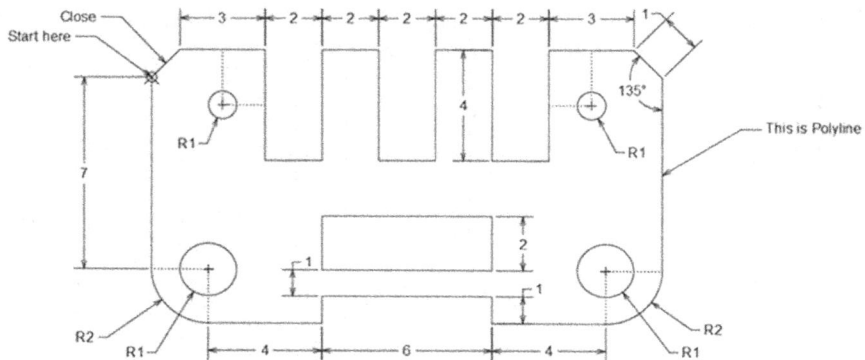

**4.** Save and close the file

## NOTES

## CHAPTER REVIEW

1. Polyline command is different from Line command in the following way(s):

   **a.** It will produce lines and arcs

   **b.** All segments drawn using the same command are considered a single object

   **c.** You can specify a starting and ending width

   **d.** All of the above

2. Perpendicular, Tangent, and Endpoint are some _____ available in AutoCAD

3. OTRACK can work by itself:

   **a.** True

   **b.** False

4. One of the following is not part of the eight pieces of information AutoCAD needs to draw an arc:

   **a.** End point of the arc

   **b.** Midpoint of the arc

   **c.** Center point of the arc

   **d.** Angle

5. To convert to lines and arcs from a polyline, use _____ command

6. If you want the Grid to follow the Snap setting, set it up to be _____ in both X and Y

7. Using the Polar Tracking dialog box, the user can only track orthogonal angles:

   **a.** True

   **b.** False

8. _____ is always required to draw an arc

9. To convert lines and arcs to polylines use:

   **a.** Polyline Edit, Convert option

   **b.** Polyline Edit, Union option

   **c.** Polyline Edit, Join option

   **d.** None of the above

## CHAPTER REVIEW ANSWERS

1. d

3. b

5. Explode

7. b

9. c

## CHAPTER 3

# MODIFYING COMMANDS PART I

## In This Chapter

- Different methods to select objects and selection cycling
- How to erase objects
- How to move and copy objects
- How to rotate and scale objects
- How to mirror and stretch
- How to lengthen and join objects
- How to use grips for editing

## 3.1  HOW TO SELECT OBJECTS IN AUTOCAD

In order to utilize any of the Modifying commands discussed in this chapter, you must select "desired objects" as the first step. Once you issue any of the modifying commands, the following prompt will appear:

```
Select objects:
```

The cursor will change to the Pick Box. At this prompt, you can work without typing anything at the keyboard, or can type few letters to activate a certain mode.

Without typing any letter, you can do the following:

- Select objects by clicking them using the pick box one by one
- Without selecting any object, click and move to the right to start Window mode, which will select all objects contained fully inside the Window
- Without selecting any object, click and move to the left to start Crossing mode, which will select all objects contained fully inside the Crossing, or touched (crossed) by it

See the following two examples:
**Window Example:**

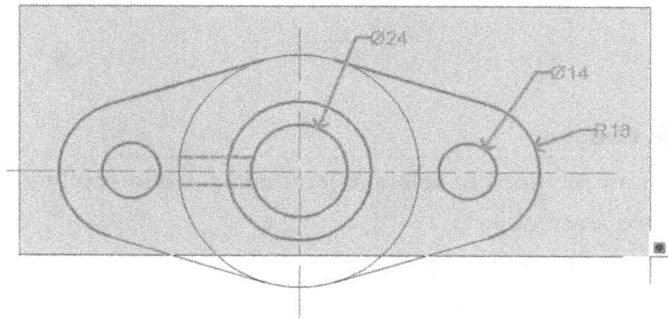

At the cursor you will see a small rectangle with blue circle inside it. All objects will be selected except the big circle and the bottom diagonal lines, but why? Because they are not fully contained inside the window.

**Crossing Example:**

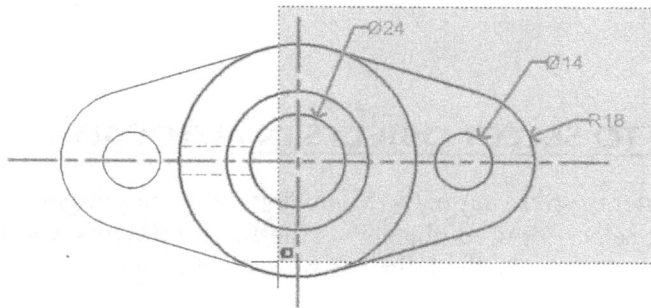

At the cursor you will see a small rectangle crossing the green circle. All objects will be selected except the two diagonal lines at the left along with the arc and circle, but why? Because they were neither contained nor crossed.

Another way to use the Select objects prompt is to type a few letters to activate a certain mode; these letters are:

### 3.1.1   Window Mode (W)

At the command prompt, typing **W** will switch the selecting mode to the **Window**, which will be available whether you go to the right or to the left.

### 3.1.2   Crossing Mode (C)

At the command prompt, typing **C** will switch the selecting mode to the **Crossing**, which will be available whether you go to the right or to the left.

### 3.1.3   Window Polygon Mode (WP)

WP mode will specify a non-rectangular window when you specify points in any fashion you like. By typing **WP** and pressing [Enter], the following prompts will appear:

```
First polygon point:
Specify endpoint of line or [Undo]:
Specify endpoint of line or [Undo]:
```

Press [Enter] to end WP mode. WP is just like W; it needs to contain the object fully in order to select it.

See the following example:

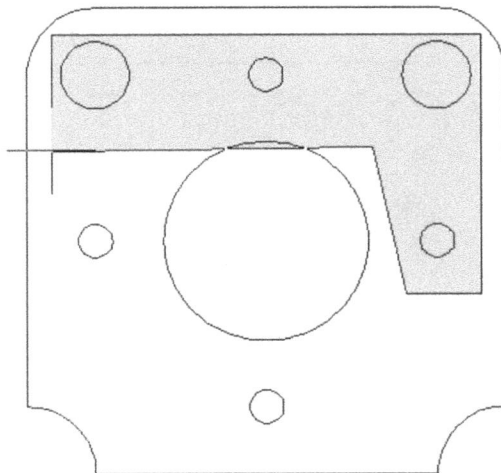

Only four circles will be selected, which are fully contained inside the WP.

### 3.1.4 Crossing Polygon Mode (CP)

Since there is W and WP, certainly there is C and CP. Crossing Polygon is just like WP, and will specify a non-rectangular shape to contain and cross objects. Refer to the following illustration:

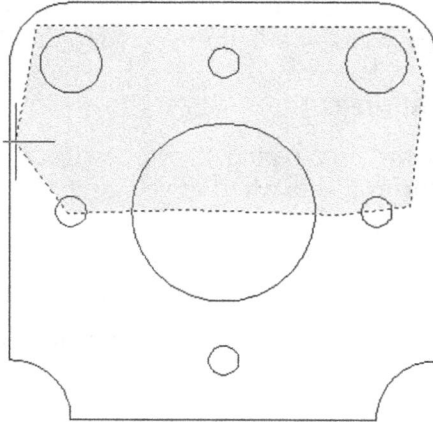

Three circles will be selected because they are fully contained inside the CP, and three more will be selected because they are crossed by the CP.

### 3.1.5 Lasso Selection

The user can combine the WP and CP modes without typing any letter. While the pick box is displayed, *click and hold*; if you go to the right, you have WP, if you go to the left, you have CP, but with an irregular shape similar the example below:

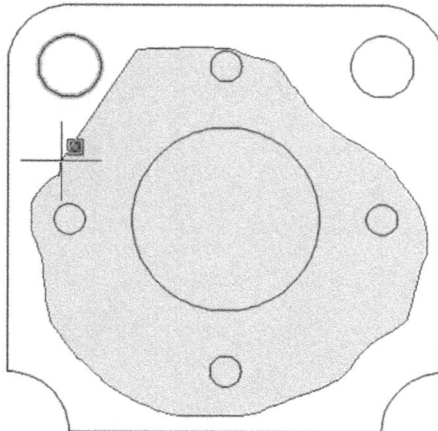

### 3.1.6   Fence Mode (F)

Fence mode will select multiple objects by crossing (touching) them. Lines in Fence mode can cross each other, contrary to WP and CP. Refer to the following example:

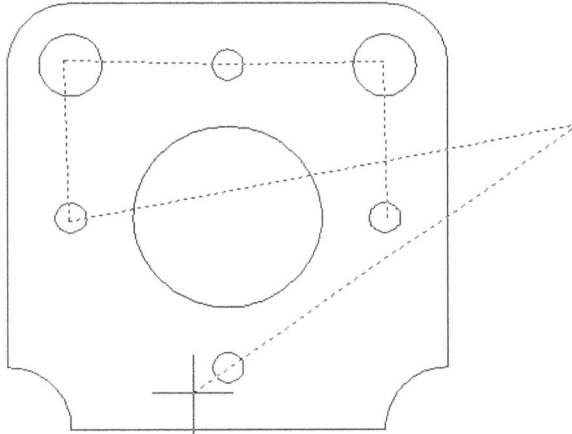

All the circles are selected because they are touched by the fence. The vertical line at the right will be selected as well. You can see that the fence lines are crossed which is 100% acceptable in AutoCAD.

### 3.1.7   Last (L), Previous (P), and All Modes

Furthermore, you can select objects using the following modes:

- **Last (L)**: To select the last object drawn
- **Previous (P)**: To select the last selected objects
- **All**: To select all objects in the drawing

**NOTE 1** *To deselect objects, hold the [Shift] key on the keyboard and click these objects. While you are in this mode, you can use window and crossing.*

**NOTE 2** *Start a selection window in one part of your drawing, and then pan and zoom to another part while keeping selection of the off-screen objects.*

### 3.1.8   Other Methods to Select Objects

There are two ways to use Modifying commands, either to issue the command and then select objects, or to select objects and then issue the command. This technique is called the **Noun/Verb** technique. Without issuing any command, you can:

- Select a single object by clicking it
- Click an empty space and go to the right to get to **Window** mode
- Click an empty space and go to the left to get to **Crossing** mode
- Click an empty space, then type W to get to **Window polygon** mode
- Click an empty space, then type C to get to **Crossing polygon** mode
- Click an empty space, then type F to get **Fence** mode

When you select objects, you can go to the **Home** tab, locate the **Modify** panel, and issue the desired command, or you can right-click to receive a shortcut menu that contains six modifying commands: Erase, Move, Copy Selection, Scale, and Rotate, as shown below:

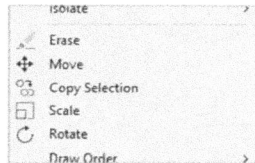

By default, Noun/Verb is working, but if you want to know how to control it, do the following steps:

- Go to the Application Menu and select the **Options** button
- Choose the **Selection** tab
- Under **Selection modes**, make sure that Noun/Verb selection is on

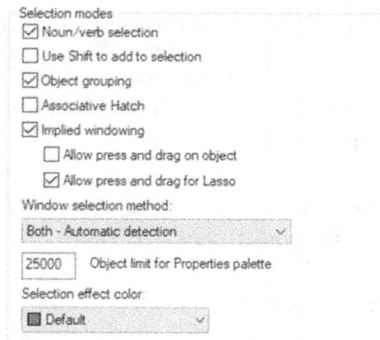

Use this part as well to do the following:

- Turn on **Allow press and drag on object** to allow you to create a window or crossing even if the cursor is not over a clean spot for picking
- Under **Window selection method**, make sure that **Both-Automatic detection** is selected so if you click then release the mouse, or if you click and drag, either method will be accepted

## 3.2   SELECTION CYCLING

While we are drafting using AutoCAD, we may unintentionally draft objects over each other, or we may click on an object using a point that is sharable by other objects. This issue used to be a problem in the past, but not anymore. AutoCAD provides us with **Selection Cycling**, which will notify us if our click touched more than one object, and will allow us the ability to choose the desired one. To activate Selection Cycling (by default it is active), using Status bar click Selection Cycling button as shown below (if this button is not displayed, simply go to the last button at the right of the Status bar; the one with the three horizontal dashes, click it to see all buttons, then select Selection Cycling):

After activating, and when you click on one object, you may see the following:

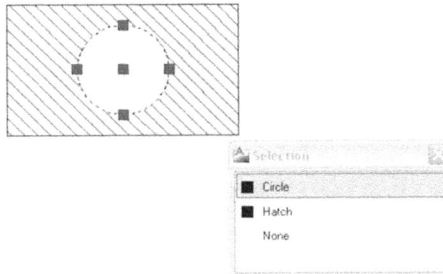

The small window will tell you there are two possible objects at the click position, Circle and Hatch, and that Circle is now selected. Using this window, if you move the mouse to Hatch, you will receive the following image:

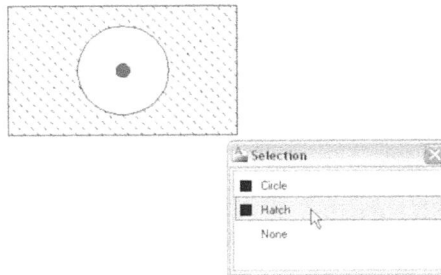

## 3.3   ERASE COMMAND

The **Erase** command will delete any object you select. To issue the command go to the **Home** tab, locate the **Modify** panel, and then select the **Erase** button:

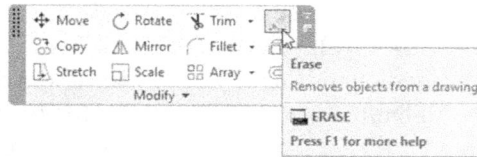

You will see the following prompt in the command window:

`Select Objects:`

If you use the one-by-one method, you will see that the cursor changes to the following shape:

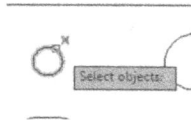

Select the desired objects. The Select Objects prompt is repetitive, so you always need to end it by pressing [Enter] or right-clicking.

You can erase using other methods:

- Click on the desired object(s), and then press [Del] key at the keyboard
- Click on the desired object(s), right-click, and select **Erase** from the shortcut menu:

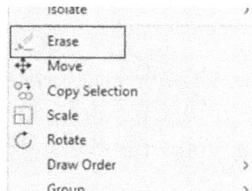

## PRACTICE 3-1   SELECTING OBJECTS AND ERASE COMMAND

**1.** Start AutoCAD 2025

**2.** Open **Practice 3-1.dwg**

**3.** Using Erase command, and the different types of selection methods we learned in the previous part, practice selecting the circles. Make sure to try all methods so you can get acquainted to them

**4.** Save and close the file

## 3.4   MOVE COMMAND

This command will move objects from one place to another in the drawing. To issue this command, go to the **Home** tab, locate the **Modify** panel, and then select the **Move** button:

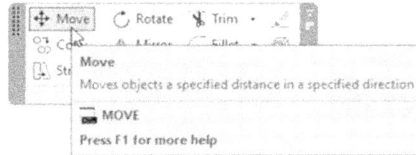

The following prompts will be shown to you:

```
Select objects:
Specify base point or [Displacement] <Displacement>:
Specify second point or <use first point as
displacement>:
```

The first step is to select objects, and then the user should select the Base Point. The base point is the point that will represent the objects; it will move a distance and angle, and the objects will follow. The main objective of the Base point is accuracy.

After you select the Base point the cursor shape will change to the following:

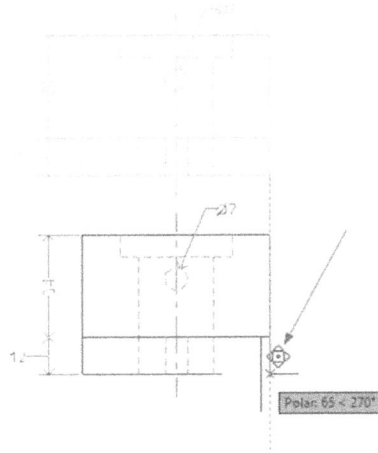

The last prompt will ask you to specify the second point or the destination of your movement.

### 3.4.1   Nudge Functionality

This function is very simple and will enable you to make an orthogonal move for selected objects. All you have to do is to select objects, hold the [Ctrl] key at the keyboard, and then use the four arrows at the keyboard; you will see objects move toward the desired direction.

## PRACTICE 3-2   MOVING OBJECTS

**1.** Start AutoCAD 2025

**2.** Open **Practice 3-2.dwg**

**3.** Using the Move command, move the three circles and the rectangle to the right places so you will get the following result (you will be needing OSNAP and OTRACK to move the rectangle accurately)

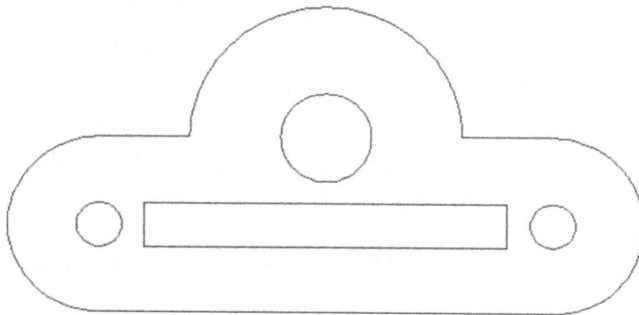

**4.** Save and close the file

## 3.5   COPY COMMAND

This command will copy objects. To issue this command, go to the **Home** tab, locate the **Modify** panel, then select the **Copy** button:

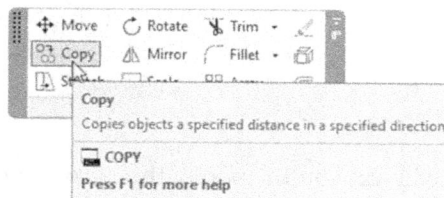

You will see the following prompts:

```
Select objects:
Current settings:   Copy mode = Multiple
```

```
Specify base point or [Displacement/mode]
<Displacement>:
Specify second point or [Array] <use first point as
displacement>:
Specify second point or [Array/Exit/Undo] <Exit>:
Specify second point or [Array/Exit/Undo] <Exit>:
```

After you select the desired objects, AutoCAD will report to you the current mode, which in our case is Multiple. This mode means the user will be able to create several copies in the same command. The other mode is Single copy. The first prompt will ask to specify the base point. AutoCAD then will ask you to specify the second point to complete a single copy process, then repeats the prompt to create another one, and so on. There are three options to use:

■ Undo, to undo the last copy
■ Exit, to end the command
■ Array, to create an array of the same object using distance and angle. Select the array option you will see the following prompts:

```
Enter number of items to array:
Specify second point or [Fit]:
```

Input the number of items in the array (including the original object); then specify the distance between the objects. The user can use the **Fit** option to specify the total distance, and AutoCAD will equally divide the distance over the number of objects.

After you select the Base point, the cursor shape will change to the following:

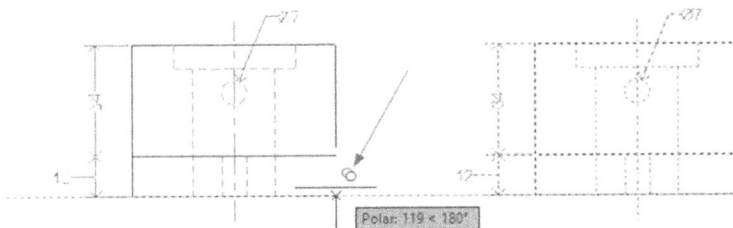

## PRACTICE 3-3   COPYING OBJECTS

**1.** Start AutoCAD 2025

**2.** Open **Practice 3-3.dwg**

**3.** Copy the door using multiple copying, and copy the toilet using the Array option (use the Midpoint OSNAP for the toilet), to achieve the following:

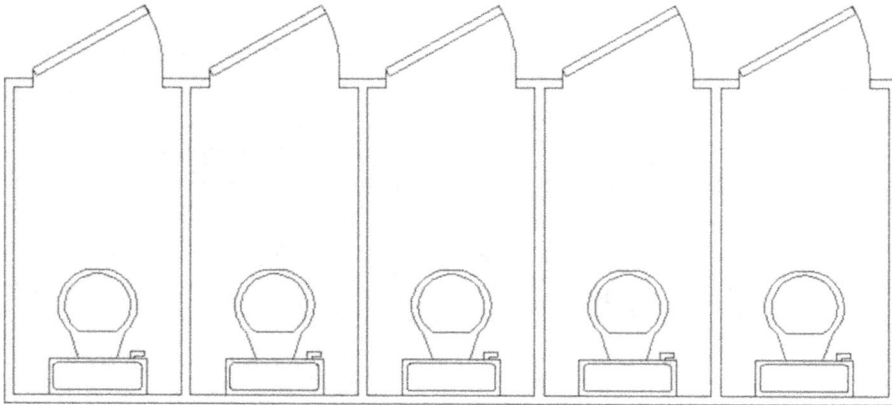

**4.** Save and close the file

## 3.6   ROTATE COMMAND

This command will rotate objects around the base point, using the rotation angle or reference. To issue the command go to the **Home** tab, locate the **Modify** panel, and select the **Rotate** button:

You will see the following prompts:

```
Current positive angle in UCS:
ANGDIR=counterclockwise  ANGBASE=0
Select objects:
Select objects:
Specify base point:
Specify rotation angle or [Copy/Reference] <0>:
```

The first message is to tell you about the current angle direction and Angle Base value. The base point here is the rotation point; all the selected objects will rotate around it.

Use the **Copy** option to make a copy of the selected objects, and then rotate them.

After you select the Base point, the cursor shape will change to the following:

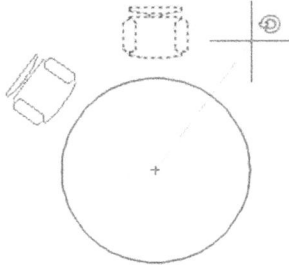

### 3.6.1 Reference Option

This option is very helpful if you do not know the rotation angle; instead, you can specify two points to indicate the current angle, and two points to input the new angle. You will see the following prompts:

```
Specify the reference angle <0>:
Specify second point:
Specify the new angle or [Points] <0>:
```

You can input the angle by typing. If you want to input angles using two points, the first point you will select will be for both angles, the current and the new. Refer to the following illustration:

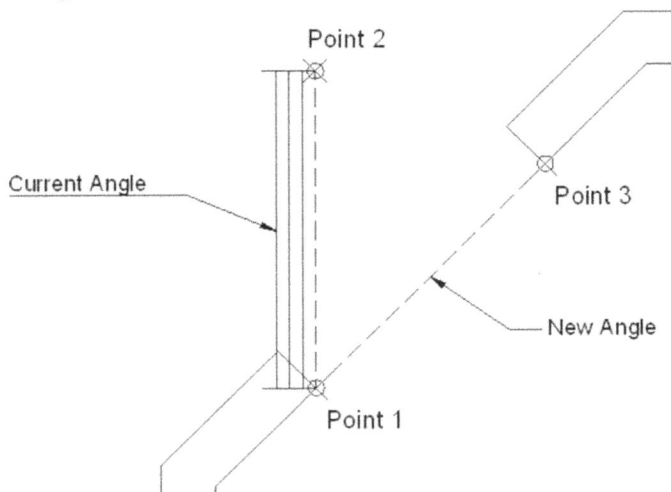

## PRACTICE 3-4    ROTATING OBJECTS

**1.** Start AutoCAD 2025

**2.** Open **Practice 3-4.dwg**

**3.** Rotate the lower window using Angle

**4.** Rotate the upper window using Reference

**5.** Rotate the chair with Copy mode to receive the following result:

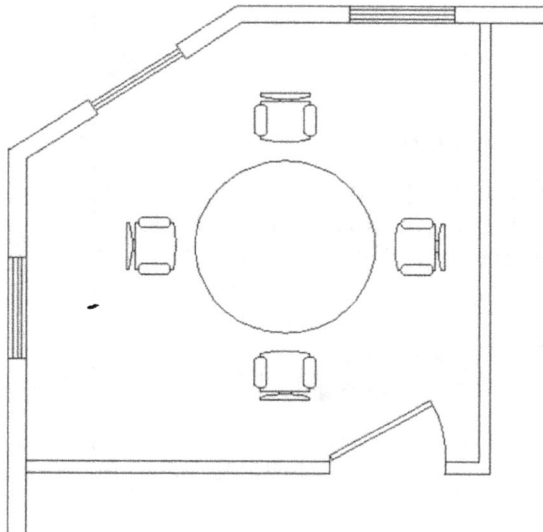

**6.** Save and close the file

## 3.7    SCALE COMMAND

This command will create bigger or smaller objects using the scale factor or the reference. To issue this command, go to the **Home** tab, locate the **Modify** panel, then select the **Scale** button:

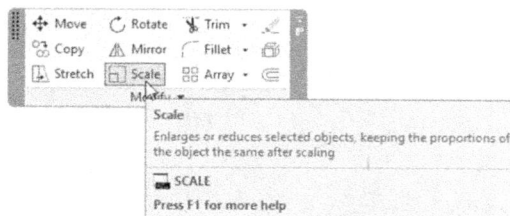

You will see the following prompts:

```
Select objects:
Specify base point:
Specify scale factor or [Copy/Reference] <1.0000>:
```

The base point here is the scaling point; all the selected objects will be bigger or smaller relative to it. Use the **Copy** option to make a copy of the selected objects, and then scale them.

After you select the Base point, the cursor shape will change to the following:

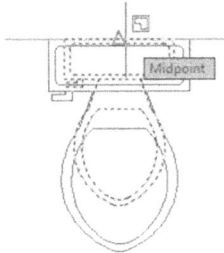

### 3.7.1 Reference Option

This option is very useful if you do not know the scaling factor as a number. Instead, you can specify two points to indicate the current length, and two points to input the new length. You will see the following prompts:

```
Specify reference length <0'-1">:
Specify second point:
Specify new length or [Points] <0'-1">:
```

You can input length by typing. If you want to input lengths using two points, the first point you will choose will be for both lengths, the current and the new. Refer to the following illustration:

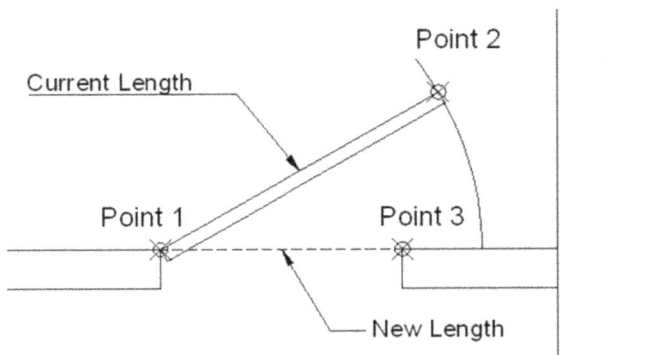

## PRACTICE 3-5   SCALING OBJECTS

**1.** Start AutoCAD 2025

**2.** Open **Practice 3-5.dwg**

**3.** Scale the toilet by scale factor = 0.9 using the midpoint of the wall

**4.** Scale the sink by scale factor = 1.2 using the quadrant of the sink

**5.** Scale the door using the Reference option to fit in the door opening

**6.** You will receive the following image:

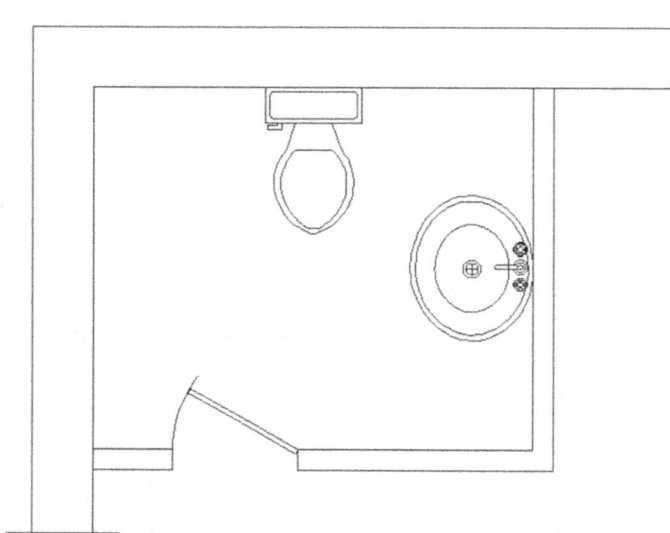

**7.** Save and close the file

## 3.8   MIRROR COMMAND

This command will create a mirror image of the selected objects using a mirror line. To issue this command, go to the **Home** tab, locate the **Modify** panel, then select the **Mirror** button:

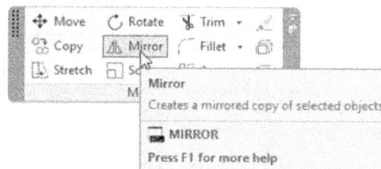

You will see the following prompt:

```
Select objects:
Specify first point of mirror line:
Specify second point of mirror line:
Erase source objects? [Yes/No] <N>:
```

After selecting objects, specify the mirror line by specifying two points (you do not need to draw a line to be able to specify these two points). The last prompt will ask you to keep or delete the original objects.

Text can be part of the selection set to be mirrored. You can tell AutoCAD what to do with it (copying or mirroring) by using the system variable MIRRTEXT. To issue this command, type it in the command window, and you will see the following:

```
Enter new value for MIRRTEXT <0>:
```

You can input either 1 (which means mirror the text just like the other objects) or 0 (which means copy the text).

See the following example:

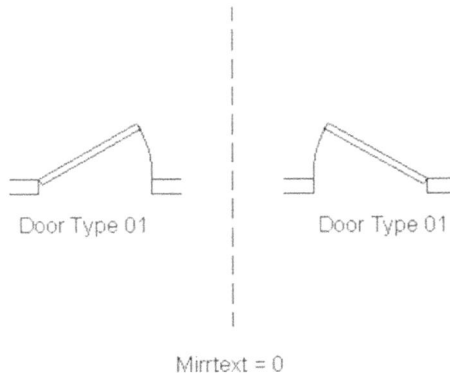

Mirrtext = 0

## PRACTICE 3-6   MIRRORING OBJECTS

1. Start AutoCAD 2025

2. Open **Practice 3-6.dwg**

3. Mirror the entrance door to open inside rather than outside, keeping the text without mirroring

4. Mirror the furniture of the room at the right with the two windows in the room at the left

**5.** You should receive the following result:

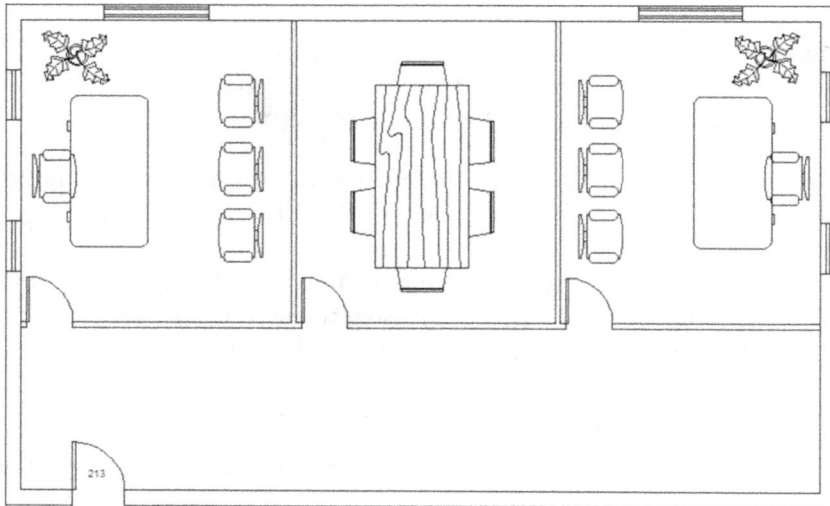

**6.** Save and close

## 3.9   STRETCH COMMAND

This command will change the length of selected objects by stretching using distance and angle. To issue this command, go to the **Home** tab, locate the **Modify** panel, and select the **Stretch** button:

You will see the following prompts:

```
Select objects to stretch by crossing-window or
crossing-polygon...
Select objects:
Specify base point or [Displacement] <Displacement>:
Specify second point or <use first point as
displacement>:
```

Stretch command is different compared to other modifying commands because it asks you to select the objects desired using C or CP modes. Why? Because Stretch will utilize both features of C and CP, which are containing and crossing. All objects contained fully inside the C or CP will be moving, whereas objects crossed will be stretching either by increasing or decreasing the length. You should then specify the base point (the same principle of the Move and Copy commands), and finally specify the second point. See the following example:

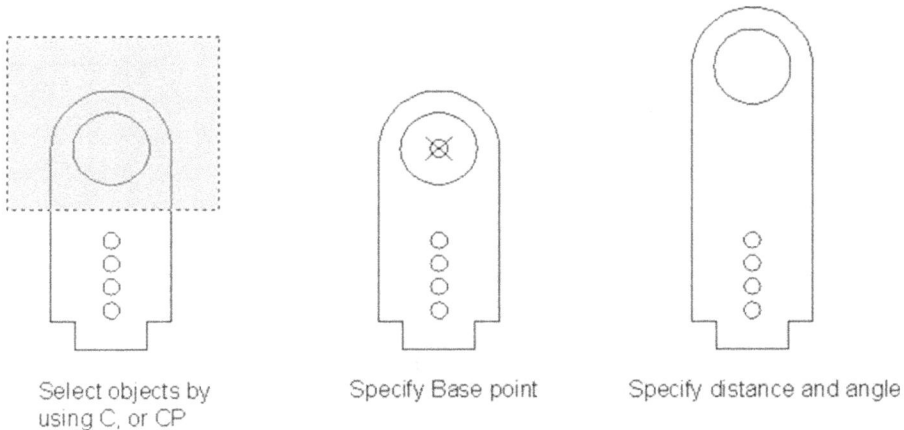

Select objects by
using C, or CP

Specify Base point

Specify distance and angle

## PRACTICE 3-7   STRETCHING OBJECTS

**1.** Start AutoCAD 2025

**2.** Open **Practice 3-7.dwg**

**3.** The vertical distance of the three rooms is not correct; it should be 1'-0" more. Use the Stretch command to do that, keeping the vertical distance of the entrance as is

**4.** The door of the room at the right is positioned incorrectly; stretch it to the right for 2'-5"

**5.** You should have the following image:

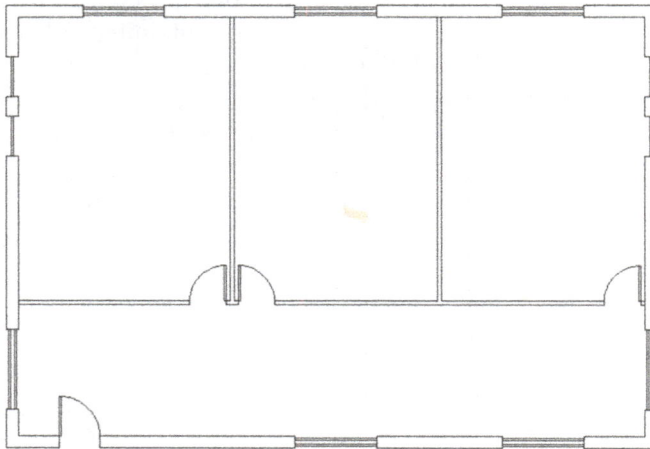

**6.** Save and close the file

## PRACTICE 3-8    STRETCHING OBJECTS

**1.** Start AutoCAD 2025

**2.** Open **Practice 3-8.dwg**

**3.** Stretch the upper part to look like the lower part by using a distance = 1.00 and crossing polygon

**4.** You should have the following image:

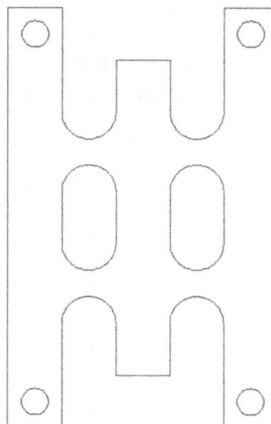

**5.** Save and close the file

## 3.10    LENGTHENING OBJECTS

With this command, you can add length or subtract length from objects using different methods. To issue this command, go to the **Home** tab, locate the **Modify** panel, and select the **Lengthen** button:

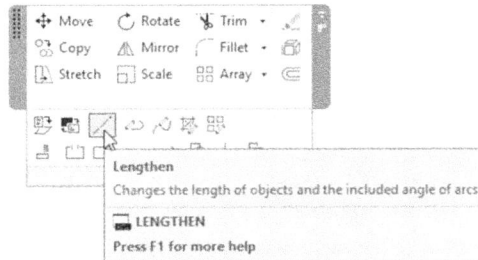

The user will see the following prompt:

```
Select an object or [DElta/Percent/Total/DYnamic]:
```

If you click an object, it will report the current length using the current length units. There are multiple methods to lengthen (or shorten) objects in AutoCAD.

The first method is called **Delta**, which will allow the user to add to or remove from the current length. Positive value means adding, and negative value means subtracting.

This is what you will see if you choose the Delta option:

```
Enter delta length or [Angle] <0.0000>:
```

The second method is **Percentage**, by which the user will be able to add to or remove from the length by specifying a percentage of the current length. To add length, input a value greater than 100; to remove length, input a value less than 100.

This is what you will see if you choose the Percentage option:

```
Enter percentage length <100.0000>:
```

The third method is **Total**, by which the user will be able to input a new total length of the object. If the new value is greater than the current value, length will be added to the object, otherwise, length will be removed.

This is what you will see if you choose the Total option:

```
Specify total length or [Angle] <1.0000)>:
```

The fourth method is **Dynamic**, which will increase/decrease the length dynamically using the mouse. You will see the following prompt:

```
Specify new end point:
```

**NOTE**    *You can lengthen/shorten a single object using a single method per command.*

## 3.11   JOINING OBJECTS

This command is a great command, as it will help you join lines to lines, arcs to arcs, and polylines to polylines. The Join command will help us do all of the above in simple and easy way.

To issue this command, go to the **Home** tab, locate the **Modify** panel, and select the **Join** button:

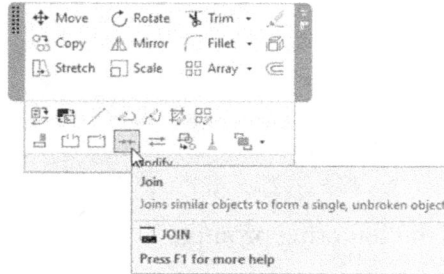

AutoCAD will show the following prompts:

```
Select source object or multiple objects to join at once:
Select lines to join to source:
Select lines to join to source:
1 line joined to source
```

The above prompts are for joining lines, and they may differ for arcs and polylines. There are some conditions for the joining to succeed:

- You can join lines, arcs, and polylines to form a single polyline, but the ends of the objects should be connected to each other
- If the lines, arcs, and polylines are not connected, each connected group will be considered as a single polyline
- If you want to join lines to form a line, they should be always collinear
- If you want to join arcs to form a single arc, they should have the same center point
- While joining arcs, there a special prompt asking if you are interested in creating a circle (this is the only command that will create a circle from an arc)

## PRACTICE 3-9   LENGTHENING AND JOINING OBJECTS

**1.** Start AutoCAD 2025

**2.** Open **Practice 3-9.dwg**

**3.** Using the Join command, join the two arcs at the top and the two collinear lines at its right

**4.** Using the Join command, convert the two arcs at the top and at the bottom to be a full circle

**5.** Using the Join command, join all objects except the four circles to form a single polyline

**6.** You should receive the following shape:

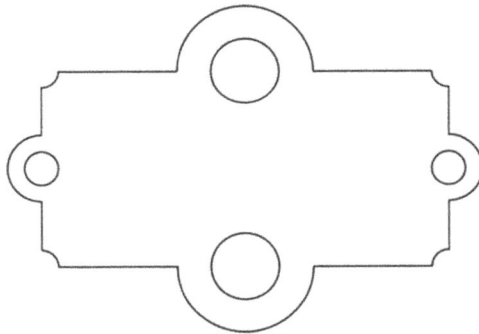

**7.** Pan to the right, until you see horizontal red lines (if you need zoom out a little bit)

**8.** Start Lengthen command, click the bottom horizontal line, what is the length of that line? _____ (7.5")

**9.** Using Delta option, input 2.5" and click the bottom line from the right end

**10.** Start Lengthen command again, select Percent, and input 200%, and click the second line from the bottom, from its right end

**11.** Start Lengthen command again, select Total, and input 10", and click the third line from the bottom, from its right end

**12.** Start Lengthen command again, select Dynamic, and drag the fourth line from the bottom, from its right end to reach the vertical line

**13.** Start Lengthen command again, select Delta, and input (-2.5"), and click the second line from the top, from its right end

**14.** Start Lengthen command again, select Total, and input 10", and click the top line, from its right end

**15.** Save and close the file

## 3.12   USING GRIPS TO EDIT OBJECTS

Grips are blue squares and rectangles appearing at certain places on each object. They will enable you to modify these objects using five modifying commands, with the selected grip as a base point. It is a clever tool for performing modifying tasks faster without compromising the accuracy of the conventional modifying commands discussed above. Depending on the type of object, grips will appear at different places, such as in the following illustration:

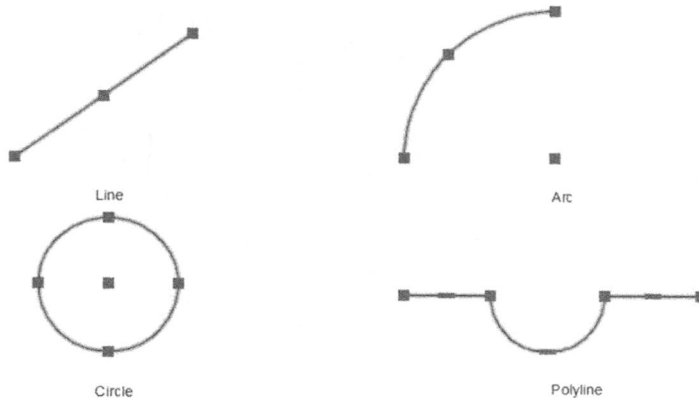

Line          Arc

Circle          Polyline

By default, the color of the grip is blue (cold); if you click it, it will turn to red (hot). Depending on the object and the location of the grip, if you right-click you will see a group of modifying commands that share one thing: the base point. For example, if you click the grip at the end of a line, then right-click, you will see the following menu:

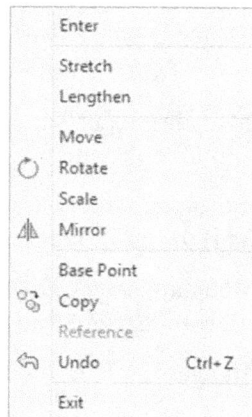

Enter

Stretch
Lengthen

Move
Rotate
Scale
Mirror

Base Point
Copy
Reference
Undo          Ctrl+Z

Exit

As you can see, there are six commands:

- Stretch
- Lengthen

- Move
- Rotate
- Scale
- Mirror

These commands (except Lengthen) share a base point (that is, if you consider the first point of the mirror line equivalent to a base point). Copy here is a mode and not a command, which means it will work with all other commands.

Holding [Shift] while selecting the grip will enable you to select more than one base point.

Holding [Ctrl] while specifying a second point (Move and Stretch), rotation angle (Rotate), specifying scale factor (Scale), and second point of mirror line (Mirror) will allow AutoCAD to remember the last input and repeat it graphically.

The base point option will enable you to select another base point other than the grip selected.

For example, select the following shape (two polylines):

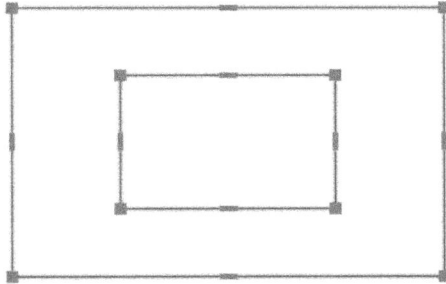

Click the upper right corner to make it hot, right-click to access the menu, and select the Rotate command:

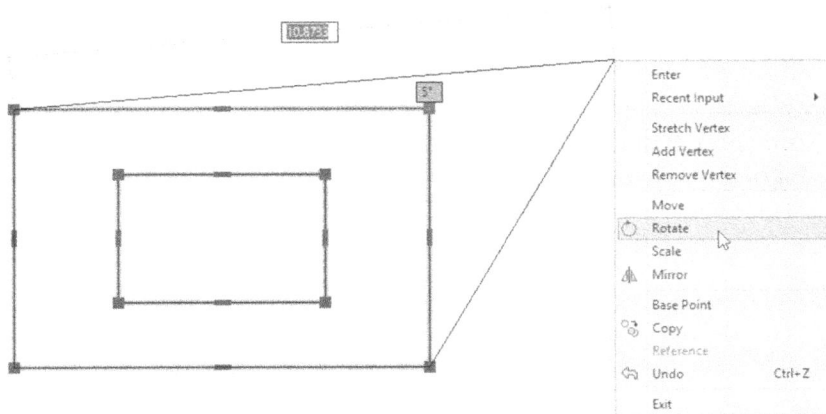

Now you can rotate the two shapes around the grip (this will be considered your base point), right-click again, and select the Base point option to select another base point, which will be the center of the two rectangles (using OSNAP and OTRACK):

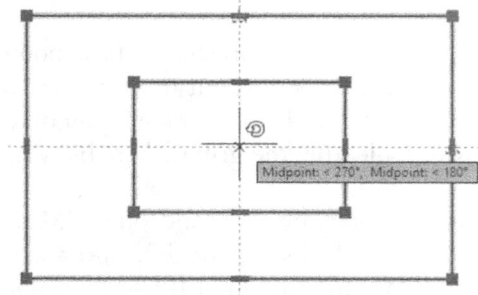

Right-click again and select the Copy option, to copy while rotating. Using Polar tracking specify an angle of 90, then press [Esc]. You will receive the following result:

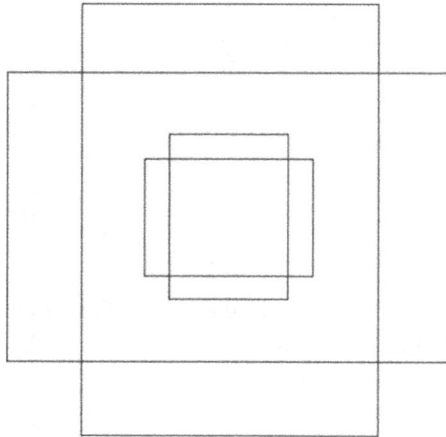

Specific objects like polylines will show more than the blue squares; they will show rectangles at the midpoint of each line and arc segment. The rectangles at the midpoint of the polyline have multiple functions. Depending on whether it is a line segment or an arc segment, AutoCAD will show a different menu. Go to the grip and stay for a second (do not click), and you will see the following:

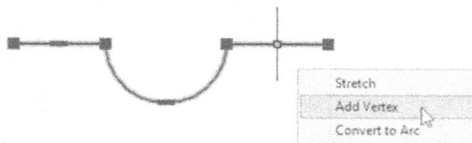

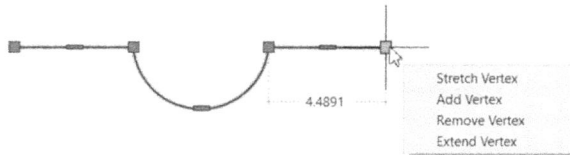

For the middle grip, you can Stretch, Adda Vertex, or Convert to Arc (if it was Line) and Convert to Line (if it was an Arc)

For the grip at the end, you can Stretch Vertex, Add Vertex, Remove Vertex, or Extend Vertex (keep the last segment and add a new segment)

When you are done with grips, click [Esc] once or twice depending on the situation you are in to end the grips mode.

## 3.13   GRIPS AND DYNAMIC INPUT

If you stay on one of the grips (without clicking), Dynamic Input along with the grips will give you information about the selected objects based on their type. (make sure that Dynamic Input in the Status bar is turned on) Check the following examples:

Using the line and one of the endpoints, you will see the length and the angle with the east along with two commands, Stretch and Lengthen:

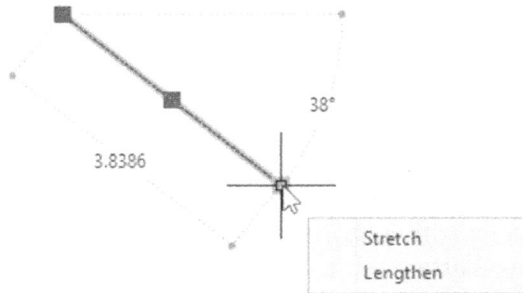

The two connected lines will show the two lengths and angles:

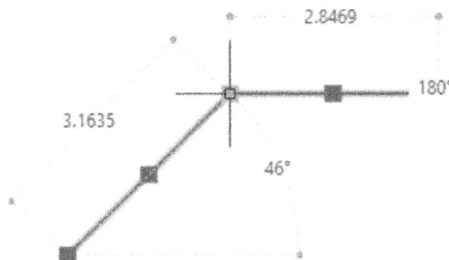

The midpoint of an arc will show the radius and the included angle along with two commands, Stretch and Radius:

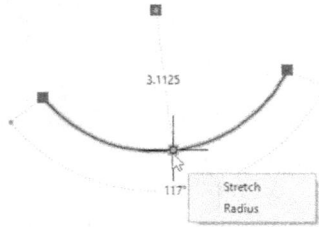

However, the endpoints of an arc will show the radius and the angle with the east, along with the Stretch and Lengthen commands:

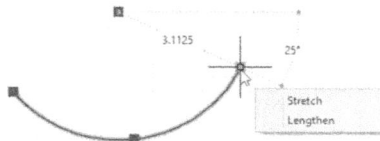

As for a circle, you will see the radius only:

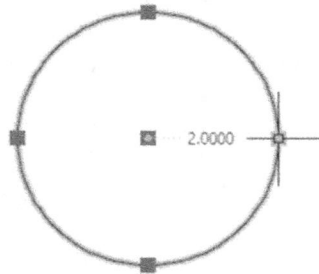

Shared endpoints between two segments of a polyline will show the length of the shared lines along with a shortcut menu which includes: Stretch Vertex, Add Vertex, and Remove Vertex.

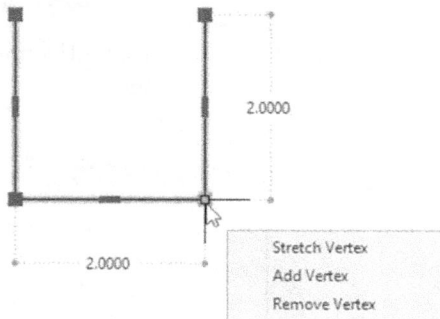

## 3.14   GRIPS AND PERPENDICULAR AND TANGENT OSNAPS

Using grips the user can specify perpendicular and tangent OSNAPs, bearing in mind that those two settings are turned on in running OSNAP. Refer to the following two examples:

**Tangent example:**

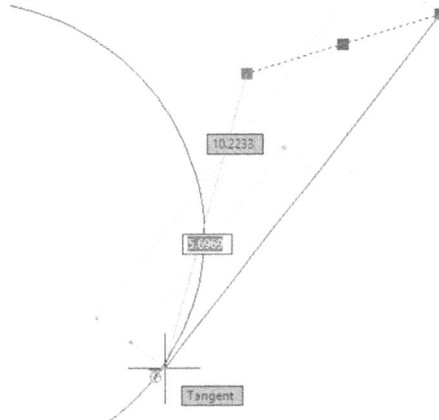

10.2233

5.6966

Tangent

**Perpendicular example:**

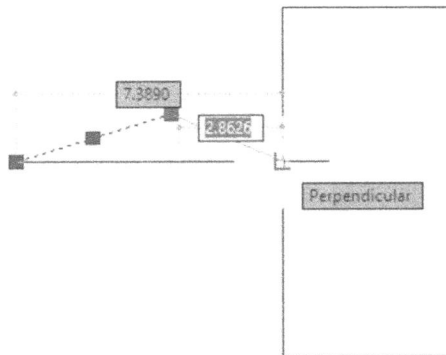

7.3890

2.8626

Perpendicular

## PRACTICE 3-10   USING GRIPS TO EDIT OBJECTS

**1.** Start AutoCAD 2025

**2.** Open **Practice 3-10.dwg** (you should solve this practice using grips techniques only)

**3.** Select the large circle, make the center hot, and scale it, copying by 1.2 as the scale factor

**4.** Select the circle at the right of the drawing and move it from its center to the center of the existing two circles

**5.** Mirror the three lines at the right to other side of the part (you should use another base point along with Copy mode)

**6.** You should by now have the following shape:

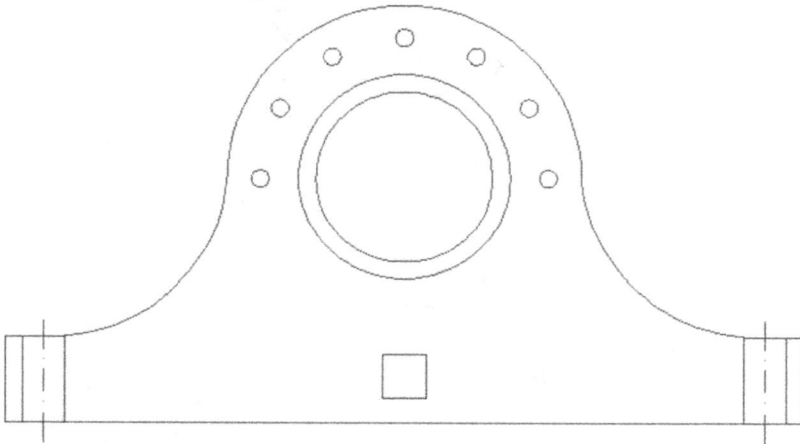

**7.** Select the polyline below the two big circles. Convert the top horizontal line to an arc by moving up distance = 0.25

**8.** Using the rectangle grip at the middle of the lower horizontal line of the polyline, stretch it downward by distance = 0.1

**9.** Using the grips and the dynamic input, what is the horizontal distance and the vertical distance of the polyline? _____, _____ (0.5,0.6)

**10.** Press [Esc] to clear the grips, reselect the polyline, then select one of the grips (any one); right-click to select the Move command, then right-click again and select Copy mode, making sure that Polar tracking is on to help you get exact angles. Now move to the right, type 2 as a distance, then press [Enter] and hold the [Ctrl] key so AutoCAD will remember this distance. Now while you are still holding [Ctrl] key, make one copy to the right and one copy to the left.

**11.** Select the lines at the right using Crossing, as in the following illustration:

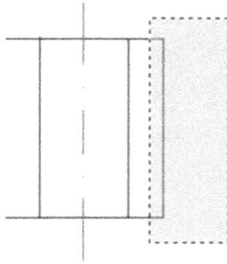

**12.** Hold the [Shift] key, select the lower right corner grip, and then select the upper right corner grip. Now release the [Shift] key, click any of the two grips, and go to the right by a distance equal to 0.25

**13.** Do the same steps to the left side of the drawing

**14.** Make sure that Tangent and Perpendicular are both on in the OSNAP dialog box

**15.** Click the right green line. Make the lower grip hot; stretch it to be perpendicular on the horizontal line. Do the same thing for the other line

**16.** Using the upper grip, make both lines tangent to the inner circle

**17.** You should have the following image:

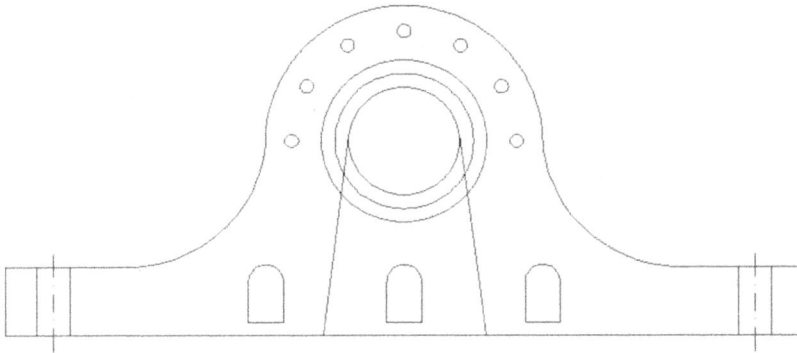

**18.** Save and close the file

# NOTES

## CHAPTER REVIEW

1. One of the following is not true about the Reference concept

    **a.** You can use it with Rotate and Scale

    **b.** In Rotate, you have to specify the current and the new angle by specifying four points

    **c.** In Scale, you have to specify two distances, the current and the new

    **d.** You cannot use it in Stretch command

2. Stretch command will ask for something that other modifying commands will not:

    **a.** True

    **b.** False

3. If you want to select the last selected set, type the letter _____ at the command window

4. In the Mirror command, you should draw a line to act as a mirror line

    **a.** True

    **b.** False

5. While using grips, holding _____ while selecting the grip will enable you to select more than one base point

6. _____ will help you control the outcome of text in Mirror command

7. One of the following commands is not among the commands for grips:

    **a.** Stretch

    **b.** Lengthen

    **c.** Join

    **d.** Rotate

**8.** While being at the Lengthen command, the option to double the current length without knowing it is:

   **a.** Delta

   **b.** Total

   **c.** Percent

   **d.** Dynamic

## CHAPTER REVIEW ANSWERS

**1.** b

**3.** P

**5.** [Shift]

**7.** c

## CHAPTER 4

# MODIFYING COMMANDS PART II

## In This Chapter

- How to offset objects
- How to fillet and chamfer objects
- How to trim and extend objects
- How to array objects using three different methods
- How to break objects

## 4.1   INTRODUCTION

These commands are modifying commands with special capabilities. They can build over the shapes you drew. Each one of them has a unique function: two of them can create objects (like Offset and Array), and others can change an existing object shape (like the Fillet, Chamfer, Trim, and Extend commands).

The following is a brief description of each command:

- **Offset command**: to create copies of an object parallel to the original
- **Fillet command**: to create a neat intersection between two objects either by extending/trimming lines or using arcs
- **Chamfer command**: to create a neat intersection but only for lines. The neat intersection is created by extending/trimming the two lines or by creating a new line showing the chamfered edge

- **Trim command**: this command will trim objects using other objects as cutting edges
- **Extend command**: this command will extend objects using other objects as boundary edges
- **Array command**: this command will create objects using three different methods: rectangular, circular, and using a path
- **Break command**: this command will break an object into two objects by specifying two points and deleting the portion between them

## 4.2    OFFSETTING OBJECTS

The Offset command will create copies of an object parallel to the original. The new object will possess the same properties of the original object. You can offset using offset distance, or by using a point the new object will pass through. You can start this command by going to the **Home** tab, locating the **Modify** panel, and selecting the **Offset** button:

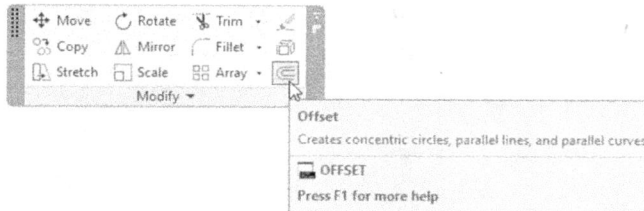

The user will see the following AutoCAD prompts:

```
Current settings: Erase source=No Layer=Source
OFFSETGAPTYPE=0
Specify offset distance or [Through/Erase/Layer]
<Through>:
```

### 4.2.1    Offsetting Using the Offset Distance Option

If you know the distance between the object and the new parallel copy, then input this value, select the original object, and finally click on the side you want the new object to go to. Going this route, you will see the following prompt:

```
Specify offset distance or [Through/Erase/Layer]
<Through>:
Select object to offset or [Exit/Undo] <Exit>:
Specify point on side to offset or [Exit/Multiple/
Undo] <Exit>:
```

This will enable you to create a single offset. To create more offsets using the same offset command, select another object and do the same steps again. Pressing [Enter] or right-clicking will end the command.

### 4.2.2 Offsetting Using the Through Option

If you do not know the offset distance, but rather you know a point in the drawing that the new parallel object will pass through, this option will help you accomplish your mission. You will see the following AutoCAD prompts:

```
Specify offset distance or [Through/Erase/Layer]
<Through>:
Select object to offset or [Exit/Undo] <Exit>:
Specify through point or [Exit/Multiple/Undo] <Exit>:
```

This will enable you to create a single offset. To create more offsets using the same offset command, select another object and do the same steps again. Pressing [Enter] or right-clicking will end the command.

Here is an example:

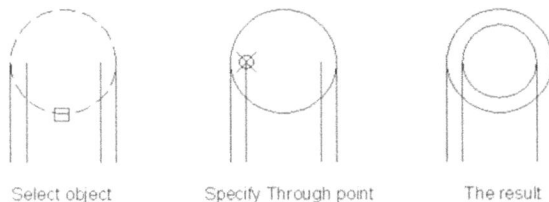

Select object          Specify Through point          The result

### 4.2.3 Using the Multiple Option

You can use the Multiple option to repeat the Offset Distance or Through option in the same command by repeatedly clicking on the side of offset, or by specifying new through point. The prompts for Multiple are:

```
Specify through point or [Exit/Multiple/Undo] <Exit>:
M Specify point on side to offset or [Exit/Undo]
<next object>:
```

While offsetting, note the following:

- If you make any mistakes, use the Undo option
- AutoCAD remembers the last offset distance used and will save it in the file
- When offsetting an arc or circle, the new arc and circle will share the same center point; hence the result will be a smaller or bigger arc or circle
- If you offset a closed polyline, the output will be smaller or bigger

- Offset has the automatic preview feature, which will show you the result before you click to accept it

## PRACTICE 4-1    OFFSETTING OBJECTS

**1.** Start AutoCAD 2025

**2.** Open **Practice 4-1.dwg**

**3.** Offset the outer polyline to the outside by distance = 0.3

**4.** Using the endpoints of the lines inside the two circles, offset the two circles to the inside using the Through option

**5.** You should have the following image:

**6.** Save and close the file

## PRACTICE 4-2    OFFSETTING OBJECTS

**1.** Start AutoCAD 2025

**2.** Open **Practice 4-2.dwg**

**3.** Offset the polyline which represents the outside edge of the outer wall by distance = 1'-0"

**4.** Offset the yellow line representing one of the stair steps ten times using distance = 1'-6", and using the Multiple option

**5.** Explode the inner polyline

**6.** Offset the vertical line at the right of exploded polyline to the left, using the Through option to pass through the inner right endpoint of the inclined line (as shown below)

**7.** Offset the newly created line to the right by distance = 6"

**8.** You will receive the following image:

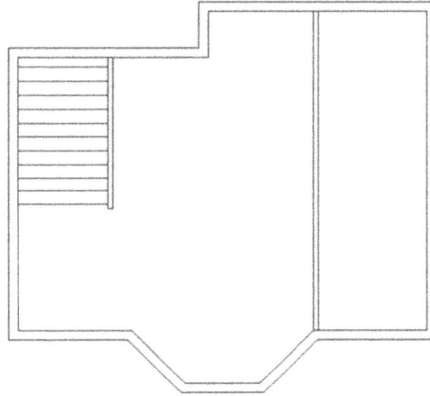

**9.** Save and close the file

## 4.3   FILLETING OBJECTS

The mission of the Fillet command is to create neat intersections. The user should set the value of the Radius as the first step. If it is 0 (zero) then you can use Fillet only between two lines, and Fillet will extend/trim the lines to the proposed intersection point. But if the value of the Radius is greater than 0 (zero), then Fillet can use lines and circles to fillet these objects with an arc.

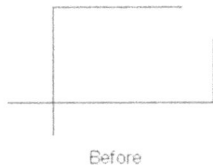

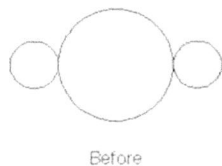

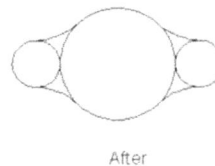

Before                        After

Before                        After

While working with a Radius greater than zero, the user can select between **Trim**, which will allow trimming to the original objects, or **No trim**, which will allow the original objects to stay as is. In both cases, the user will be able to see the arc when hovering over the second object, and can make sure that the value is correct!

Here is an example:

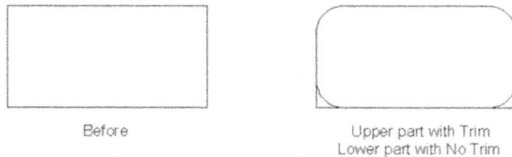

Before

Upper part with Trim
Lower part with No Trim

To start this command, go to the **Home** tab, locate the **Modify** panel, then select the **Fillet** button:

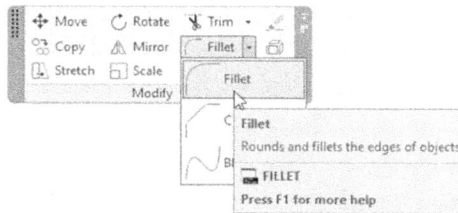

The user will see the following AutoCAD prompts:

```
Current settings: Mode = TRIM, Radius = 0.0000
Select first object or [Undo/Polyline/Radius/Trim/
Multiple]:
```

The user should always check the first line of this command, because it will report the current value of the **Radius**; accordingly, you will decide whether to keep it or change it. Type **r** or right-click and select the Radius option to set the new value of the Radius. You will see the following prompt:

```
Specify fillet radius <0.0000>:
```

Type **t** or right-click and select **Trim** to change the mode to **Trim** or **No trim**. You will see the following prompt:

```
Enter Trim mode option [Trim/No trim] <Trim>:
```

The Fillet command will allow only a single fillet per command; to make multiple fillets in the same command, simply change the value to **Multiple** mode. If you make an error, use the **Undo** option to undo the last action. To end the command press [Enter].

The user can do two significant things while using the Fillet command:

- Fillet will fillet two parallel lines regardless of the current Radius value
- By holding the [Shift] key, Fillet will fillet any two lines with radius = 0 regardless of the current value of the Radius

NOTE *If you start the Multiple option, you can use different radius values in the same command.*

## PRACTICE 4-3   FILLETING OBJECTS

**1.** Start AutoCAD 2025

**2.** Open **Practice 4-3.dwg**

**3.** Using the Fillet command, try to get the following final result:

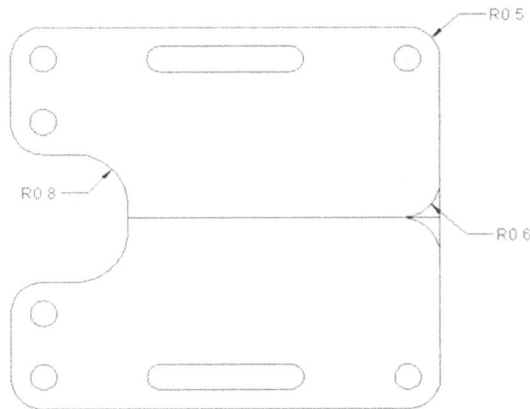

**4.** Save and close the file

## 4.4   CHAMFERING OBJECTS

The Chamfer command will create neat intersections as well. But here the user should set the value of Distance (or Distance and Angle) as the first step. If it is 0 (zero) then you can use Chamfer to extend/trim the lines to the proposed intersection point. If the value of the Distance, however, is greater than 0 (zero), then Chamfer will create a sloped edge between the two lines. While working with a distance greater than zero, the user can select between **Trim**, which will allow trimming of the original objects, and **No trim**, which will allow the original objects to stay as is. In both cases, the user will be able to see the chamfer line when hovering over the second object, and can make sure that the value is correct.

To create a sloped edge, use one of two available methods:

- Distance (two distances)
- Distance and Angle

### 4.4.1 Chamfering Using the Distance Option

To chamfer using the Distance option, you will see the two following prompts:

```
Specify first chamfer distance <0.0000>:
Specify second chamfer distance <0.0000>:
```

There will be two different cases, which are:

Distances are equal          Distances are not equal

### 4.4.2 Chamfering Using Distance and Angle

To chamfer using Distance and Angle, you will see the two following prompts:

```
Specify chamfer length on the first line <0.0000>:
Specify chamfer angle from the first line <0>:
```

Set the length (which will be cut from the first object selected) and an angle. See the following illustration:

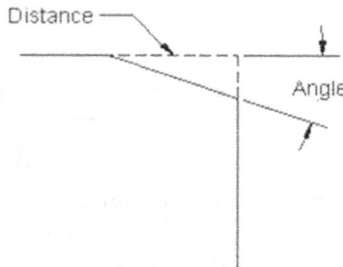

Distance

Angle

Distance and angle

To start this command, go to the **Home** tab, locate the **Modify** panel, then select the **Chamfer** button:

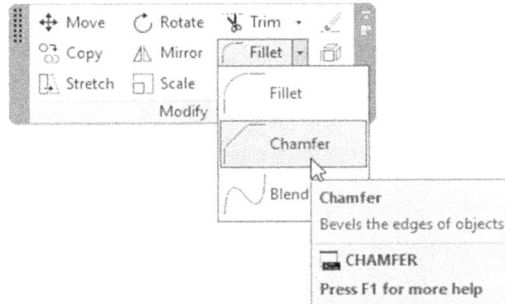

The user will see the following prompts:

```
(TRIM mode) Current chamfer Dist1 = 0.0000,
Dist2 = 0.0000
Select first line or [Undo/Polyline/Distance/Angle /
Trim/Method/Multiple]:
```

The user should always check the first line of this command, because it will report the current value of the method used (whether Distance or Distance and Angle), and the current values. Accordingly, you will decide whether to keep them or change them.

Other options like Multiple, Trim, and Undo are identical to what we learned in the Fillet command. The Method option selects the default method to be used in the chamfering process.

Another similarity to the Fillet command is that holding the [Shift] key will chamfer by extending/trimming the two lines regardless of the current Distance values.

**NOTE**  *Trim and Untrim in Chamfer will affect the Fillet command and vice versa..*

## PRACTICE 4-4   CHAMFERING OBJECTS

**1.** Start AutoCAD 2025

**2.** Open **Practice 4-4.dwg**

**3.** Use the Chamfer command to make the shape look similar to the following:

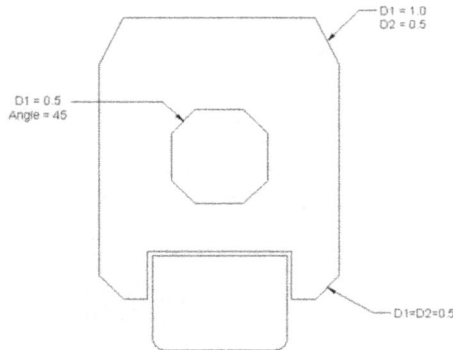

**4.** Save and close the file

## 4.5    TRIMMING OBJECTS

This command will remove part of an object based on cutting edges. The default is the **Quick** mode which considers all objects in the drawing are cutting edges, so, all you have to do is to click on the part of the object which you want to remove.

The following example will illustrate the process of trimming:

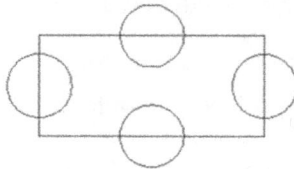

Select Cutting Edges

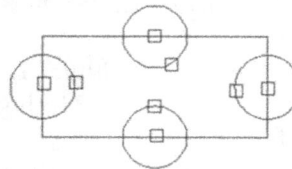

Select objects to trim

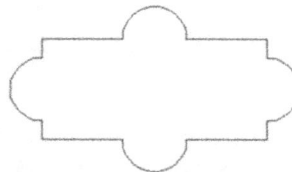

The result

To issue this command go to the **Home** tab, locate the **Modify** panel, and then select the **Trim** button:

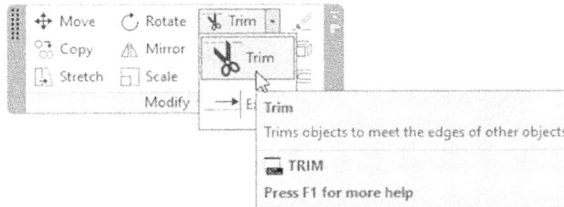

You will see the following prompt:

```
Current settings: Projection=UCS, Edge=None, Mode=Quick
Select object to trim or shift-select to extend or
[cuTting edges/Crossing/mOde/Project/eRase]:
```

The first line is a message from AutoCAD telling you the current settings. (Check that the current Mode = Quick). The second line is asking you to click on the part of the object which you want to remove or trim, using clicking, Fence, or you can choose Crossing option

While you are trimming, you may receive an orphan object as a result; to remove it, issue the eRase option (use the letter r).

The Mode options available are:

- If you want to go back to the Standard method, type O (for mode) and you will see the following prompt:

```
Enter a trim mode option [Quick/Standard] <Quick>: S
```

- The next time you will use Trim command, you will see the following prompt:

```
Current settings: Projection=UCS, Edge=None,
Mode=Standard
Select cutting edges ...
Select objects or [mOde] <select all>:
```

- This means you have to specify the Cutting Edges first, or press [Enter] to select all objects to be cutting edges
- The Standard mode will be the default until you change it to Quick again
- If you type T (for Cutting Edges) in both modes you will be allowed to specify you cutting edges as you wish

## PRACTICE 4-5   TRIMMING OBJECTS

**1.** Start AutoCAD 2025

**2.** Open **Practice 4-5.dwg**

**3.** Using the Trim command, try to make a shape that looks similar to the following:

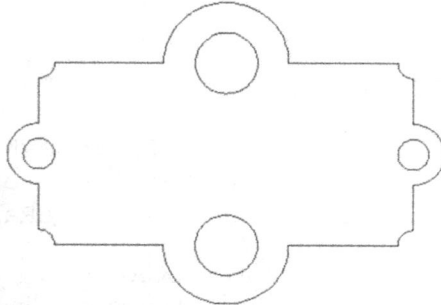

**4.** Save and close the file

## 4.6   EXTENDING OBJECTS

This command will extend an object to a boundary edge. The default is the **Quick** mode which considers all objects in the drawing are boundary edges, so, all you have to do is to click on the part of the object which you want to extend.
See the following illustration:

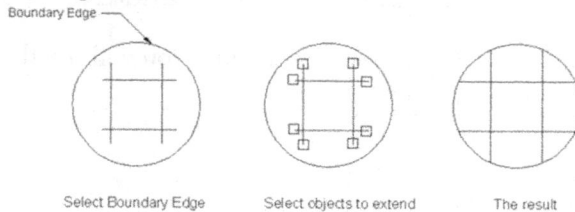

Select Boundary Edge      Select objects to extend      The result

To issue this command, go to the **Home** tab, locate the **Modify** panel, then select the **Extend** button:

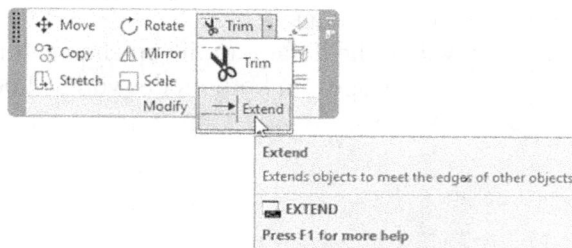

All options and prompts are already discussed in the Trim command

The last feature in both the Trim and Extend commands is the ability to use each command while you are using the other. See the following example:

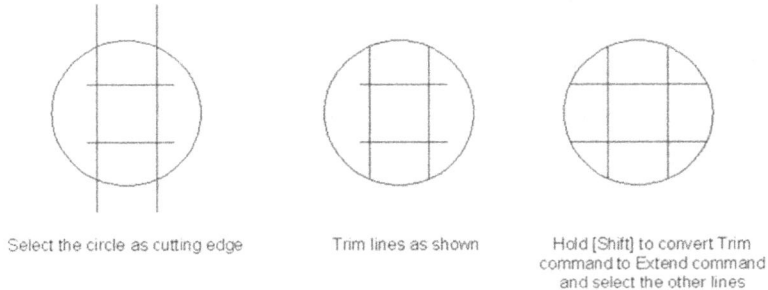

Select the circle as cutting edge        Trim lines as shown        Hold [Shift] to convert Trim
command to Extend command
and select the other lines

## PRACTICE 4-6    EXTENDING OBJECTS

1. Start AutoCAD 2025

2. Open **Practice 4-6.dwg**

3. Using the Extend command (and Trim if needed), correct the architectural plan to look similar to the following:

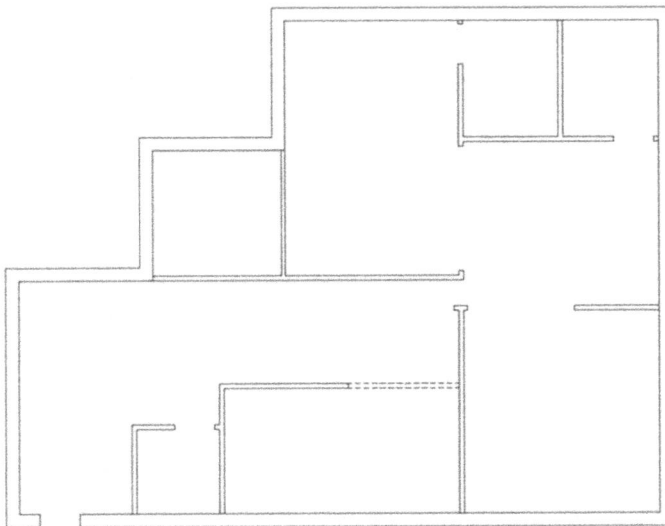

4. Save and close

## 4.7 ARRAYING OBJECTS – RECTANGULAR ARRAY

This command will create replicates in matrix fashion using rows and columns. The resultant shape will be one object that can be edited.

### 4.7.1 The First Step

To issue this command, go to the **Home** tab, select the **Modify** panel, and select the **Rectangular Array** button:

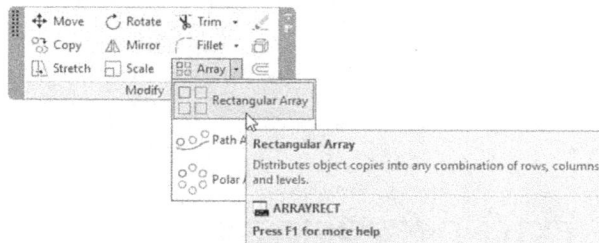

The following prompt will be shown:

```
Select objects: 1 found
```

The first prompt will ask you to select the desired objects; once done, press [Enter] and AutoCAD will immediately add a three-row, four-column grid showing grips as in the following:

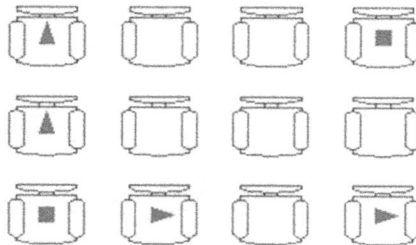

Subsequently, you will see the following context tab titled Array Creation:

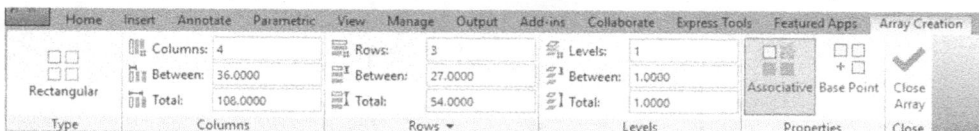

The following prompts will appear at the command window:

```
Type = Rectangular Associative = Yes
Select grip to edit array or [ASsociative/Base point/
COUnt/Spacing/COLumns/Rows/Levels/eXit]<eXit>:
```

AutoCAD is prompting you to select a suitable grip to edit the array. The user is invited to use two things to reach the desired result:

- Use the Array Creation context tab to do the above mentioned things
- Use grips to set the number of rows, columns, distances between columns, distances between rows, and direction of arraying (downward or upward, right or left)

### 4.7.2 Using the Array Creation Context Tab

Using the Array Creation context tab is more convenient if you know the numbers and distances. So, we will use this method to start:

- Use the Column panel to input two of three pieces of information: Columns (number of columns), Between (Distance between Columns), and Total (Total distance the columns will occupy). The user should be consistent, taking the same reference point to measure distances (from left to left or from center to center). Also, the user should consider the direction of the arraying, and that positive distances mean upward, and negative distances mean downward
- Use the Rows panel to input two of three pieces of information: Rows (number of rows), Between (Distance between Rows), and Total (Total distance the rows will occupy). The user should be consistent, taking the same reference point to measure distances (from top to top or from center to center). Also, the user should consider the direction of the arraying, and that positive distances mean going to the right, and negative distances mean going to the left
- Ignore Levels for 2D, because this is only for 3D
- Associative means all objects resultant from the array will be considered a single object holding all the information used to build it up
- By default, the first object will be considered the base point for the array, but AutoCAD allows you to select a different one
- Click the Close Array button to end the command

### 4.7.3 Editing a Rectangular Array Using Grips

After you insert a rectangular array, click on any object in order to edit it. Grips will appear on the objects as in the following:

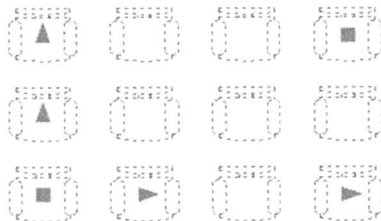

Each one of these has a function:

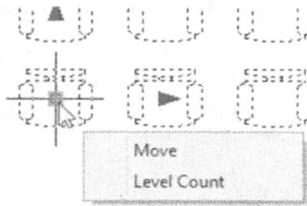

This grip is for Moving the whole array object or controlling the Level count (3D only). Use [Ctrl] to browse between these options.

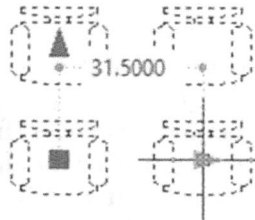

This grip is for changing the column spacing alone:

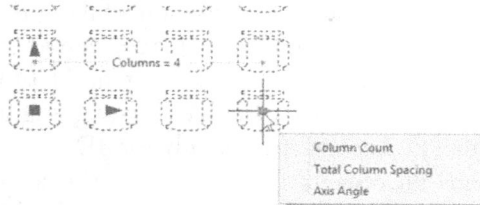

This grip is for changing the Column count, Total Column Spacing, or Axis Angle (to specify another angle other than horizontal and vertical). Use [Ctrl] to browse between these options.

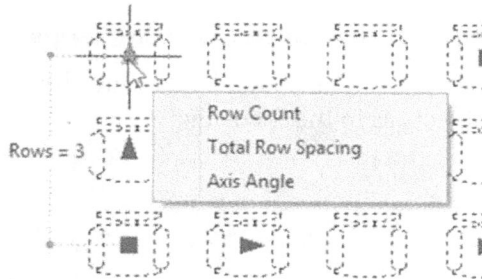

This grip is for changing the Row Count, Row Spacing, or Axis Angle. Use [Ctrl] to browse between these options.

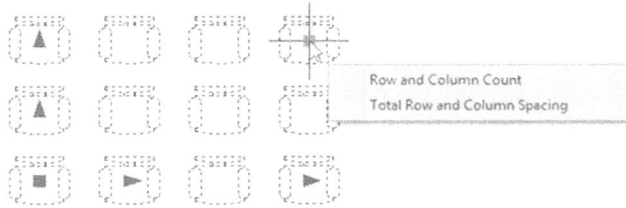

This grip is for changing the Row and Column Count or the Total Row and Column Spacing. Use [Ctrl] to browse between these options.

### 4.7.4 Editing a Rectangular Array Using the Context Tab

When you click a rectangular array, a context tab named Array will appear, as in the following:

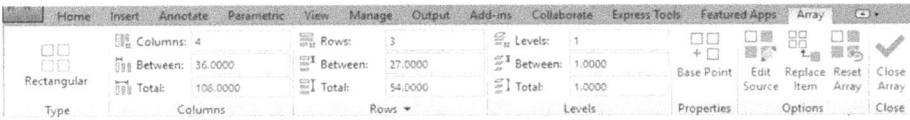

This tab is almost identical to the Array Creation context tab, except for the Options panel, which includes three buttons:

- The Edit Source command will make changes to one of the objects arrayed, which will reflect to all other arrayed objects
- The Replace Item command will replace one or more of the arrayed shapes with another shape
- The Reset Array command will reverse the effects of the Replace Items command

### 4.7.5 Editing a Rectangular Array Using Quick Properties

When you click a rectangular array, Quick Properties will be shown as in the following:

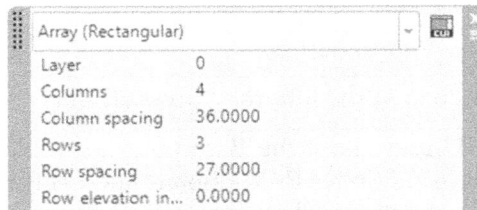

You can change the Columns number, Columns spacing, Rows number, and Rows spacing.

# PRACTICE 4-7   ARRAYING OBJECTS USING RECTANGULAR ARRAY

**1.** Start AutoCAD 2025

**2.** Open **Practice 4-7.dwg**

**3.** Using the chair, try to create a rectangular array as in the following:

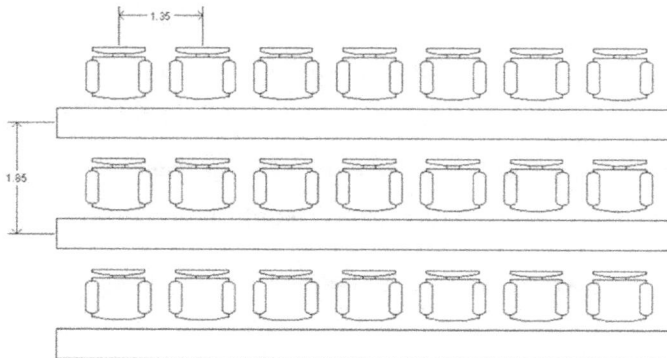

**4.** Exit the Array command

**5.** Click the rectangular array, using grips to make the number of columns = 5 and distance between columns = 2

**6.** Click the rectangular array to select it, click the Edit Source button, and select the upper right chair (you can select any other one if you want); click OK for the message that appears, and you will notice that only the chair you selected is highlighted; zoom to it and add two diagonal lines

**7.** A small panel titled Edit Array is displayed at the right; click it and select the Save Changes button, and you can see that all the other chairs are holding the new changes

**8.** Zoom out until another chair appears at the left

**9.** Select the rectangular array

**10.** Select the Replace Item button

**11.** Select the new chair at the left, then press [Enter]

**12.** When AutoCAD asks about the Base point for replacement objects, set the new base point using OSNAP (Midpoint) and OTRACK to locate the center of the rectangle of the new chair

**13.** Click the all chairs at the row at the bottom, press [Enter] twice, then close the array command

**14.** The final look should be as in the following:

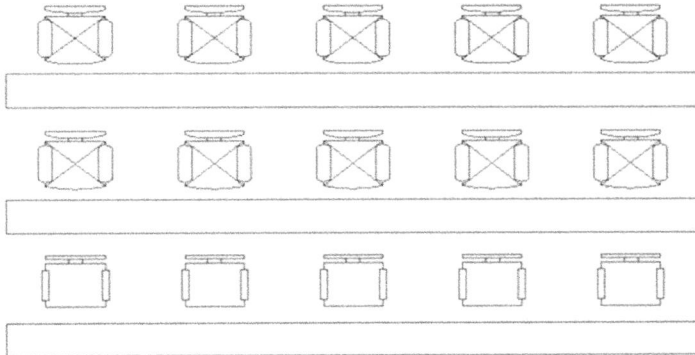

**15.** Save and close the file

## 4.8   ARRAYING OBJECTS – PATH ARRAY

This command will create an array using an object like a polyline, spline, arc, etc. To issue this command, go to the **Home** tab, locate the **Modify** panel, and select the **Path Array** button:

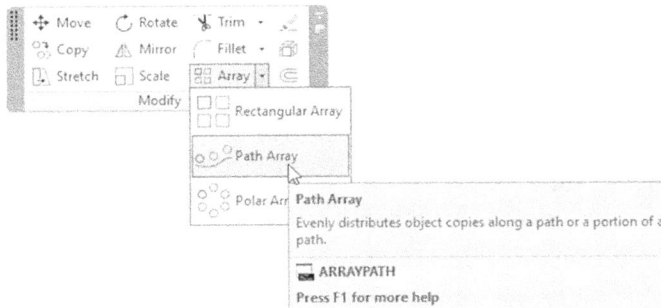

The following prompts will be shown:

```
Select objects:
Type = Path Associative = Yes
Select path curve:
```

The first prompt will ask you to select the desired objects; once done, press [Enter]. The second prompt will show a message that the type of array is path and the associativity is on. AutoCAD then asks the user to select the path curve; once selected, AutoCAD will show ten objects arrayed using the path. The following prompt will appear:

```
Select grip to edit array or [ASsociative/Method/
Base point/Tangent direction/Items/Rows/Levels/Align
items/Z direction/eXit]<eXit>:
```

Meanwhile, you will see the Array Creation context tab, which looks as the following:

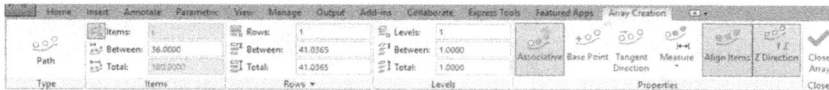

When you think about arraying an object using a path, three things should come to your mind:

- Base point
- Aligning objects with the path
- Measure or Divide

    Let's see how AutoCAD will treat the following case:

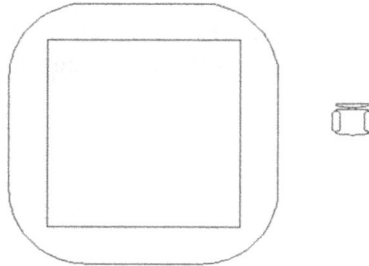

We will select the chair as the object to be arrayed and the outer polyline to be the path. See how AutoCAD will handle this:

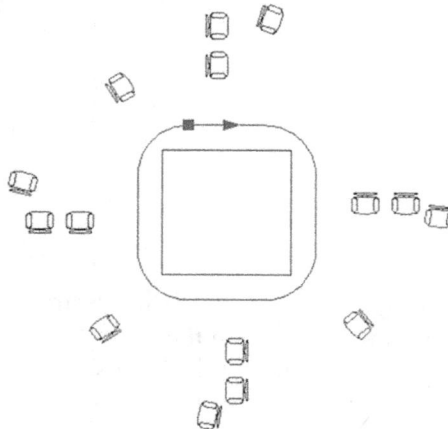

Now we will go to the context tab and select the Base point option to specify a new base point (which will be the center of the chair):

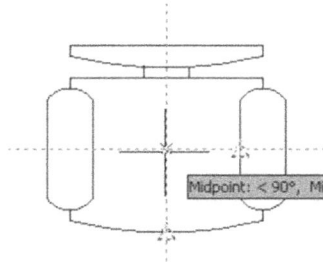

The following is the result after specifying the new base point:

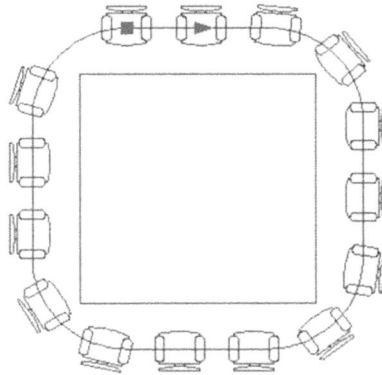

Of course, the Align button is on by default; if you turn it off, this is what you will receive:

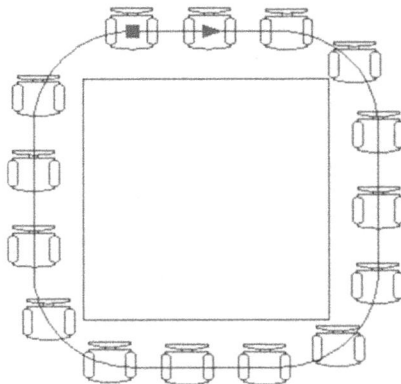

If you select the Divide option, AutoCAD will divide the path equally for all the objects, but if you select Measure, you should specify the distance between each object. The previous was produced by using Measure, the following by using Divide:

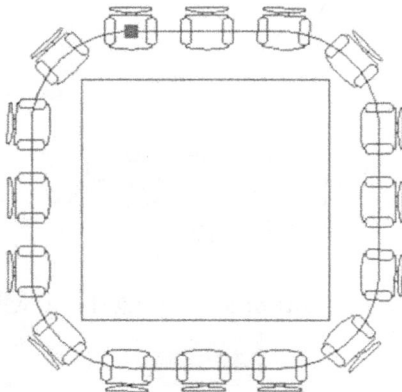

After you specify the three factors, you will be able to select Items (number of items), Between (the distance, which will be off in case of Divide), and Total (the total distance). Also, you can add rows to receive the following:

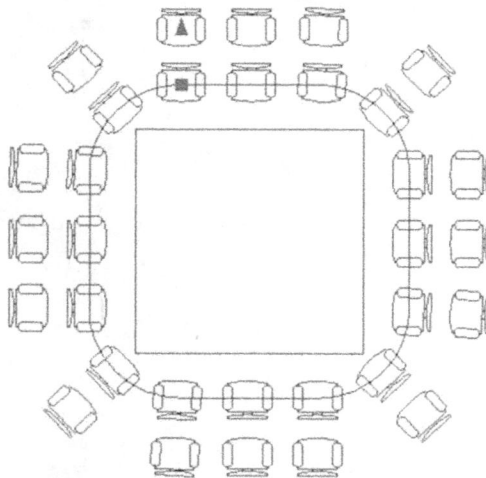

After you create the path array, you can select the array for editing. You will see the following three grips (this will be true only for Divide; in Measure, you will see an extra arrow). The first one is:

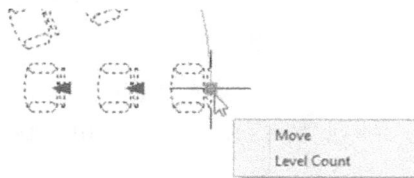

This will control the movement and the level count.

The second one is:

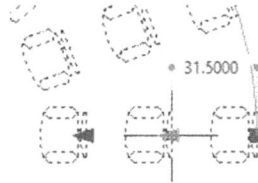

This grip will specify distance between rows. The third one is:

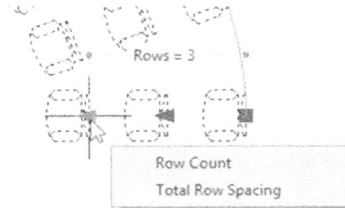

This will control the Row Count and the Total Row Spacing.

Meanwhile, a new context tab called Array will appear; it appears as the following:

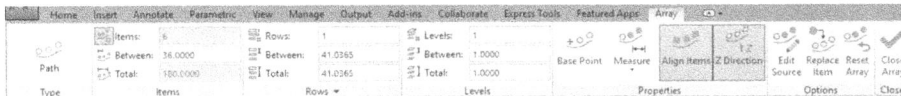

This is identical to what we discussed in the rectangular array.

Also, when you select a path array, you will see Quick Properties:

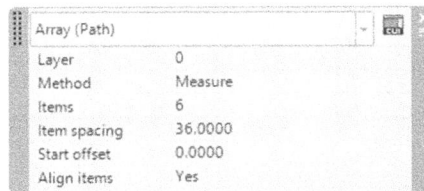

In Quick Properties you can change the Method (Divide or Measure), the number of items, the item spacing, the Start Offset (not to start from the first point of the path), and finally whether to align items or not while arraying them.

## PRACTICE 4-8    ARRAYING OBJECTS USING PATH

**1.** Start AutoCAD 2025

**2.** Open **Practice 4-8.dwg**

**3.** Use Path Array to create an array, bearing in mind the following:

    **a.** Number of items = 22

    **b.** Base point = Center of the Chair (use OSNAP + OTRACK)

    **c.** Divide

    **d.** Align

    **e.** Number of rows = 2

    **f.** Distance between rows = 1.5

**4.** Erase the outer polyline

**5.** You should obtain the following result:

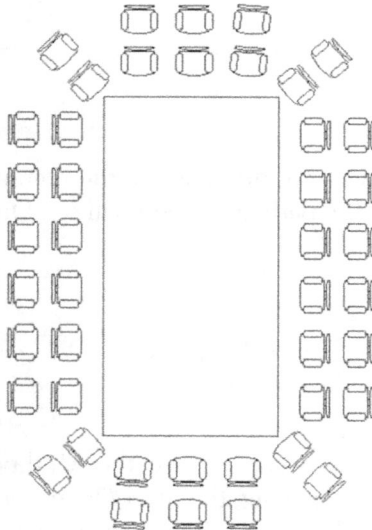

**6.** Using grips you can make the number of rows = 3, and 4, and back to 2

**7.** Save and close the file

## 4.9 ARRAYING OBJECTS – POLAR ARRAY

This command will duplicate objects in a circular fashion. To issue this command go to the **Home** tab, locate the **Modify** panel, then select the **Polar Array** button:

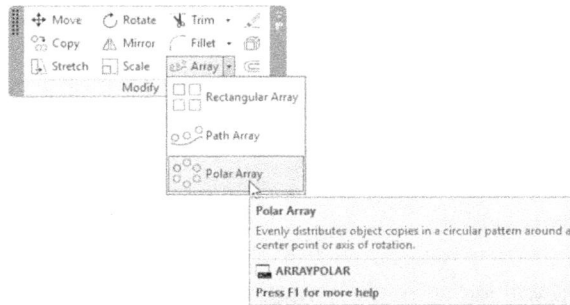

The following prompts will be shown:

```
Select objects:
Type = Polar Associative = Yes
Specify center point of array or [Base point/Axis of
rotation]:
```

The first prompt will ask you to select the desired objects; once done, press [Enter]. The second prompt will show a message that the type of array is polar and the associativity is on. The third line is asking you to specify the center point of array; once you specify the center point, a polar array with six objects filling 360o will be created. Meanwhile, you will see a new context tab titled Array Creation, which appears as the following:

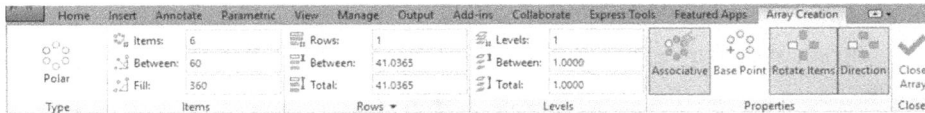

Using the context tab, the user can specify the number of items, the angle between items, and the Angle to fill. You are also invited to specify the number of rows and the distance between rows. Using the Properties panel, input if this is going to be an associative array or not, specify a new base point for the object to be arrayed, specify whether to rotate items as you are copying them, and finally specify the direction of arraying (CW or CCW).

AutoCAD allows you to select the polar arraying for further editing. Once you select it, you will receive an image similar to the following:

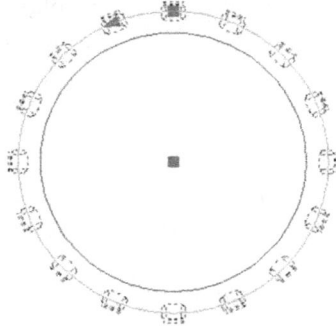

The first grip will show the following:

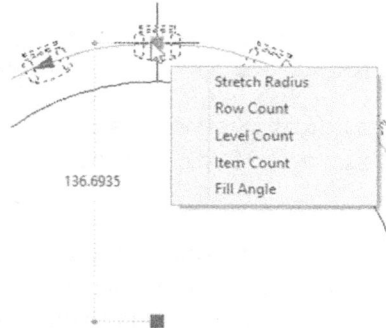

You will see the current radius value along with a menu to help you stretch the radius, change the row count, change the level count (for 3D only), change the item count, and finally change the fill angle. The following is an example of changing the row count:

The second arrow grip will show the current angle between the first and second items.

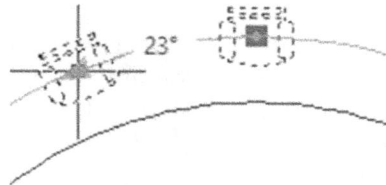

Meanwhile, a new context tab called Array will appear:

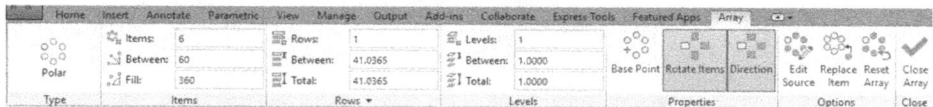

This is identical to what we discussed in the rectangular array.

Also, when you click polar array you will see the following Quick Properties:

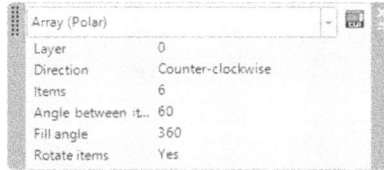

| Array (Polar) | |
| --- | --- |
| Layer | 0 |
| Direction | Counter-clockwise |
| Items | 6 |
| Angle between it... | 60 |
| Fill angle | 360 |
| Rotate items | Yes |

In Quick Properties the user can change the direction of the array (CW or CCW), change the number of items and angle between items, change the total fill angle, and choose whether or not to rotate items while they are copied.

## PRACTICE 4-9    ARRAYING OBJECTS USING POLAR ARRAY

**1.** Start AutoCAD 2025

**2.** Open **Practice 4-9.dwg**

**3.** Using Polar Array create the following shape:

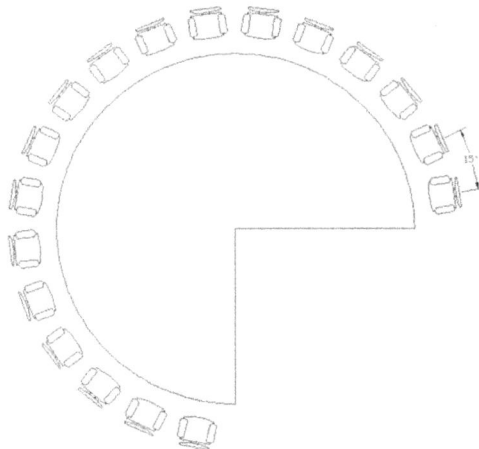

**4.** Save and close the file

## 4.10   BREAK COMMAND

This command will help you break any object into two objects by removing the portion between two specified points. To issue this command, go to the **Home** tab, locate the **Modify** panel, and select the **Break** button:

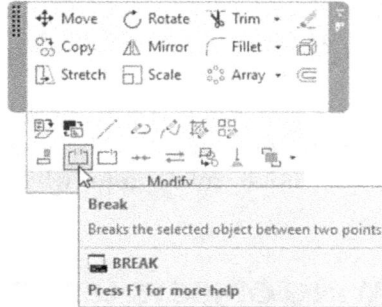

You will see the following prompts:

```
Select object:
Specify second break point or [First point]:
```

If you consider selecting the object is – as well – specifying the first point, then select the second point. This will end the command. But, if you consider selecting is purely selecting, and you did not specify the first point, then type **F** so you can see the following two prompts:

```
Specify first break point:
Specify second break point:
```

Using these two prompts, specify two points on the object, and the Break command will end.

AutoCAD offers another command using a different technique which is breaking on the same point. To issue this command, go to the **Home** tab, locate the **Modify** panel, then select the **Break at Point** button:

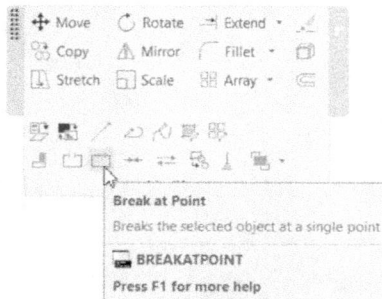

You will see the following prompts:

```
Select object: Specify break point:
```

All you have to do is to select an object, then specify a point to break at.

*If the object to break is a circle, be careful to specify the two points counterclockwise. Check this example:*

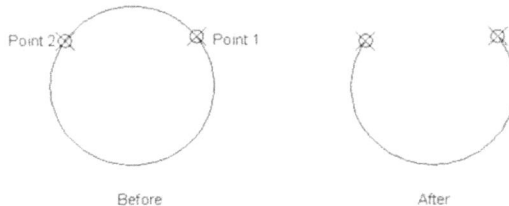

Before                          After

## PRACTICE 4-10    BREAKING OBJECTS

**1.** Start AutoCAD 2025

**2.** Open **Practice 4-10.dwg**

**3.** In order to locate the meeting table exactly at the center of the room horizontally and vertically, we need to break the upper horizontal line and the left vertical line

**4.** Using the Break command break the upper horizontal line from the points shown below:

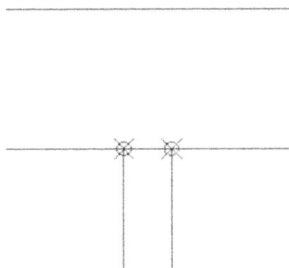

**5.** Using the Break command break the left vertical line using one point as shown below:

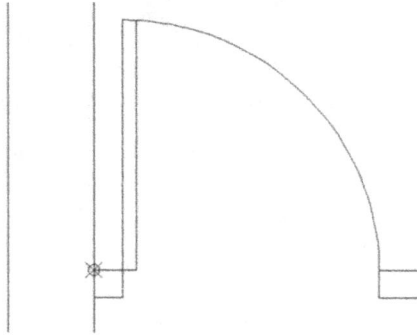

**6.** Using the Move command move the meeting table at the center of the room, using OSNAP and OTRACK

**7.** You will receive the following image:

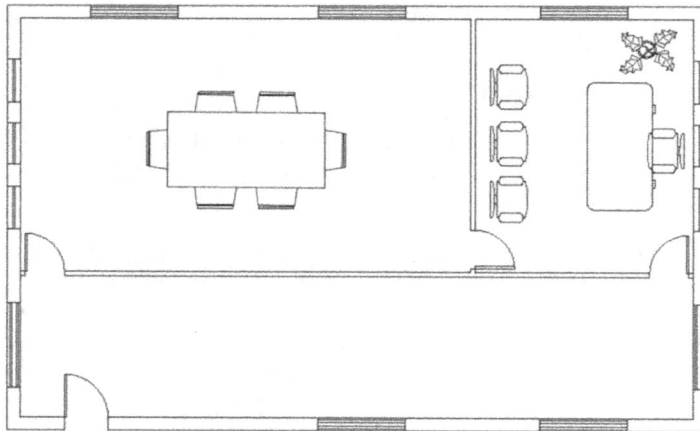

**8.** Save and close the file

# NOTES

## CHAPTER REVIEW

**1.** You can fillet using more than one fillet radius using the same command:

　　**a.** True

　　**b.** False

**2.** There are three types of arrays:

　　**a.** True

　　**b.** False

**3.** If you hold _____ while filleting you will get radius = 0.0 regardless of the current radius

**4.** One of the following is not related to the Chamfer command:

　　**a.** Distance 1 and Distance 2

　　**b.** Distance and Angle

　　**c.** Distance and Radius

　　**d.** Trim and No trim

**5.** You can offset using offset distance and _____

**6.** In the Trim command the default mode is _____

**7.** While you are at the Extend command, if you hold _____ you will convert the command to Trim:

　　**a.** [Ctrl]

　　**b.** [Ctrl] + [Shift]

　　**c.** [Shift]

　　**d.** [Alt] + [Ctrl]

## CHAPTER REVIEW ANSWERS

**1.** a

**3.** [Shift]

**5.** Through

**7.** c

## CHAPTER 5

# LAYERS AND INQUIRY COMMANDS

**In This Chapter**

- What are layers in AutoCAD?
- How to create and set layer properties
- What are layer controls?
- How to use Quick Properties and Properties
- What are the inquiry commands and how are they used?

## 5.1 LAYERS CONCEPT IN AUTOCAD

Layers are the most important way to organize and control your AutoCAD drawings. Managing layers means managing the drawing. So, what are layers in AutoCAD? Layers are a simulation of a transparent piece of paper in which you will draw part of the drawing using a certain color, linetype, and lineweight. Each object on a layer will hold the properties of that layer, meaning the object will have the same color, linetype, and lineweight of the layer it resides in. This setting is called BYLAYER, which means we will control the drawing through controlling layers rather than controlling objects.

Each layer should have a name, which will be considered as the first step of the creation process. Proper naming should adhere to the following rules:

- Name length should not exceed 255 characters
- Use all letters (small or capital)

- Use all numbers
- Use (-) hyphen, (_) underscore, and ($) dollar sign

A unique layer will exist in all AutoCAD drawings called 0 (zero). This specific layer can't be deleted or renamed. Other layers can be deleted and renamed.

The layer at the top of the pile is the only layer we can draw on; we call it the current layer. So as a rule of thumb, make the desired layer current first, and then start drawing. To start building up your layers, go to the **Home** tab, locate the **Layers** panel, and then click the **Layer Properties** button:

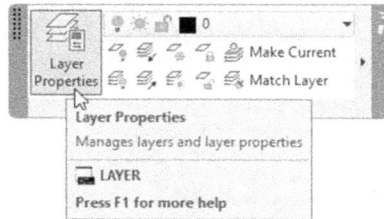

You will see the following palette which is called the **Layer Properties Manager**:

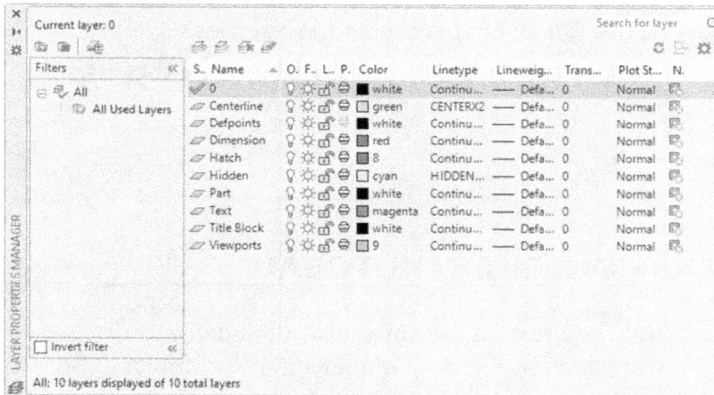

Palettes are much better than the normal dialog box:

- A dialog box has two states, either displayed on the screen or closed. But a palette can be displayed but hidden, which will spare you the continuous effort of opening and closing

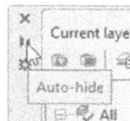

- Almost all of the dialog boxes cannot be resized, as opposed to palettes, which can be resized horizontally, vertically, and diagonally

- Dialog boxes cannot be docked but palettes can be docked at the four sides of the screen

## 5.2    CREATING AND SETTING LAYER PROPERTIES

We will learn how to accomplish the following:

- How to create a new layer
- How to set a color for a layer
- How to set the linetype for a layer
- How to set a lineweight for a layer
- How to set the current layer

### 5.2.1    How to Create a New Layer

This command will add a new layer to the current drawing. Using the **Layer Properties Manager**, click the **New Layer** button:

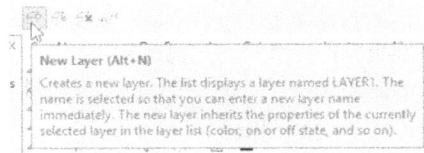

There will a new layer with temporary name *Layer1*. The **Name** field will be highlighted. Type the desired name of the layer. You should always stick to good naming convention; a layer containing doors should be named Door.

**NOTE** *All the following settings require you to select layers in the Layer Properties Manager. Selecting layers in AutoCAD is just like any other software running under Windows OS. You can hold the [Ctrl] key and/or [Shift] to select multiple layers.*

### 5.2.2    How to Set a Color for a Layer

You can use one of 256 colors available in AutoCAD. The first seven colors can be set using the name or its number (from color 1 to color 7). These are:

- Red (1)

- Yellow (2)
- Green (3)
- Cyan (4)
- Blue (5)
- Magenta (6)
- Black/White (7)

Other colors should be set using only their numbers. To set the color for a layer, the following steps should be done:

- Using the **Layer Properties Manager**, select the desired layers
- Using the **Color** field, click the icon of the color, and the following dialog box should appear:

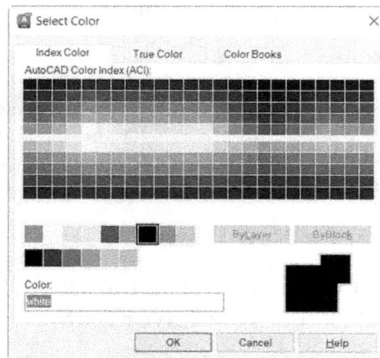

- Select the desired color (or you can type the name/number of the color in the **Color** field), then click the **OK** button to end this action

Another way to set up (or modify) a layer's color is by using the pop-up list in the **Layers** panel, as shown below:

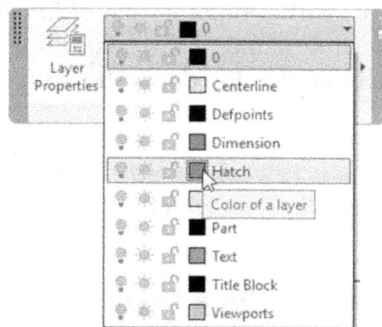

### 5.2.3 How to Set the Linetype for Layers

Two linetype files come with AutoCAD 2025: acad.lin and acadiso.lin. These two linetype files are not adequate for all types of engineering designing and drafting, so you should consider buying more linetype files available on the market.

Contrary to colors, linetypes are not loaded into the current file. Accordingly, the user should load the desired linetypes when needed. To set the linetype for a layer, do the following steps:

- Using the **Layer Properties Manager**, select the desired layers
- Using the **Linetype** field, click the name of the linetype, and you will receive the following dialog box:

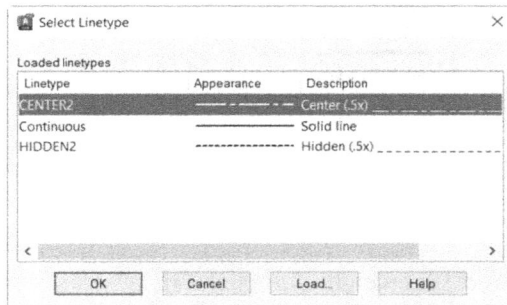

- If the desired linetype is listed, then select it. If not, you need to load it. Click the **Load** button, and the following dialog box will appear:

- Browse for your desired linetype, select it to be loaded, then click **OK**. Now the linetype is loaded; select it and then click **OK**.

### 5.2.4 How to Set a Lineweight for Layers

This option will set the lineweight for layers. All objects in AutoCAD (except polyline with width) have a lineweight of Default, which is 0 (zero), but you can set the lineweight for objects through their layers. The following steps should be done:

- Using the **Layer Properties Manager**, select the desired layers
- Under the field **Lineweight**, click the lineweight icon and the following dialog box will appear:

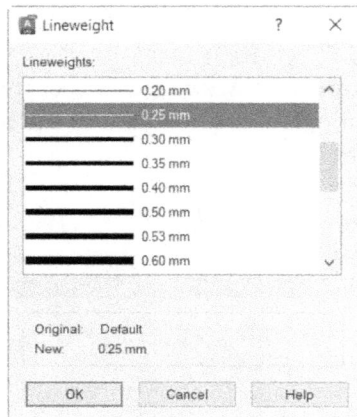

- Select the desired Lineweight and click **OK**
- To see the lineweight on the screen, use the status bar and click the **Show/ Hide Lineweight** button on.

### 5.2.5  How to Set the Current Layer

There are several ways to make a layer the current layer.

The simplest way is to use the layer pop-up list in the Layers panel, as shown in the following:

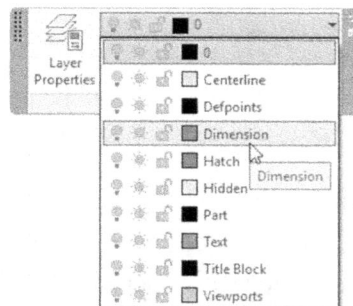

Another way is to use the **Layer Properties Manager** palette, then double-click the Status of the desired layer's name.

The longest way is to use the **Layer Properties Manager** palette, select the desired layer, and click the **Set Current** button:

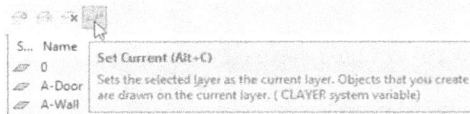

## PRACTICE 5-1    CREATING AND SETTING LAYER PROPERTIES

1. Start AutoCAD 2025

2. Open **Practice 5-1.dwg**

3. Create a new layer and call it Centerlines, with the color yellow and the linetype Center; make it current, and draw two lines, horizontal and vertical, using the centerline of the circle and OSNAP and OTRACK

4. Create another layer and call it Hidden, with the color 9 and the linetype Dashed; make it current and draw a circle using the same center point of the other circles with R = 1.5

5. You will receive the following shape:

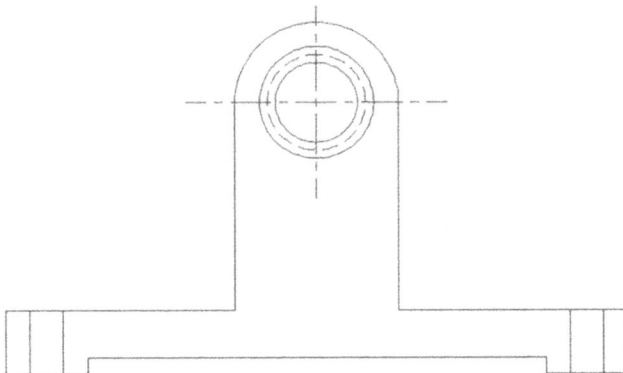

6. Save and close the file

## 5.3   LAYER CONTROLS

The commands discussed here will allow you to have full control over layers. We will learn how to control the visibility of layers, locking layers, plotting layers, deletion of layers, renaming of layers, etc.

### 5.3.1   Controlling Layer Visibility, Locking, and Plotting

AutoCAD provides controls to show/hide layers (Freeze and Off), lock and unlock layers, and plot and not plot layers. Refer to the two examples below:
The first example is layer "Part":

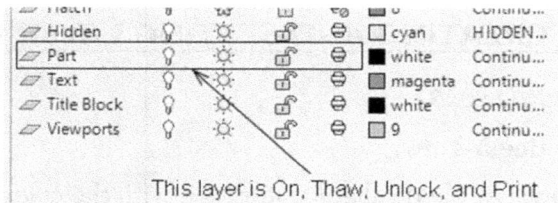

This layer is On, Thaw, Unlock, and Print

The second example is layer "Hatch":

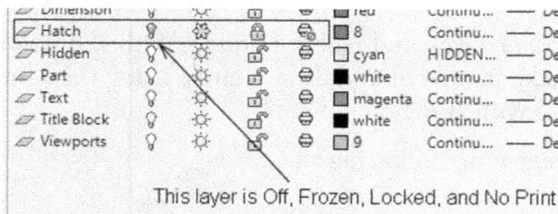

This layer is Off, Frozen, Locked, and No Print

When a new layer is created, the settings will be On, Thaw, Unlock, and Plot. You can hide the contents of layers by turning them off or freezing. Freeze has a deeper effect than off, as objects in a frozen layer will not be considered in the drawing; hence, drawing size will be less temporarily (we use Freeze to lessen the drawing size if the drawing becomes slow).

If you try to freeze the current layer, AutoCAD will display the following message:

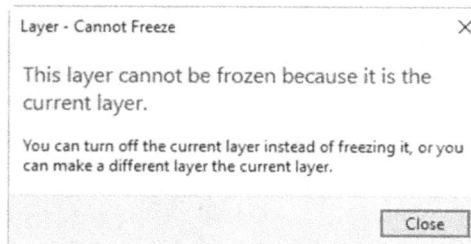

Layer - Cannot Freeze                                    ✕

This layer cannot be frozen because it is the current layer.

You can turn off the current layer instead of freezing it, or you can make a different layer the current layer.

Close

If you try to switch off the current layer, however, you will receive the following message:

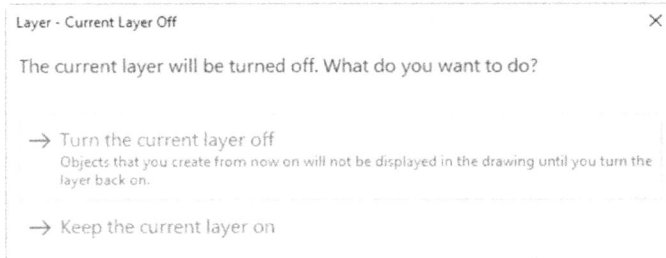

If you lock a layer, then objects reside in it and will not be selected for any modifying commands. Objects in a locked layer will be faded away, and when you get close to them a small lock icon appears to tell you that the layer is locked. Refer to the following illustration:

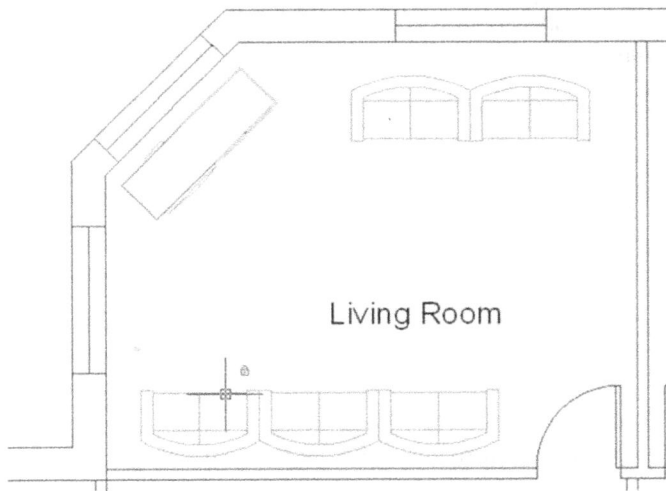

If you turned a layer to be No Plot, then objects in this layer will be displayed but not plotted.

On/Off, Thaw/Freeze, and Lock/Unlock can be controlled using the pop-up list in the Layer panel and the Layer Properties Manager, whereas Plot/No Plot can be controlled only in the Layer Properties Manager.

### 5.3.2 Deleting and Renaming Layers

AutoCAD will not delete a layer that contains objects; it will delete only empty layers. In order to delete layers, these steps should be followed:

- In the **Layer Properties Manager** palette select the desired layer(s).
- Press [Del] at the keyboard or click the **Delete Layer** button

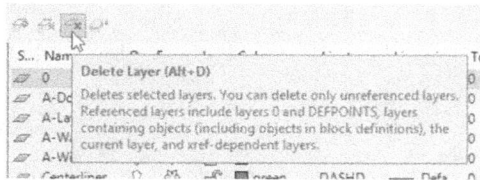

In the **Layer Properties Manager**, the user can rename layers as well. In order to do that, follow these simple steps:

- In the **Layer Properties Manager** palette, select the desired layer
- Click the layer name once and the name will be highlighted for editing; type the new name, then press [Enter]

If you try to delete a layer that contains objects, you will receive the following message:

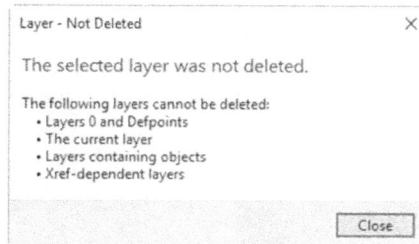

To rename a layer, you need to do the following steps:

- Select the desired layer
- Click the name with a single click and you will see the following:

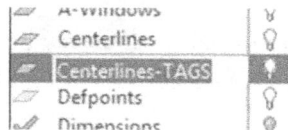

- The name will become editable; type in the new name and press [Enter]

### 5.3.3  How to Make an Object's Layer the Current Layer

This is the fastest way to make a layer the current layer. After issuing the command, simply select an object that resides in that layer. Follow these steps:

- Go to the **Home** tab, locate the **Layers** panel, then click the **Make Current** button:

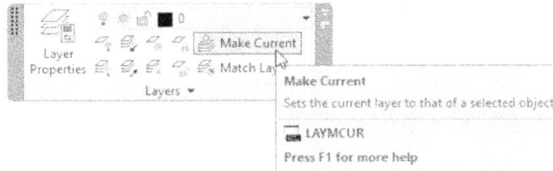

- You will see the following prompt:

```
Select object whose layer will become current:
```

- Select the desired object. Check the current layer and you will find it has become the object's layer (even without you knowing the object's layer)

### 5.3.4  How to Undo Only Layer Actions

The function to undo the layer actions only is called **Layer Previous**. This function will help you restore the previous states of the layers (like Freeze, Thaw, On, Off, etc.) without affecting other drawing or modifying commands. Follow these steps:

- Change the layer states as needed
- Go to the **Home** tab, locate the **Layers** panel, then click the **Previous** button

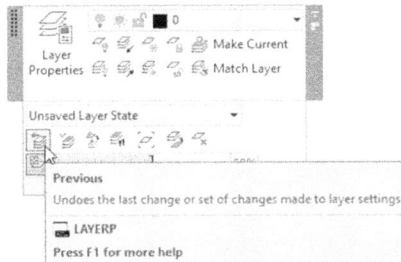

In the command window, you will see the following message:

```
Restored previous layer status
```

### 5.3.5   Moving Objects from One Layer to Another

All similar objects must reside in the same layer but mistakes may occur. If you draw on the wrong layer and you want to shift the objects to the right layer, you can use the **Match** command. The following needs to be done:

- Go to the **Home** tab, locate the **Layers** panel, then click the **Match** button:

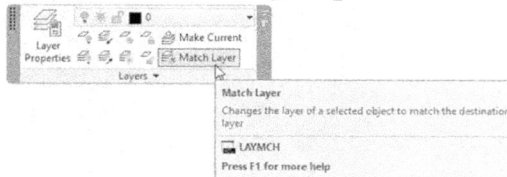

The following prompt will appear:

```
Select objects to be changed:
Select object on destination layer or [Name]:
```

This command contains two prompts; using the first you will select the object mistakenly drawn in the wrong layer, then at the second prompt, either you select an object resides in the right layer, or simply type its name.

At the end you will see the following at the command window:

```
8 objects changed to layer "Dimensions"
```

## 5.4   WHILE YOU ARE AT THE LAYER PROPERTIES MANAGER

While you are in the Layer Properties Manager palette, you can do several actions; for example, if you select a layer and then right-click, you will see a menu as shown in the following:

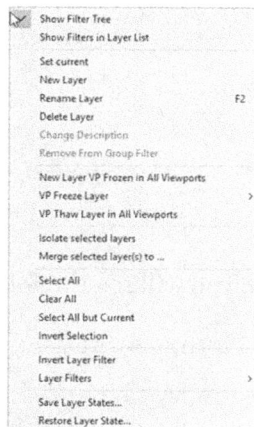

In this menu, you can do all or any of the following:

- Set the current layer
- Create a new layer
- Delete a layer
- Select All layers
- Clear the selection
- Select All but Current
- Invert Selection (make the selected unselected and vice versa)

One of these things is to show or hide the Filter Tree:

- Show Filter Tree (by default it is turned on)
- Show Filters in the Layer List (by default it is turned off)

If you turned off **Show Filter Tree**, you will be allowed to see more information about your layers, as shown in the following:

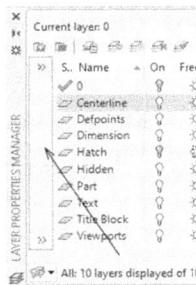

Another way to do the same thing is by clicking the small arrows at the top right part of the Filter Tree pane:

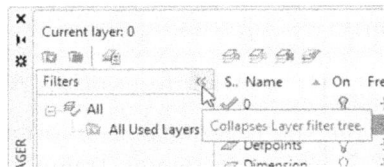

# PRACTICE 5-2    LAYER CONTROLS

**1.** Start AutoCAD 2025

**2.** Open **Practice 5-2.dwg**

**3.** Make Layer 0 the current layer

**4.** Freeze layers: Centerlines and Hatch

**5.** Lock layers: Furniture. What happened to the color? _____

**6.** Get closer to any object in the Furniture layer, and check how the icon appears when you get closer

**7.** Try to erase one of the objects in layer Furniture and type down the message: _____

**8.** Using the Layer Properties Manager, try to freeze layer 0 (the current layer) and type down the message from AutoCAD: _____

**9.** Try to rename layer 0 and type down the message from AutoCAD:

_____

**10.** Rename layer Partition to become Inside Wall

**11.** Using the Match command, select the two doors at the bottom (their color is black or white) to match one of the blue doors

**12.** Using the Layer Properties Manager, select all layers, then change their color to black or white. Then close the Layer Properties Manager

**13.** Click Layer Previous; what happened? _____

**14.** Click the current button for Make Object's layer and click one of the yellow lines; what is the current layer now? _____

**15.** Save and close the file

## 5.5 CHANGING AN OBJECT'S LAYER, QUICK PROPERTIES, AND PROPERTIES

We will learn to control an object's properties.

### 5.5.1 Reading Instantaneous Information About an Object

AutoCAD provides you with instantaneous information about any object when your mouse hovers over the object, as demonstrated in the following illustration:

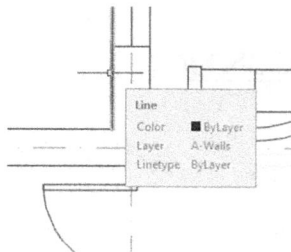

AutoCAD is showing the type of object (Line), its color, its layer, and its linetype.

### 5.5.2   How to Move an Object From a Layer to Another Layer

Any object in AutoCAD should reside in a layer. To move an object from one layer to another, follow these steps:

- Click the desired object(s)
- Go to the **Home** tab, locate the **Layers** panel, and check the layer name that the pop-up list is displaying. This part may sometimes be blank; this happens when your desired objects reside in different layers. Click the layer's pop-up list and select the new layer name
- Press [Esc] once to deselect all the selected objects

### 5.5.3   What is Quick Properties?

This is the first of two commands that will enable you to change the properties of selected objects. An object's properties differ depending on the object type; line properties are different comparing to arcs, circles, or polylines.

Quick Properties will appear on the screen whenever you select objects without issuing a command (using grips selection); if it does not appear, click the following button on the Status bar to switch it on/off:

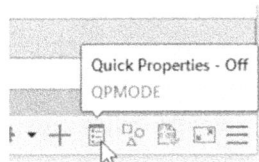

When you select an object, you will see the following (using circle as an example):

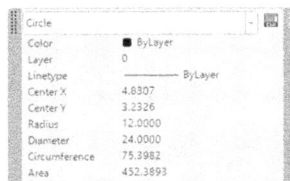

You will see information such as:

- Color
- Layer
- Linetype
- Coordinates of Center point (**X** & **Y**)
- Radius and Diameter
- Circumference and Area

If you select more than one object from the same type, you will see the following:

But if you select more than one object from different types, you will receive the following:

You will see All(number) which means you are seeing nine objects (as our example) from different types, but you can see a breakdown of the selected objects by clicking the pop-up list, as shown below:

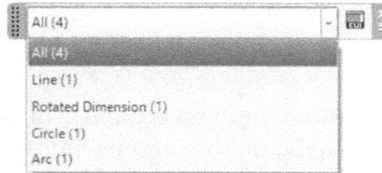

When you select a single object type, you can change the general and specific object's properties.

The user should note that while at Quick Properties, whenever you select a different layer, color, linetype, etc., you will see the effect of your change concurrently. This applies to the next command as well.

### 5.5.4 What Is Properties?

As the name indicates, Quick Properties will be your fast way to change the properties of objects. However, the Properties command is more detailed. All the rules we discussed for Quick Properties are applicable here as well.

To issue the **Properties** command, do the following:

- Select the desired object(s)
- Right-click and select the **Properties** option

Another way is to double-click any object to get the Properties palette. This way has two drawbacks: first, it will be for a single object only; second, some objects like polylines, blocks, hatch patterns, and text will interpret this action as an edit. Either way, you will receive the following:

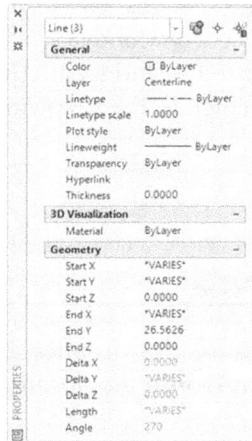

As you can see there is much more information displayed here than in the Quick Properties, hence, providing you the power to make more changes to the selected objects.

**NOTE** *Some of the information is shaded which means it is Read-Only.*

## PRACTICE 5-3    CHANGING OBJECT'S LAYER, QUICK PROPERTIES, AND PROPERTIES

1. Start AutoCAD 2025

2. Open **Practice 5-3.dwg**

3. Hover over one of the circles of the centerlines and type down the name of the layer: _____

4. Select all circles and text inside them and move them to layer Centerlines

5. Change the properties of the circles to be Continuous

6. Delete layer Centerlines-TAGS

7. Zoom to the lower door and you will find two red lines at the right and at the left; move them from layer Dimensions to layer A-Walls

8. Save and close the file

## 5.6  INQUIRY COMMANDS – INTRODUCTION

The main purpose of this set of commands is to measure lengths between two points, inquire the radius of a circle or arc, measure the angle, measure the area, or measure the volume for 3D objects. The user will use this set of commands to make sure that the drawing is correct and according to the design intent. To issue these functions go to the **Home** tab and locate the **Utilities** panel. For all of these commands, the cursor will change to the following:

## 5.7  MEASURING DISTANCE

This command will measure the distance between two selected points. To issue this command, go to the **Home** tab, locate the **Utilities** panel, and click the **Distance** button:

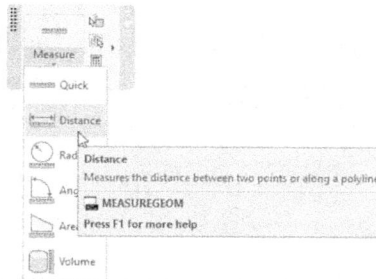

You will see the following prompt:

```
Specify first point:
Specify second point or [Multiple points]:
```

Either select the desired two points, or input m for Multiple which will allow you to select as many points as you wish, hence measuring multiple lines. AutoCAD will display the following on the screen:

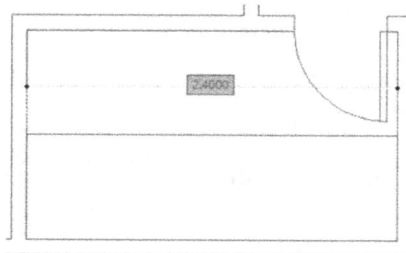

And you will see the following result on the command prompt:

```
Distance = 2.4, Angle in XY Plane = 0,
Angle from XY Plane = 0 Delta X = 2.4,
Delta Y = 0.0000, Delta Z = 0.0000
```

If you select Multiple, you will see the following result (as an example):

```
Specify next point or [Arc/Length/Undo/Total] <Total>:
Distance = 18.0000
Specify next point or [Arc/Close/Length/Undo/Total]
<Total>:
Distance = 24.0000
Specify next point or [Arc/Close/Length/Undo/Total]
<Total>:
Distance = 32.0000
Specify next point or [Arc/Close/Length/Undo/Total]
<Total>:
Distance = 32.0000
```

## 5.8   INQUIRING RADIUS

This command will check the radius (and diameter as well) of an existing circle or arc. To issue this command, go to the **Home** tab, locate the **Utilities** panel, and select the **Radius** button:

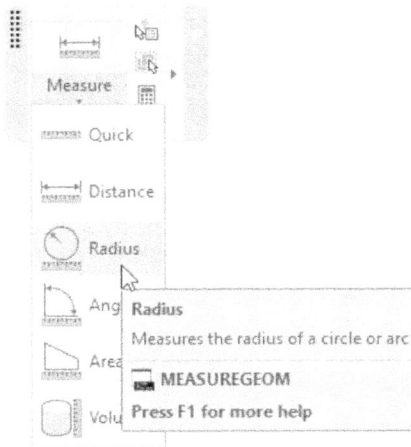

AutoCAD will display the following prompt:

```
Select arc or circle:
```

Select the desired arc or circle and you will receive the following on the screen:

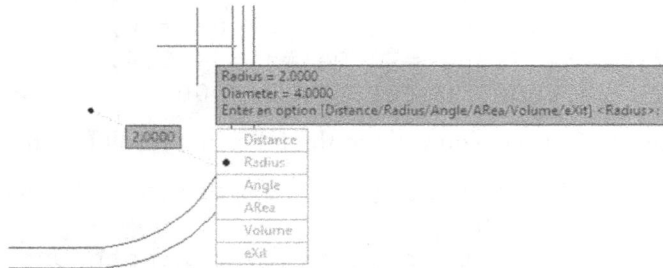

You will see the following on the command window:

```
Radius = 2.000
Diameter = 4.000
```

## 5.9   MEASURING ANGLE

This command will measure an angle (between two lines, the included angle of an arc, or two points and the center of the circle). To issue this command go to the **Home** tab, locate the **Utilities** panel, then select the **Angle** button:

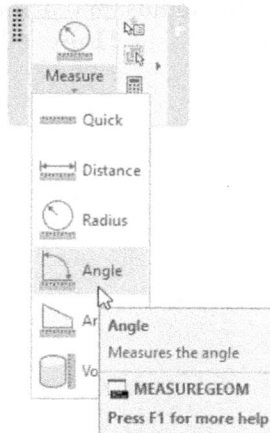

You will see the following prompt:

```
Select arc, circle, line, or <Specify vertex>:
```

Select the desired objects (whether two lines, an arc, or points on a circle) and AutoCAD will display the following on the screen:

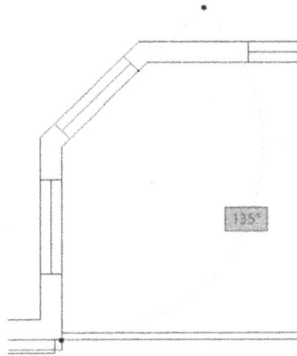

Also, you will see the following in the command window:

```
Angle = 135°
```

## 5.10   MEASURING AREA

This command will measure areas, whether simple area (area that has no islands inside), or complex area (areas that have islands inside). AutoCAD can measure areas between points (assuming lines and arcs connect them), or objects (like circles, closed polylines, etc.). To issue this command, go to the **Home** tab, locate the **Utilities** panel, then select the **Area** button:

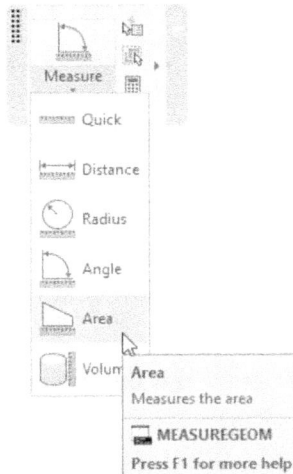

The user will see the following prompt:

```
Specify first corner point or [Object/Add area/
Subtract area/eXit] <Object>:
```

### 5.10.1 How to Calculate Simple Area

The definition of simple area is any closed area without any objects (islands) inside it. AutoCAD will assume you want to measure a simple area if you start by specifying points or selecting objects. If you start specifying points, AutoCAD will assume there are either lines or arcs connecting them. For lines, you will see the following prompts:

```
Specify next point or [Arc/Length/Undo]:
Specify next point or [Arc/Length/Undo]:
Specify next point or [Arc/Length/Undo/Total] <Total>:
Specify next point or [Arc/Length/Undo/Total] <Total>:
```

If you see an arc in your area, simply change the mode to Arc, and you will see prompts identical to the Polyline command. Keep specifying points of lines or arcs until you press [Enter]. You will see the Total value of measured area in the command window and you will receive the following on the screen:

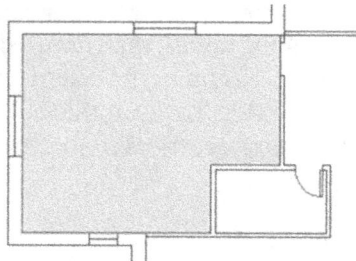

Also, you will see the following displayed in the command window:

```
Area = 18.8366, Perimeter = 17.0416
```

If you have a simple area and the parameter is a single object, you can select the object rather than specifying points. Either right-click and select the **Object** option, or type **o** at the command window, and you will see the following prompt:

```
Select objects:
```

Select the object whose area you want to measure, then press [Enter].

### 5.10.2   How to Calculate Complex Area

The definition of a complex area is any closed area with objects (islands) inside it. To tell AutoCAD you want to calculate a complex area, you *have* to start with either **Add area** or **Subtract area**.

If you start with Add area or Subtract area, AutoCAD will start with area = 0, and will add areas and subtract areas as needed. For the Add area mode you will see the following prompt:

```
Specify first corner point or [Object/Subtract area/
eXit]:
```

Specify the area using the same methods discussed above (points or object), then switch to Subtract area mode; you will see the following prompts:

```
Specify first corner point or [Object/Add area/eXit]:
```

AutoCAD will offer you a subtotal each time after adding or subtracting areas. When you are done, press [Enter] twice to end the command and receive the final net area.

You will receive the following image:

## 5.11   QUICK MEASURE

Measuring has become much faster with Quick option. With this option, you can quickly review the dimensions, distances, angles, and areas within a 2D drawing.

Start this option, you will see two yellow cross hairs, once you move them you will see all nearby measurements, both inside and outside the nearest parts of a drawing.

To issue this command go to the **Home** tab, locate the **Utilities** panel, then select the **Quick** button:

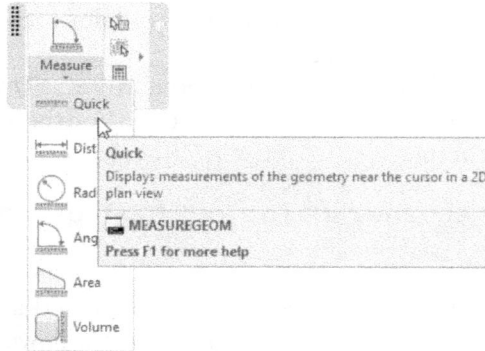

The user will see the following prompt:

```
Move cursor or [Distance/Radius/Angle/ARea/Volume/
Quick/Mode/eXit] <eXit>:
```

Start moving your cursor over the 2D geometry, you will receive the following:

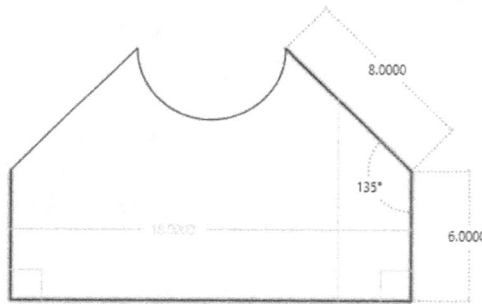

Keep moving and you will see more dimensions. The two rectangles at bottom right and left means 90° angles.

If you click inside a closed area with islands, AutoCAD will calculate the area and the perimeter of this area, and display the results both on the screen, and command window:

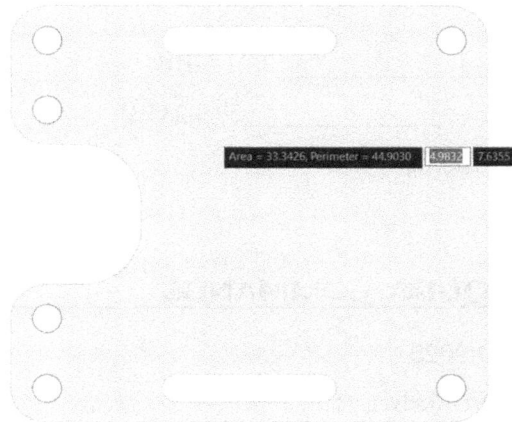

Area = 33.3426, Perimeter = 44.9030   4.9832   7.6355

## PRACTICE 5-4   INQUIRY COMMANDS

**1.** Start AutoCAD 2025

**2.** Open **Practice 5-4.dwg**

**3.** Freeze all layers except A-Walls (make layer A-Walls current, then select all layers except current, and freeze them)

**4.** Measure the length of the slanted wall from the inside and type the information given by AutoCAD:

  **a.** Length = _____ (1.6971)

  **b.** Angle in XY plane = _____ (45, or 135)

  **c.** Delta X = _____ (1.2000)

  **d.** Delta Y = _____ (1.2000)

**5.** Thaw layer Partition

**6.** Measure the horizontal and vertical lengths of the room at the upper right part (you can use Nearest and Perpendicular osnaps) and type them down:

  **a.** Horizontal Distance = _____ (5.05)

  **b.** Vertical Distance = _____ (3.9)

7. Start Quick option, and move it over the different lines, and arcs, and see how measurements are displayed automatically

8. Measure the inside area of the room at the lower right part of the plan and type it down:

   **a.** Area = _____ (21.8016)

   **b.** Parameter = _____ (18.2416)

9. Save and close the file

## PRACTICE 5-5  INQUIRY COMMANDS

1. Start AutoCAD 2025

2. Open **Practice 5-5.dwg**

3. Calculate the net area of the shape using the standard method, or the quick method without all the inside objects and type it down: Area = _____ (33.3426)

4. Save and close the file

# NOTES

## CHAPTER REVIEW

**1.** One of the following is not true about layer name:

    **a.** Should not exceed 256 characters

    **b.** Space is allowed

    **c.** $ is allowed

    **d.** $ is not allowed

**2.** You cannot _____ the current layer

**3.** You can _____ the current layer

**4.** You can undo layer actions only

    **a.** True

    **b.** False

**5.** Area command can calculate only areas without any islands inside it

    **a.** True

    **b.** False

**6.** The following are facts about layer 0 (zero) except one:

    **a.** User cannot rename it

    **b.** User cannot set a new color for it

    **c.** User cannot delete it

    **d.** It is in all AutoCAD files

**7.** Delta X is one example of information available in _____ command

**8.** Most objects will respond to _____ to display Properties palette

## CHAPTER REVIEW ANSWERS

**1.** d

**3.** Turn off

**5.** b

**7.** Measure distance

## CHAPTER 6

# BLOCKS AND HATCH

## In This Chapter

- What are blocks and how are they defined?
- How to use (insert) blocks
- How to explode and convert blocks
- How to hatch in AutoCAD
- How to control hatch in AutoCAD
- How to edit hatch

## 6.1 WHY DO WE NEED BLOCKS?

In your daily work there will be shapes you will need to use repeatedly. You have two options: either draw it each time; or draw it once and save it as a block (the block will be a single object) that can be used (inserted) as many times as you wish in the current file, and in other files as well.

Many benefits will be gained if blocks are used:

- The file size will be less, due to the fact that each block will be counted as a single object
- Standardization for the same company
- Speed of completing a drawing
- Using of Design Center and Tool Palettes

## 6.2  HOW TO CREATE A BLOCK

In order to create a block, the following steps must be completed:

- Draft the shape that you want to create a block from in layer 0 (zero). Layer 0 will enable the block to inherit the properties (color, linetype, lineweight) of the layer which will reside.
- Make sure to control the "Block unit," which will enable AutoCAD to automatically scale the block to appear in the right size in any other drawing
- Draft the shape that you want to create a block from in its real-life dimensions

Let's assume the following shape is drawn:

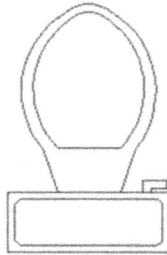

Now you are ready to issue the command: go to the **Insert** tab, locate the **Block Definition** panel, and select the **Create Block** button:

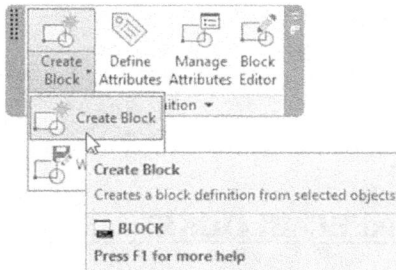

You will receive the following dialog box:

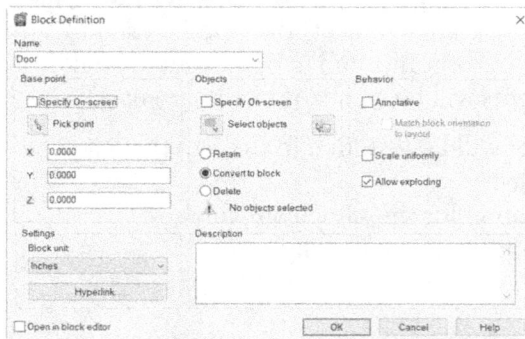

Now do the following:

- Type the block's name (it should not exceed 255 characters, use only numbers, letters, -, _, $, and space)
- Specify **Base point**, either by typing X Y Z coordinates, or select the **Pick point** button to input the base point graphically
- Click the **Select objects** button to select the desired objects
- AutoCAD will create from the drawn objects the needed block, but what should be done with the objects afterward? The user should select one of the three choices available, either to Retain (leave) objects as they are, Convert them to a block, or Delete them

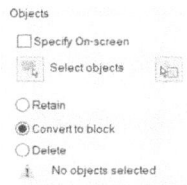

- AutoCAD will create from the drawn objects the needed block, but what should be done with the objects afterward? The user should select one of the three choices available, either to Retain (leave) objects as they are, Convert them to a block, or Delete them

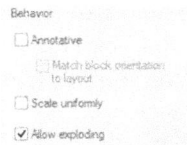

- Select the **Block unit**. This will tell AutoCAD what each AutoCAD unit used in this block will equal; this will help AutoCAD in the Automatic scaling feature

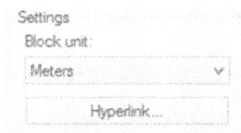

- Type block description
- Turn the checkbox "Open in a block editor" off, because this is an advanced feature which is used for creating dynamic blocks
- Once you finished inputting all the above data, click **OK**

Later on, you will use (insert) the block; you will insert only a copy of the block definition and the original block definition will stay intact.

## 6.3  HOW TO USE (INSERT) BLOCKS

After block creation, the user is ready to use (insert) the block in the current drawing. Nonetheless, you should make sure that you are in the right layer, and that the drawing is ready to accept the block (make sure the door openings are made before the insertion of the door, for example).

Now you are ready to issue the command: go to the **Insert** tab, locate the **Block** panel, then click the **Insert** button:

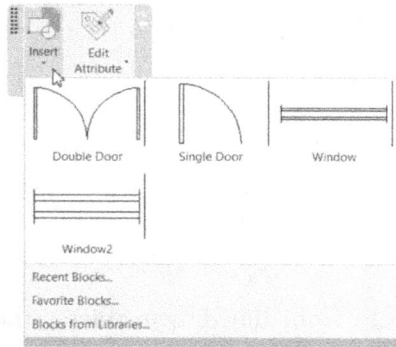

You will see a list of the current blocks available in your drawing. Using this method, you cannot change the scale or rotation angle. This is a rapid way to insert blocks in your drawing. Conversely, if you prefer to customize the insertion process, you can select either Recent Blocks option, Favorite Blocks, or Blocks from Libraries option, you will see the following palette:

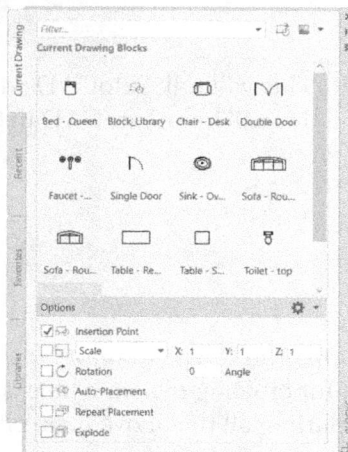

In the Current Drawing tab, you will see the blocks available in this drawing.

The user should specify the **Insertion point**, the **Scale**, the **Rotation** angle, whether you want **Auto-Placement** or not, whether you want **Repeat Placement** or not, and finally whether you want to insert the block exploded or as one unit, either by using the Specify On-Screen checkbox or by typing the needed value.

If you right-click any block, you will see the following menu:

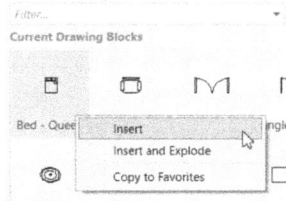

In this menu, you can select to Insert the block, Insert it and explode it, or Copy it to Favorites.

If you click the **Recent** tab, it will display all the most recently inserted blocks (just like the below figure) regardless of the current drawing. These preserve between drawings and sessions. You can remove a block from this tab by right clicking it and choosing Remove from Recent List:

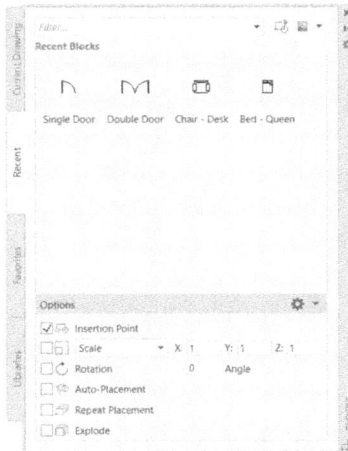

Go to **Favorites** tab, you will see the blocks you chose to be favorites for your future use, you will see the following:

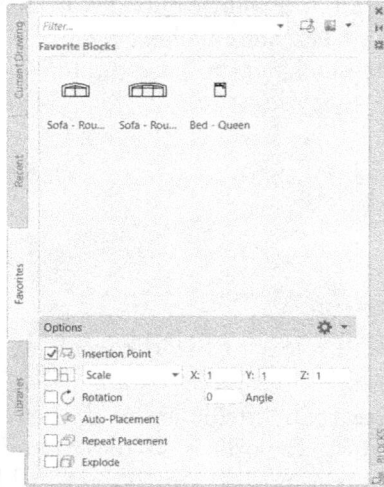

Click on **Libraries** tab, you will see the following:

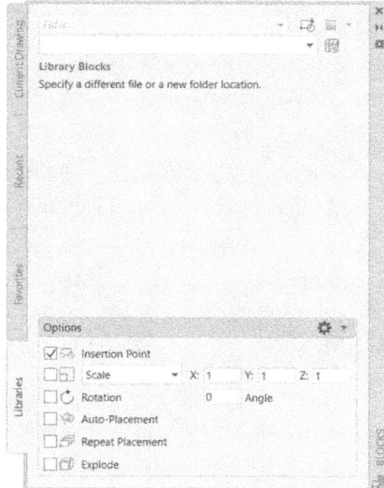

Click the shape of the books at the top right part and specify either a file contains your desired blocks, or, select a folder which contains multiple files containing your desired blocks. You will see the following:

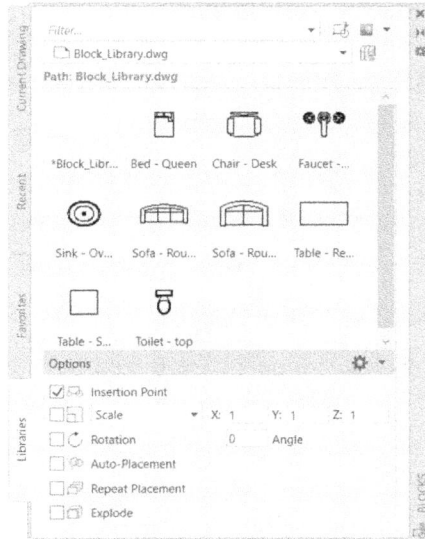

### Scale

While using the **Scale** you can insert mirror images of the block by using negative values. Refer to the following illustration:

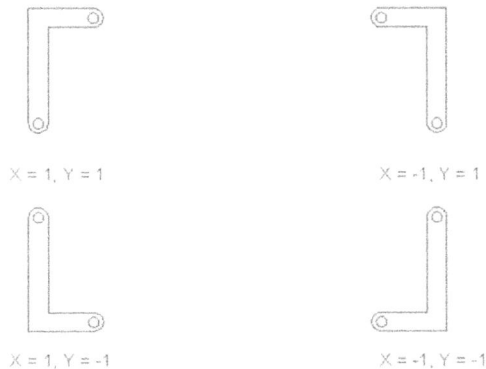

While you are using the Rotation angle, remember the CCW is always positive.

**Auto-Placement**

Auto-Placement option will allow AutoCAD to offer placement suggestions based on where you have placed that block before in the drawing.

AutoCAD learns how the existing block instances are placed in your drawing to suggest the next placement of the same block.

**Repeat Placement**

If Repeat Placement checkbox is on, right-click the desired block and select Insert option, then this block will be inserted as many times as you wish in your drawing in the same command. This will save you lots of minutes in the insertion process.

### 6.3.1   Block Insertion Point OSNAP

After inserting a block incidence, click it to see the following:

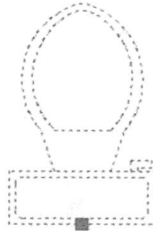

The whole block is one unit, and only one grip, the insertion point, is highlighted. There is a specific OSNAP to snap at this point called Insertion (or insert depending on where you are looking). Refer to the following illustration:

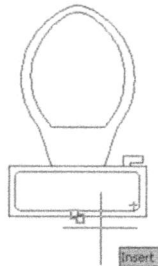

### 6.3.2   Block Count

This command will quickly and precisely count the instances of blocks in the current drawing.

It will allow you to zoom and navigate the instances, of any given block.

Go to the **View** tab, locate the **Palettes** panel, then click the **Count** button:

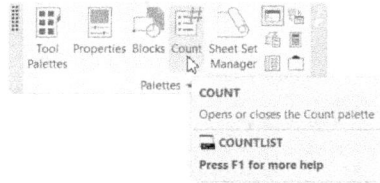

You will see the following palette:

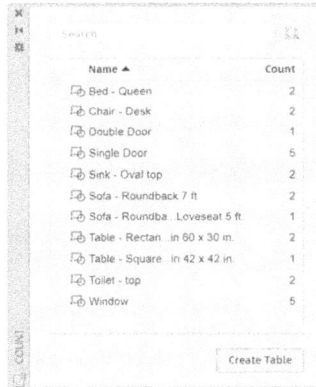

You can use Count command in the whole drawing, or you can specify a window (click the green upper right button) in the drawing you want to count the blocks inside it.

Once you click one of the blocks, a blue frame around the screen will appear, the block will be highlighted in green, and the Count toolbar will appear:

Use this toolbar for the following:

- Use the left and right arrow to zoom in for the block instances
- You can specify a window within the drawing to limit your zooming
- You can select the instances
- You can insert a count field

At the bottom of the Count tool palette, click Create Table button, you will see the following:

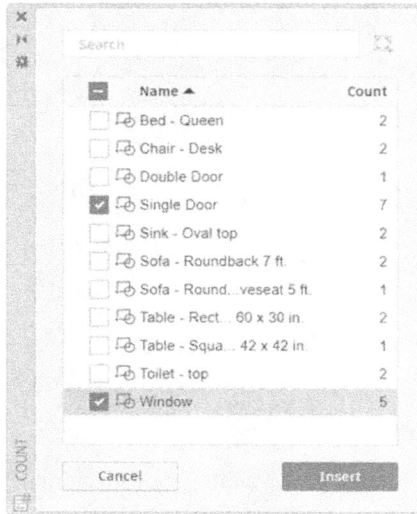

Select the desired blocks you want to create a table to (it can be one or more), then click Insert button to insert the table, you will receive the following:

| Item | Count |
|---|---|
| Single Door | 7 |
| Window | 5 |

### 6.3.3   Block Replace

This command will replace specified block references with a different block.

You can replace one or more blocks with another block you specify from a drawing, or from a list of recent or suggested blocks.

Go to the **Insert** tab, locate the **Block** panel, then click the **Replace** button:

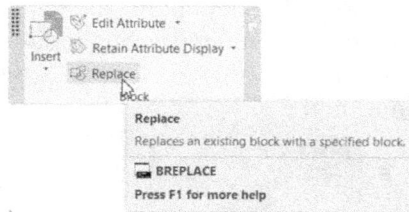

You will see the following prompt:

```
Select one or more blocks to replace:
```

Select the desired block to replace

Select the desired block (or blocks, you cannot select different blocks, they should have the same name) to replace, you will see the following palette:

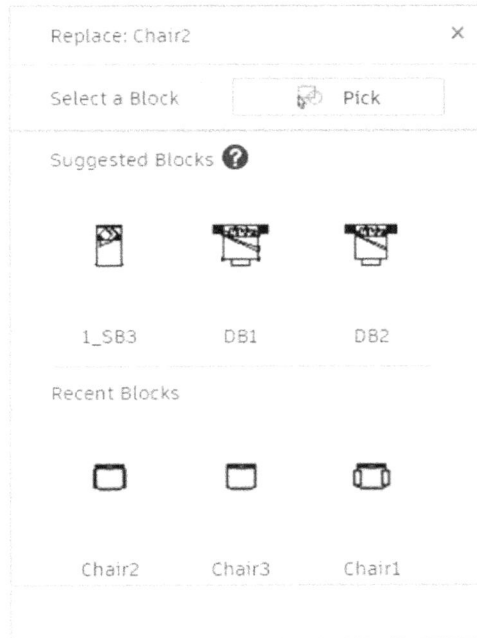

You can choose one of three options:

- Select an existing block in the current drawing
- Select one of the suggested blocks
- Select one of the recent blocks

## PRACTICE 6-1A   CREATING AND INSERTING BLOCKS

**1.** Start AutoCAD 2025

**2.** Open **Practice 6-1a.dwg**

**3.** There are three shapes at the left drawn in layer 0. Create from these shapes three blocks, naming them Window, Single Door, and Double Door, leaving the block unit to default value, the Base point is always the lower left point of the shape.

4. Use the Insert command to insert the three blocks in the proper places using the proper layers, as in the following picture. Also, use the Other Drawing tab, to load the file in your exercise folder called Block_Library.dwg to bring in the furniture blocks and input them in Furniture layer:

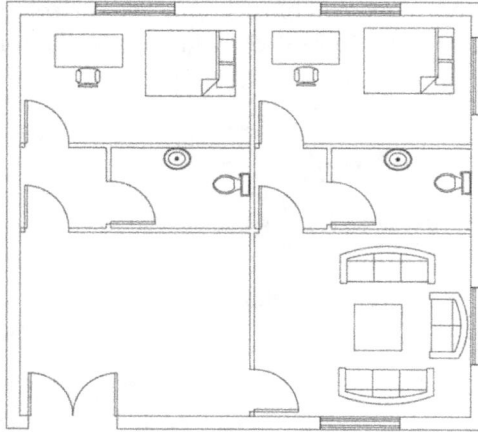

5. Start Count Command, how many Single Door blocks are inserted? _____ (7).

6. How many Window blocks are inserted? _____(5). Use Zoom in to identify the 5 instances of the Window (you can specify a window look at the upper two windows only

7. Make Layer 0 current

8. Create a table of the Single Door and Window blocks and insert it beside the drawing

9. Save and close the file

## PRACTICE 6-1B    AUTO-PLACEMENT AND REPLACE

1. Start AutoCAD 2025

2. Open **Practice 6-1b.dwg**

3. Go to **Insert** tab, locate **Block** panel, click the drop-down list of **Insert**, and select **Recent Blocks** option. Blocks palette will be displayed, select Current Drawing tab, make sure Auto-Placement is turn on, select Desk block, go to the top right corner of the plan, the block will align itself to the corner.

**4.** Repeat the process for the lower left corner of the room, and lower right corner of the room (you may need some movement after the insertion process)

**5.** Repeat the process for the upper right corner of the room (zooming out sometimes will help)

**6.** Using Auto-Placement add Chair2, and Chair3, for the new added desks

**7.** Using Replace command, select Chair2, and replace them with Chair1 (if a message comes out, do not redefine the block)

**8.** Using Replace command, select Chair3, and replace them with Chair1

**9.** Save and close the file

## 6.4   EXPLODING BLOCKS AND CONVERTING THEM TO FILES

### 6.4.1   Exploding Blocks

As with polylines when we explode them into lines and arcs, we can do the same thing with blocks. The explode command will bring them back to their original objects. This practice is definitely not recommended, as we always advocate keeping blocks as one object and not exploding them.

To issue the command, go to the **Home** tab, locate the **Modify** panel, and select the **Explode** button:

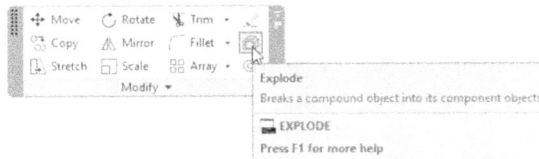

You will see the following prompt:

```
Select objects:
```

Select the desired blocks and press [Enter] to end the command.

### 6.4.2   Converting Blocks to Files

This is an old practice in which we used to convert all of our blocks to files in order to use them in other files. This practice was eclipsed by the emergence of the Design Center and Tool Palettes.

To issue the command, go to the **Insert** tab and locate **Block Definition**, then click the **Write Block** button:

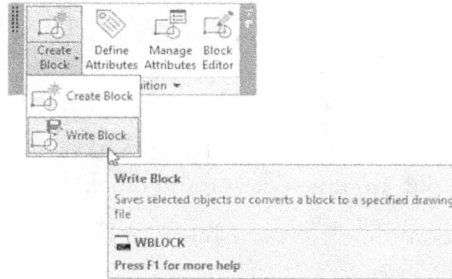

You will receive the following dialog box:

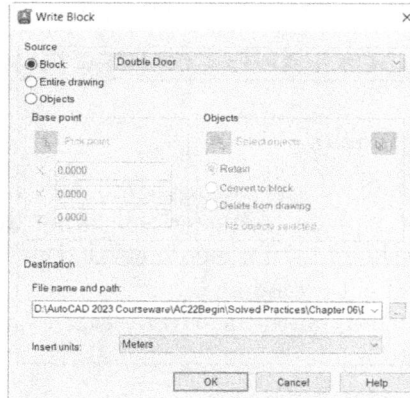

Select the **Block** option under **Source**. Select the name of the block. Under Destination input the file name and path and specify the insert unit. You can use the same dialog box to create a file from the entire drawing or from some of the objects in the current drawing.

## PRACTICE 6-2    EXPLODING, CONVERTING

**1.** Start AutoCAD 2025

**2.** Open **Practice 6-2.dwg**

**3.** Start the Write Block command and select the Single Door block to convert it to a file and save it in your practice folder

**4.** Explode one of the single door insertions. Using Quick Properties, select one of the objects resulting from the exploding process. In which layer does it reside? _____ Why?_____

**5.** Save and close the file

## 6.5 HATCHING IN AUTOCAD

AutoCAD can hatch closed areas and non-closed areas (with maximum distance defined by you).

There are two hatch pattern files that come with AutoCAD, and they are: *acad. pat* & *acadiso.pat* (the hatch pattern file's extension is *.PAT).

The following stated are four pattern types:

- **Solid** (a single pattern covers the area with a single solid color)
- **Gradient** (two gradient colors are mixed together in several fashions)
- **Pattern** (several predefined patterns)
- **User defined** (simplest pattern: parallel lines)

## 6.6 HATCH COMMAND: FIRST STEP

This command will put hatches in the drawing and control all of its properties. The preview is instantaneous. To issue the Hatch command, go to the **Home** tab, locate the **Draw** panel, then click the **Hatch** button:

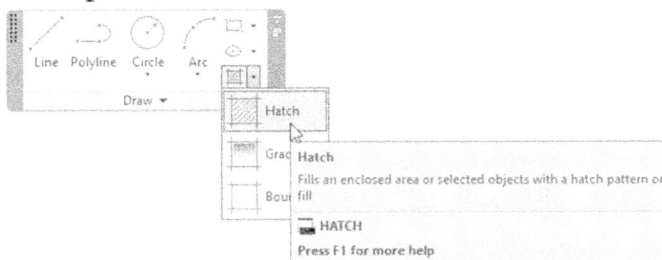

You will see a new context tab added to ribbons called **Hatch Creation**. You will see several panels (they be discussed once we come to each one). Your first step should be locating the **Properties** panel; at the top left part, select the **Hatch Type** as shown below:

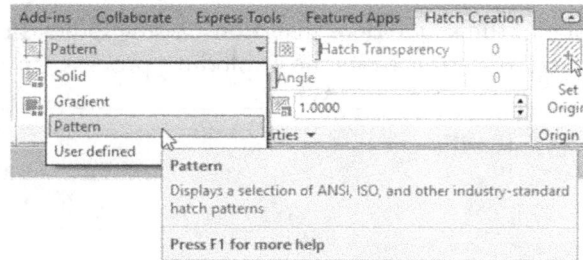

Once you select the Hatch Type, locate the **Pattern** panel; AutoCAD will take you to first pattern in the selected type. For instance, if you select **Gradient** in the Hatch Type, the first pattern in the Pattern panel will be GR_LINEAR, which is the first pattern in the gradient patterns:

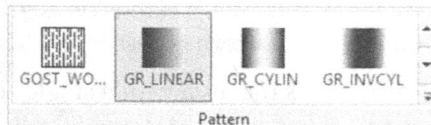

Now, simply go (without any clicking) to the desired area to be hatched, and you will see the area filled. At this moment there are two choices:

- If you like the result, click to pick the area, then go to the **Close** panel and click **Close Hatch Creation**
- If the result did not satisfy you, you need to control the properties of the hatch in an attempt to rectify the result

## 6.7 CONTROLLING HATCH PROPERTIES

If you clicked inside the area and you do not like the result, you need to alter the properties of the hatch. All of these functions exist in the **Properties** panel. They are:

- **Hatch Color**: Specify the color of the hatch or leave it to "Use Current"

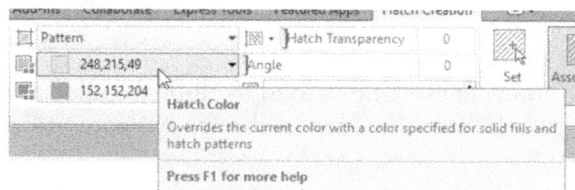

- **Background Color:** Specify the color of the background or leave it to "None"

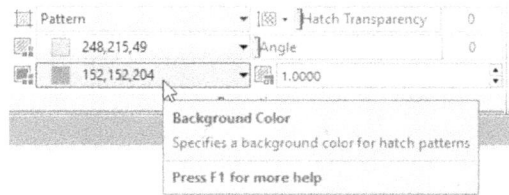

- **Transparency:** By default all colors will be with their normal colors, but you can increase the value of transparency (maximum 90) to decrease the intensity of the color (hatch color and background color)

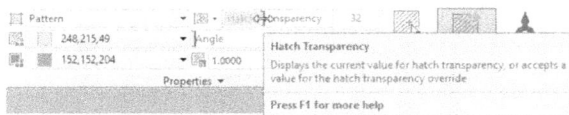

- **Angle:** Specify the angle of the hatch pattern (this has no effect with Solid hatching)

- **Scale:** (If you chose a User-defined type, this will be called Spacing): Input the scale or spacing for the selected hatch pattern

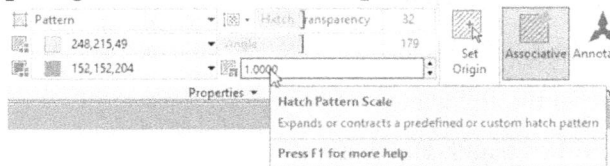

- **Hatch Layer Override:** By default, the hatch will reside in the current layer; using this function, you can specify the layer you want the hatch to reside in, regardless of the current layer

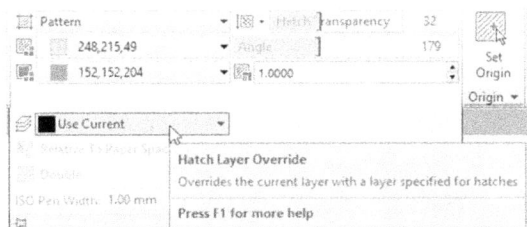

---

**NOTE** *You can use the system variable HPLAYER, to create a new layer, and set it as the default layer for hatch.*

---

■ **Double**: This option is only valid if the hatch type is User-defined. It controls whether the lines are in one direction or crosshatched

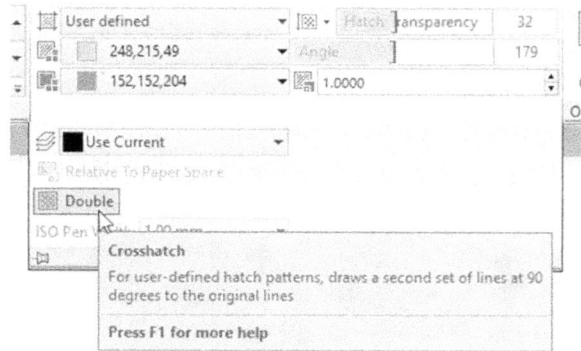

## PRACTICE 6-3A    INPUTTING HATCH AND CONTROLLING HATCH PROPERTIES

**1.** Start AutoCAD 2025

**2.** Open **Practice 6-3.dwg**

**3.** Make layer Hatch the current layer

**4.** Start the Hatch command

**5.** Make sure of the following:

At Properties panel, Hatch Type is Pattern

At Pattern panel, ANSI31

**6.** Hover over any part of the drawing and you will see the preview of the hatch. You will notice that the Scale is a little bit small, so increase it to become equal to 2.

**7.** Select Hatch background = Yellow

**8.** Then select the following areas to hatch:

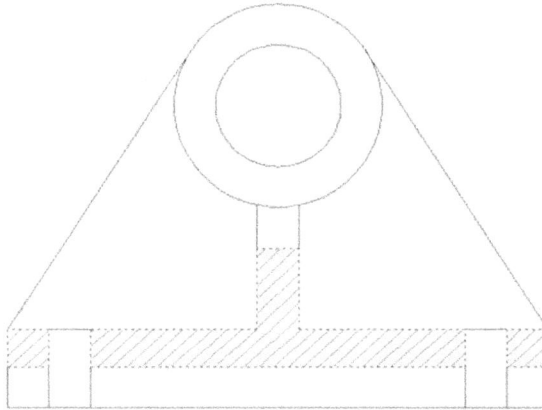

**9.** Click the Close Hatch Creation button at the right

**10.** Start the Hatch command again and you will notice that all options used by the previous hatch are still valid. Change the angle to 90o and apply it to the lower portion of the shape as shown below; click the Close Hatch Creation button at the right

**11.** You will have the following result:

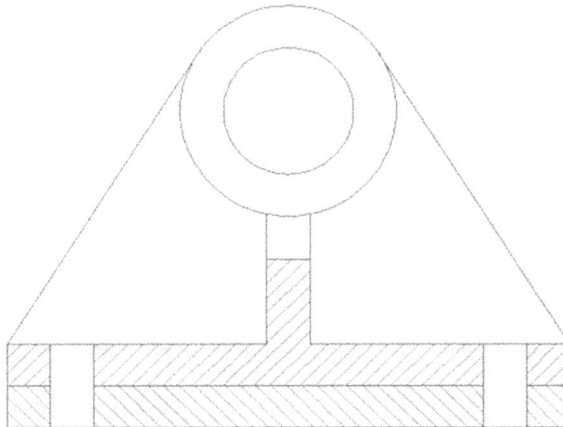

**12.** Pan to the right until you see a circle, and rectangle

**13.** Start Gradient command, from Pattern panel, select GR_SPHER pattern, click inside the circle

**14.** Change the two gradient colors as you wish, then finish the command

**15.** Start Gradient command again, from Pattern panel, select GR_CYLIN pattern, click inside the rectangle

**16.** Change the Angle to be 90, see how the gradient hatch reacted to your change. Finish the command

**17.** Save and close the file

## 6.8 HATCH COMMAND USING COMMAND WINDOW

Another method of using Hatch command, is to start the command, and utilize the options in the Command Window instead of using the ribbon. Start the Hatch command as we did previously, if you see the following prompt:

```
Pick internal point or [Select objects/Draw/Undo
/seTtings]:
```

Select the Draw option. Look at the Command Window, you will see the following:

```
Specify start point or [picK internal point/Select
objects/Rectangle/Circle/Mode /seTtings]:
```

You may receive this prompt immediately depending on the last time this command was used in this file. If you select Mode, you will see the following:

```
Specify draw hatch mode [Area/Path] <Area>:
```

You have two options:
- **Area**, which will allow you to draw an area filled with the current hatch pattern. The area shape can be any shape you draw, or a specific shape like Rectangle or Circle
- **Path**, which will allow you to draw a path filled with hatch pattern. The path shape can be any shape you draw, or a specific shape like Rectangle or Circle
- If you choose Path, you will see the following:

```
Specify start point or [picK internal point/
Select objects/Rectangle/Circle/ALignment/Width/Mode/
seTtings]:
```

- You can specify the Alignment of the Path (you have three choices: Center, Inside, and Outside)
- You can specify the Width of the path
- All hatch shapes produced from these options do not have any boundary

## PRACTICE 6-3B   HATCH USING COMMAND WINDOW

1. Start AutoCAD 2025

2. Open **Practice 6-3b.dwg**

3. Make layer Hatch the current layer

4. Start the Hatch command. Make sure you are at the Draw mode

5. Select Mode option and set it to Area. From the ribbon set the current pattern to ANSI32

6. From Command Window, draw the shape at the right (once done specifying points, press [Enter])

7. From Command Window, select Rectangle option, and fill the exiting rectangle. Do the same thing for the circle using Circle option

8. Set the current pattern to be Earth, and set Scale = 2

9. Select the Mode, and change it to Path

10. Set Alignment to Inside, and set the Width = 1

11. Draw a path around the lower part of the shape press [Enter])

12. You should receive the following result:

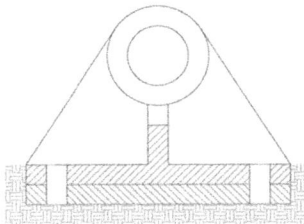

13. Save and close the file

## 6.9    SPECIFYING HATCH ORIGIN

When you want to hatch an area and want the pattern to start from a certain point and not to abide by the default settings of AutoCAD, then you need to manually set the Hatch Origin. By default, AutoCAD uses 0,0 as the starting point for any hatch, which means you will never know for certain if your hatch will be displayed correctly or not. In order to control Hatch Origin, make sure you are still in the **Hatch Creation** context tab, locate the **Origin** panel, and you will see the following:

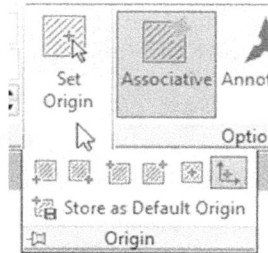

The apparent button is **Set Origin**, which will ask you to:

```
Specify origin point:
```

Specify the desired point or use the other predefined points (lower left corner, upper left corner, etc.). You have the ability to save the point you picked for future use instead of using 0,0.

See the following example:

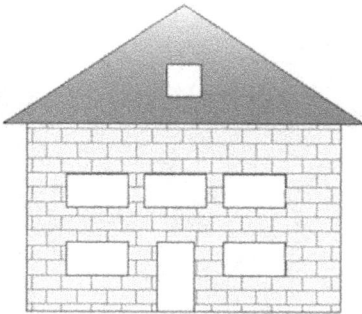

Default Origin

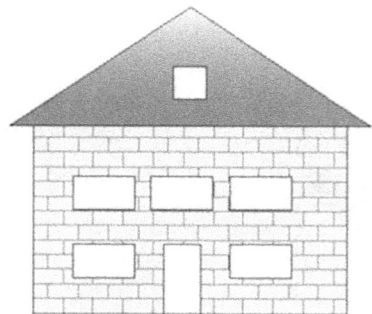

Origin is the lower left corner

## 6.10    CONTROLLING HATCH OPTIONS

These options will control the outcome of the hatching process; using these options you will be able to hatch an open area (Gap Tolerance) or create separate hatches, to mention only a few:

### 6.10.1    How to Create Associative Hatching

Hatch in AutoCAD is associative, which means hatch understands the boundary it fills. When this boundary changes, hatch will respond correctly.
See the illustration below:

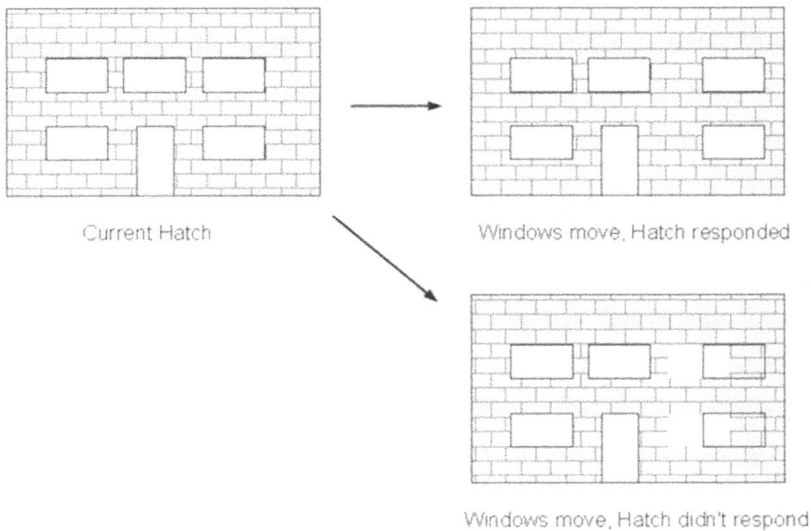

Current Hatch

Windows move, Hatch responded

Windows move, Hatch didn't respond

### 6.10.2    How to Make Your Hatch Annotative

Annotative is an advanced feature related to printing will be discussed later in this book.

### 6.10.3 Using Match Properties to Create Identical Hatches

This option will create an identical hatch from an existing hatch (it will reside in the same layer; also, it will have the same angle, scale, transparency, etc.).

This option has two-button associated with it: Use current origin or Use source hatch origin; both options are self-descriptive.

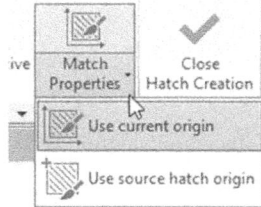

Select the **Match Properties** button, and you will see the following prompt:

```
Select hatch object:
```

Select the hatch object that you desire to mimic, and you will see the following prompt:

```
Pick internal point or [Select objects/seTtings]:
```

Click inside the desired area. Keep selecting the area, and when done press [Enter] to end the command.

### 6.10.4 Hatching an Open Area

By default, AutoCAD will hatch only closed areas. However, you can ask AutoCAD to hatch an area with an opening. To tell AutoCAD to allow hatching in an open area, simply set the **Gap Tolerance**, which is the maximum allowable opening; any area with an opening bigger than this value will not be hatched.

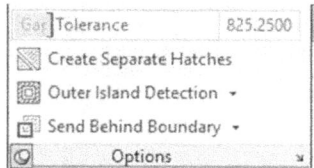

When you want to hatch this area, the preview will not be displayed; accordingly, you need to click inside the opened area, and you will see the following warning message:

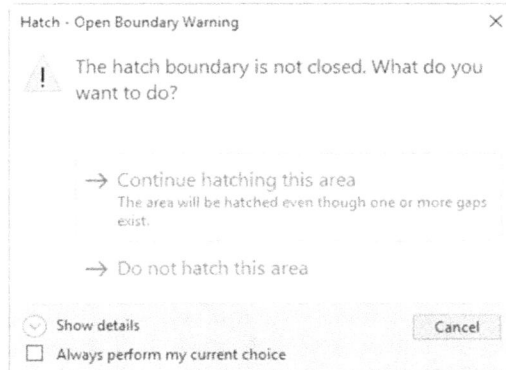

Hatch - Open Boundary Warning ✕

⚠ The hatch boundary is not closed. What do you want to do?

→ Continue hatching this area
The area will be hatched even though one or more gaps exist.

→ Do not hatch this area

⌄ Show details                    Cancel
☐ Always perform my current choice

Either continue hatching the open area or skip this operation.

### 6.10.5 Creating Separate Hatches in the Same Command

Using the same command, if you hatched several separated areas, AutoCAD will consider them a single hatch (single object).

You can override this default setting by telling AutoCAD that you want separate hatches for separate areas: simply click this button on.

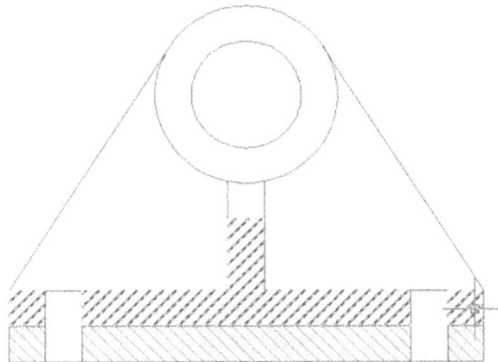

Single Object

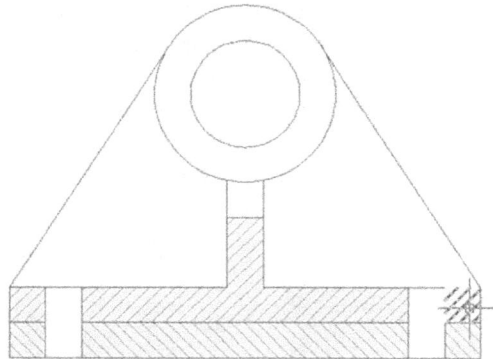

Seperate Hatches

### 6.10.6 Island Detection

When you are hatching an area that contains several areas (islands), these inside areas may contain more islands. We want to know how AutoCAD will treat these islands.

There are four different ways to do this:

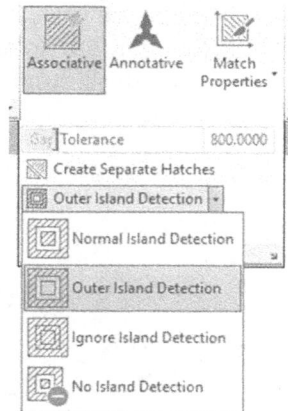

- **Normal Island Detection:** AutoCAD will hatch the first area (the outer one), then leave the second one, hatching the third, and so on
- **Outer Island Detection:** AutoCAD will hatch the outer area only
- **Ignore Island Detection:** AutoCAD will ignore all of the inner islands, and hatch the outer area, as if there were no areas inside
- **No Islands Detection:** this option will turn off the island detection feature, which is the same result as the Ignore Island Detection option

### 6.10.7   Set Hatch Draw Order

Hatch like any other object in AutoCAD: you can set the draw order for it, relative to the other objects. You have five choices to choose from:

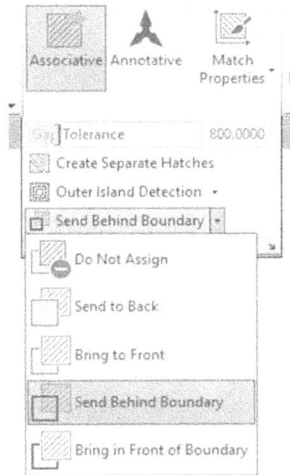

The five cases are:

- Do Not Assign, use the default
- Send to Back
- Bring to Front
- Send Behind Boundary
- Bring in Front of Boundary

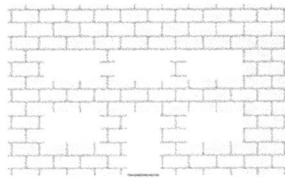

See the following example:

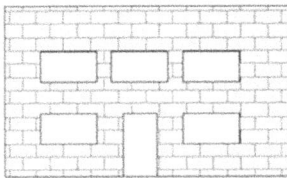

Send Behind Boundary                    Bring in Front of Boundary

# PRACTICE 6-4  HATCH ORIGIN AND OPTIONS

1. Start AutoCAD 2025

2. Open **Practice 6-4.dwg**

3. Start Hatch command

4. Using the Options panel, click the Match Properties button and select the hatch of the Toilet, and then apply it to the Kitchen. Press [Enter] to end the command

5. Start the Hatch command again, set background color = 40, scale = 4, and transparency = 0; hover over the Study room and change the Origin point to the lower left corner of the room. Click inside the Study and end the command

6. Zoom to the lower right corner of the Living Room, and you will see the area is opened. Start the Hatch command again and set the Gap Tolerance = 0.3 (the opening in this drawing is 0.2, so 0.3 is enough), then set to create separate hatches. Click inside the Living Room and you will see a warning message; select the **Continue hatching this area** option, then click inside the Sitting Room and end the command

7. Thaw layer **A-Doors**, and notice that the two doors of the Living Room and Study are not shown properly. To solve this problem, select the hatch, right-click, select the **Draw Order** option, and select the **Send to Back** option.

8. Start the Hatch command for the fourth time, and from Hatch Type select Solid and hatch the outside walls.

9. Zoom to any window of the Kitchen and move it a small distance; what happened to the hatch? _____ Why? _____

10. Save and close the file

## 6.11  HATCH BOUNDARY

If we were discussing older versions of AutoCAD, this panel (Boundary panel) would be the first panel to discuss, but because of the instantaneous display of the hatch once you are inside the hatch area, this part is not important in this version; hence, we delayed discussing it until the end.

Depending on what you are doing—creating a new hatch or editing an existing one—some of the buttons will be turned off and some of them will be valid.

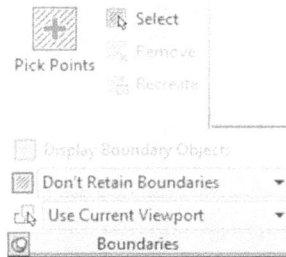

The **Pick Points** button is always on, which will allow you to pick the areas to hatch. **Select** and **Remove** will add/remove more objects to be included in the hatch boundaries.

If you are editing a hatch (by clicking it), you will see the **Recreate** button on. This button will help you to recreate the boundary if (for any reason) the boundary was deleted. Simply click the hatch without its boundary, then click the **Recreate** button, and you will see the following prompts:

```
Enter type of boundary object [Region/Polyline] <Polyline>:
Reassociate hatch with new boundary? [Yes/No] <N>:
```

The first prompt will ask you to select the desired boundary type, then ask to reassociate the boundary with the hatch.

If you select any hatch, AutoCAD will enable you to highlight (**Display**) **Boundary Objects**, so you can edit the boundary. See the following illustration:

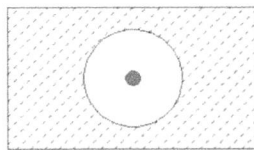

Display Boundary objects = Off          Display Boundary objects = On

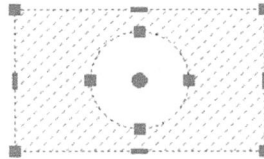

When you create a hatch, AutoCAD normally creates a polyline (or region) that fits the boundary exactly; once the command ends, AutoCAD will delete it. Using the **Retain Boundary** pop-up list, you can ask AutoCAD to keep it as a polyline or as a region, or not to keep it at all.

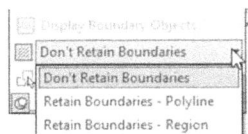

In order for the Hatch command to work, it needs to analyze all the objects in the current viewport (in model space this means the area you are seeing right now), which may take very long time (depending on the number of objects). You can ask AutoCAD to analyze only the relative objects rather than all objects. Locate the **Select New Boundary Set** button:

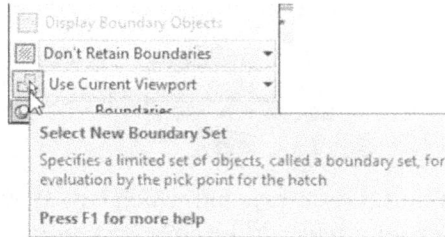

You will see the following prompt:

```
Select objects:
```

Select the desired objects, then press [Enter]; the pop-up list will read this time as **Use Boundary Set** instead of the default prompt.

## 6.12    EDITING HATCH

Editing hatches in AutoCAD has never been easier. There are two ways to edit a hatch: by clicking (single-click) or by using Properties (double-click).

If you click a hatch with a single click, three things will happen:

- You will see a grip (small dot) at the center of the area
- The context tab **Hatch Editor** will appear, which includes the same panels Hatch Creation includes.
- The Quick Properties palette will appear

If you move your mouse to the grip (without clicking) you will see the following menu:

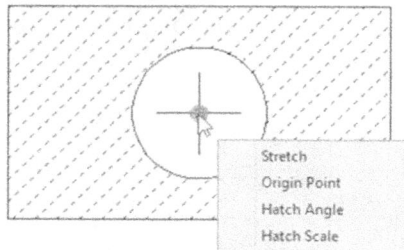

You can do any or all of the following:

- Stretch (which should be stretching the hatch, but will move the hatch instead). In order to stretch the hatch, it is preferable to display boundary objects as discussed above
- Modify the Origin Point on the spot
- Modify the Hatch Angle on the spot
- Modify the Hatch Scale on the spot

Also, the context tab Hatch Editor will make all the necessary modifications because it contains the same panels as the Hatch Creation tab.

The **Quick Properties** palette will appear as well, and hence, the user can make the edits needed:

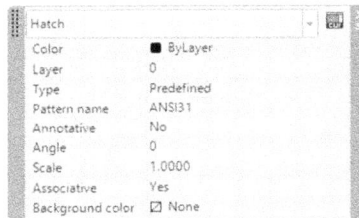

| Hatch | |
|---|---|
| Color | ■ ByLayer |
| Layer | 0 |
| Type | Predefined |
| Pattern name | ANSI31 |
| Annotative | No |
| Angle | 0 |
| Scale | 1.0000 |
| Associative | Yes |
| Background color | ☑ None |

Alternatively, double-clicking the desired hatch, or selecting the hatch, right-clicking, and selecting the **Properties** option will display the **Properties** palette:

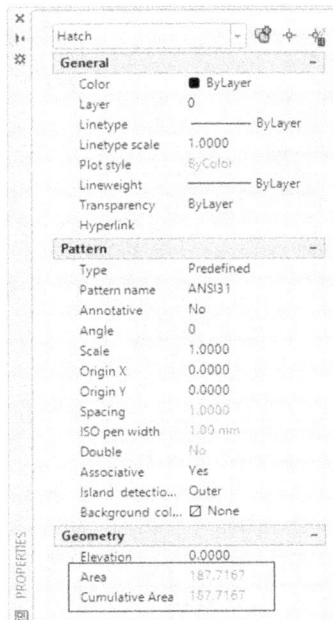

| Hatch | |
|---|---|
| **General** | |
| Color | ■ ByLayer |
| Layer | 0 |
| Linetype | ——— ByLayer |
| Linetype scale | 1.0000 |
| Plot style | ByColor |
| Lineweight | ——— ByLayer |
| Transparency | ByLayer |
| Hyperlink | |
| **Pattern** | |
| Type | Predefined |
| Pattern name | ANSI31 |
| Annotative | No |
| Angle | 0 |
| Scale | 1.0000 |
| Origin X | 0.0000 |
| Origin Y | 0.0000 |
| Spacing | 1.0000 |
| ISO pen width | 1.00 mm |
| Double | No |
| Associative | Yes |
| Island detectio... | Outer |
| Background col... | ☑ None |
| **Geometry** | |
| Elevation | 0.0000 |
| Area | 187.7167 |
| Cumulative Area | 187.7167 |

The data valid for editing is everything related to the hatch properties, options, and boundary. Moreover, the Properties palette provides a single piece of information that other methods do not, which is the Area of the hatch. If you select a single hatch (whether single area, or multiple areas hatched as a single area), the Area field and the Cumulative Area will be the same. If you select multiple areas created using different commands, however, you will see only Cumulative Area filled, and the Area field showing Varies as a value.

If you select the hatch and right-click, you will see some Hatch-related editing commands such as:

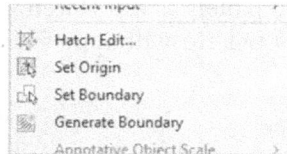

Which will do all or any of the following:

- **Hatch Edit**, which will show the old Hatch dialog box
- **Set Origin**, which will help you set a new origin
- **Set Boundary**, which will help you set a new boundary
- **Generate Boundary**, which will regenerate a new boundary for this hatch

## PRACTICE 6-5    HATCH BOUNDARY AND HATCH EDITING

**1.** Start AutoCAD 2025

**2.** Open **Practice 6-5.dwg**

**3.** Select the existing hatch at the top left of the drawing

**4.** From the **Boundaries** panel, click **Recreate**, then select **Polyline** and **Yes**. You can see that AutoCAD recreated a new boundary for this orphan hatch

**5.** Using the existing hatch at the middle left, select the hatch so you can see the special grip, then change the angle to 90°

**6.** Using the existing hatch at the bottom left, select the hatch so you can see the special grip, then change the origin point to be the lower left corner and the scale to be = 0.1

**7.** Using the existing hatch at the upper right, select, right-click, then select the **Properties** option; change the scale to be 0.5 and the background color to be Magenta. What is the total area of the hatch? _____

**8.** Start the Hatch command, select pattern name = steel, and set the scale = 2.0; now go to the Options panel and change Island detection to Normal. Try Hatching the shape at the middle right. Change the Island detection to Outer and try it, then Ignore and try it. Finally, return it back to Normal and finish the command

**9.** Save and close the file

## NOTES

## CHAPTER REVIEW

1. The command to convert a block to a file is:

   **a.** Makeblock

   **b.** Createblock

   **c.** Wblock

   **d.** None of the above

2. There are _____ hatch types in AutoCAD

3. When you create a block in a drawing, you cannot use it in other drawings:

   **a.** True

   **b.** False

4. Hatch grips are like the normal objects:

   **a.** True

   **b.** False

5. In the Insert command, you can see the blocks available in the current drawing:

   **a.** True

   **b.** False

## CHAPTER REVIEW ANSWERS

**1.** c

**3.** a

**5.** a

# WRITING TEXT

## In This Chapter

- How to write using single line text and multiline text
- How to edit text

## 7.1 WRITING TEXT USING SINGLE LINE TEXT

This command will create lines of text; each line is an independent object. To issue the command, go to the **Annotate** tab, locate the **Text** panel, then click the **Single Line** button:

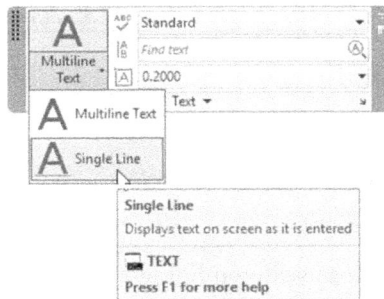

You will see the following prompt:

```
Current text style: "Standard" Text height: 0.2000
Annotative: No Justify:
Left Specify start point of text or [Justify/Style]:
Specify rotation angle of text <0>:
```

The first prompt is to provide you with some information about the current settings. The report shows the current style (in our example Standard), the current height (in our example = 0.20), whether this text will be annotative or not, and the justification of the text.

At the first prompt you can change the current Justification (discussed in the Multiline text) and Style settings by typing J or **S**, or you can specify the start point of the baseline of the text, then specify the rotation angle (the default value is 0 (zero); once you press [Enter] you will see a blinking cursor on the screen, ready for you to type any text you want. To finish any line, press [Enter]; to end the command press [Enter] twice.

Another way of setting the current text style is to go to the **Annotate** tab, then locate the **Text** panel; at the top right, the current text style can be changed to any desired text style:

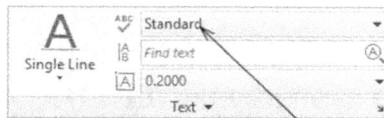

## PRACTICE 7-1    CREATING TEXT STYLE AND SINGLE LINE TEXT

**1.** Start AutoCAD 2025

**2.** Open **Practice 7-1.dwg**

**3.** Make Room Names text style current

**4.** Make layer Text current

**5.** Type the room names as shown below

**6.** Make the Standard text style current (in this text style the height = 0, hence you should set it every time you want to use this style)

**7.** Make layer Centerlines current

**8.** Zoom to the upper left centerline, check that the letter A is missing

**9.** Start Single Line text, right-click, and select the Justify option; from the list pick MC (Middle Center), select the center of the circle as the Start point, set the height to be 0.25 and the rotation angle = 0, then type A and press [Enter] twice

**10.** You should receive the following results:

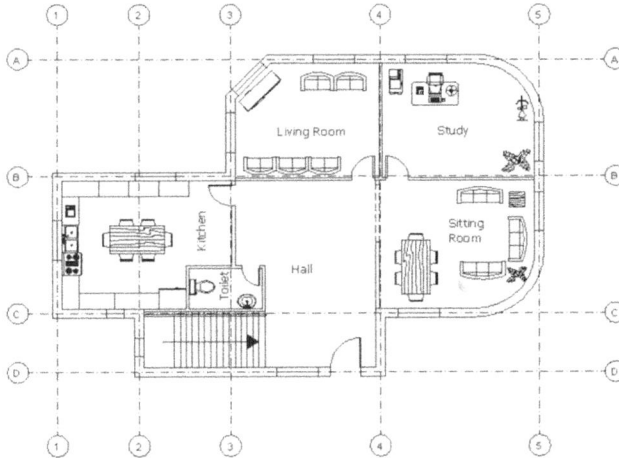

**11.** Save and close the file

## 7.2 WRITING TEXT USING MULTILINE TEXT

This command will enable you to type text in a MS Word-similar environment. To issue this command, go to the Annotate tab, locate the **Text** panel, then click the **Multiline Text** button:

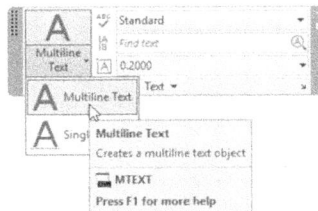

You will see the following in the command window:

```
Current text style: "TNR_05" Text height: 0.500
Annotative No
Specify first corner:
```

```
Specify opposite corner or [Height/Justify/
Line spacing/Rotation/Style/Width]:
```

The first prompt is to provide you with some information about the current settings. The report shows the current style (in our example TNR_05) and the current height (in our example = 0.50) and whether this text will be annotative or not. You will need to specify an area to write in, so after the prompts, the cursor will change to:

At the command window, you will see the following prompt:

```
Specify first corner:
```

Specify the first corner and you will see the following:

At the command window you will see the following prompt:

```
Specify opposite corner or [Height/Justify/
Line spacing/Rotation/Style/Width/ Columns]:
```

Specify the other corner or select one of the options; accordingly, a text editor with a ruler will appear:

AutoCAD will show a context tab called **Text Editor**, which appears as the following:

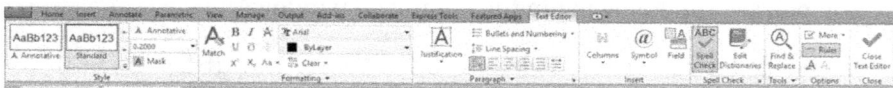

Using these panels will allow you to do many things. Here is a list of the panels available:

### 7.2.1 Style Panel

Below is a picture of the Style panel:

Use the Style panel to select the current text style you want to use and set the height (this value will overwrite the text style height, so be careful). Check **Background Mask**, for the selected text, you will see the following dialog box:

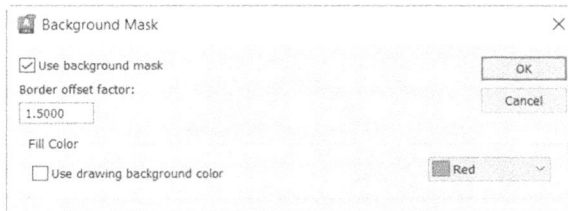

### 7.2.2 Formatting Panel

Below is a picture of the Formatting Panel:

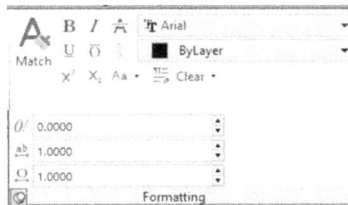

Use the **Formatting** panel to do all or any of the following:

- Match the selected text properties and apply them to other text
- Make the selected text **Bold**, **Italic**, **Underlined**, **Overlined**, or **Strike-through**
- Change the **Font** and the **Color** of the selected text
- Change the selected text to be **Superscript** or **Subscript**
- Change capital letters to small letters, and vice versa
- Clear the text formatting
- Change the **Oblique Angle** of the selected text
- Change the **Tracking** (to increase or decrease the spaces between letters). If the value is greater than 1, there will be more spaces between letters, and vice versa)
- Change the **Width Factor**

### 7.2.3 Paragraph Panel

Below is a picture of the Paragraph panel:

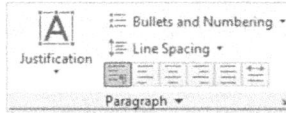

Use the Paragraph panel to do all or any of the following:

- Change the **Justification** of the text related to the text area selected by choosing one of the following options:

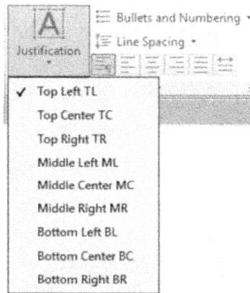

Below is an illustration of the nine points available related to the area:

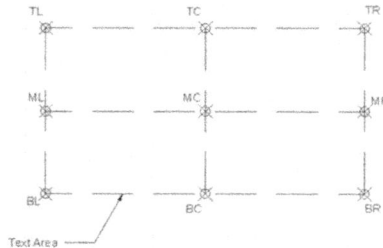

- Change the text selected to use **Bullets and Numbering**; there are three choices listed below (Numbered, Lettered, and Bulleted):

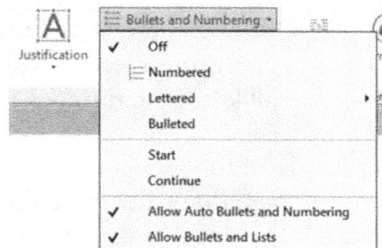

- Change the **Line Spacing** of the paragraph; the choices are listed below:

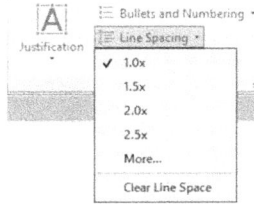

- Change the horizontal justification of the text, by using one of the six buttons as shown below:

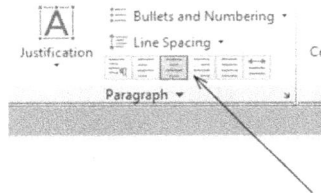

### 7.2.4 Insert Panel

Below is a picture of the Insert Panel:

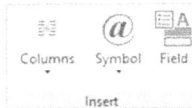

In Insert Panel, you can do all or any of the following to the selected text:

- Convert text to two **Columns** or more. If you click the **Columns** button, the following menu will appear:

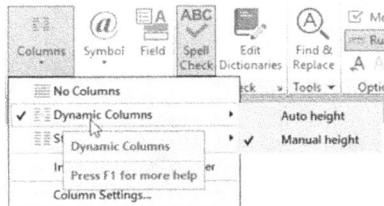

- The above shows **Dynamic Columns**, which will select whether you want AutoCAD to specify the height (Auto height) or you want to set the height (Manual height). If you want static columns then select the **Static Columns** option to specify the number of columns:

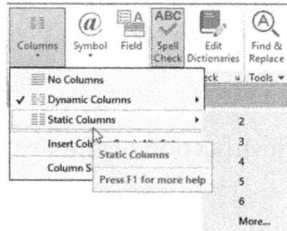

- Select the **Insert Column Break Alt+Enter** option to insert a column break at a certain line, which means the rest of the column will go to the next column
- Select the **Column Settings** option to show the **Column Settings** dialog box, which will allow you to do all of the above settings as well as set the **Column** width and the **Gutter** distance:

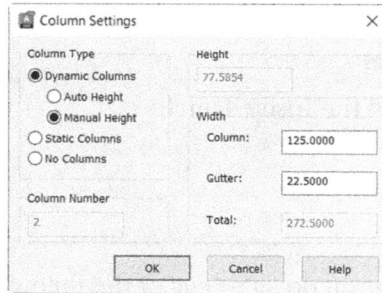

- This is an example of columns:

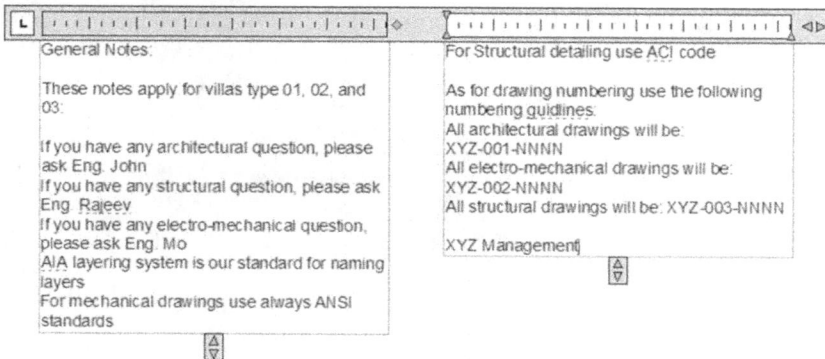

■ Select **Symbols** to add scientific characters to your text that are not available at the keyboard. You will see the twenty available symbols:

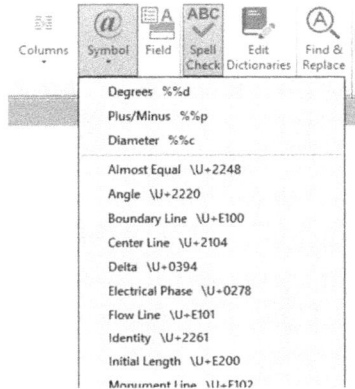

### 7.2.5  Spell Check Panel

Below is a picture of the Spell Check Panel:

■ As long as you are in the Text editor, you can leave the **Spell Check** button on to catch any misspelled words; a dotted red line will appear underneath them, as shown in the following example:

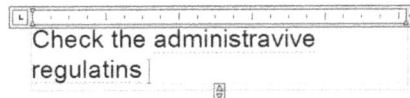

■ To obtain suggestions for the correct word, move to the word, right-click, and you will see the following:

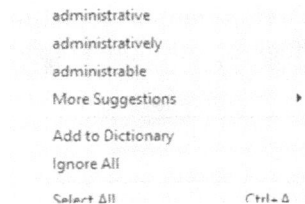

■ Select the **Edit Dictionaries** button, which will enable you to select another dictionary other than the default.

### 7.2.6   Tools Panel

Below is a picture of the Tools Panel:

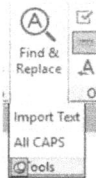

In the **Tools** panel, you can do all or any of the following:

- Click the **Find & Replace** button to search for a word and then replace it with another word, as you will see in the following dialog box:

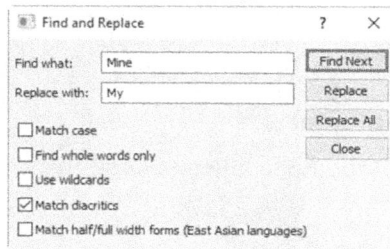

- Use the **Import Text** button to import text from text files
- Use the **AutoCAPS** button to write in the Text editor using capital letters

### 7.2.7   Options Panel

Below is the picture of the Options Panel:

Using the Options panel, you can do all or any of the following:

- Change the Character set, Remove formatting of the selected text change, change the Editor Settings, or get more help about Multiline text:

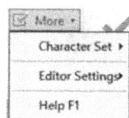

- Show the ruler (by default it is displayed) or hide it
- Undo and redo text actions

### 7.2.8   Close Panel

Below is a picture of the Close panel, which contains a single button to close the Text editor; as a consequence, the Text editor context tab with all its panels will close:

### 7.2.9   Text Editor

While you are in the text editor you can do the following things:

■ The user can use the ruler to set the First Line indent and Paragraph indent:

■ If you right-click you will see the following menu, which includes all the functions discussed above; hence, you can reach these commands either using the panels and buttons or by this method:

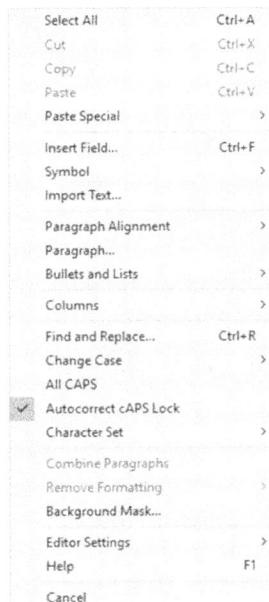

▪ You can change the width and height of the area using the following controls:

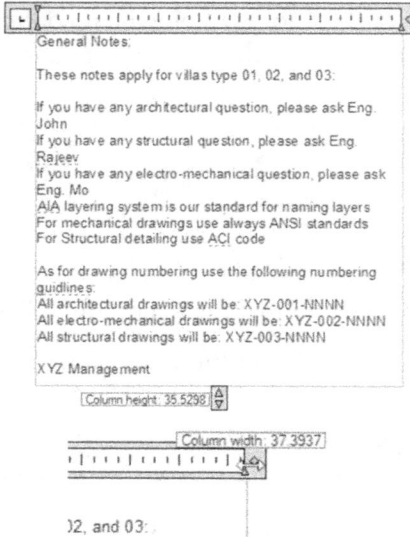

General Notes:

These notes apply for villas type 01, 02, and 03:

If you have any architectural question, please ask Eng. John
If you have any structural question, please ask Eng. Rajeev
If you have any electro-mechanical question, please ask Eng. Mo
AIA layering system is our standard for naming layers
For mechanical drawings use always ANSI standards
For Structural detailing use ACI code

As for drawing numbering use the following numbering guidlines:
All architectural drawings will be: XYZ-001-NNNN
All electro-mechanical drawings will be: XYZ-002-NNNN
All structural drawings will be: XYZ-003-NNNN

XYZ Management

Column height: 35.5298

Column width: 37.3937

)2, and 03:

# PRACTICE 7-2   WRITING USING MULTILINE TEXT

1. Start AutoCAD 2025

2. Open **Practice 7-2.dwg**

3. You are now in a layout called "Cover"

4. Make sure that the current layer is Text

5. Make text style Title the current text style

6. Start Multiline text and specify the two corners of the rectangle as your text area

7. Before you write anything, change the Justification to MC

**8.** Type the following as shown:

Villa Type 03
Ground Floor Plan
Architectural Plan
Architectural Details

**9.** Erase the rectangle

**10.** Go to layout called ISO A1 – Overall

**11.** Make Notes text style the current text style

**12.** Using the rectangle at the right, specify the two corners of your text area

**13.** Using the **Tools** panel, select the **Import Text** command and select the Notes.text file

**14.** Select General Notes, and make it Bold, Underlined, and size = 5.0

**15.** Stretch the width a little bit so "03" will be in the same line

**16.** Select the text from the first "If you have…" to "ACI code," then select Bullets and Numbering and select Numbered

**17.** Select the text starting from "As for…" until the last "NNNN," and make them Bulleted with line spacing = 1.5 x

**18.** Select the last line "XYZ Management" and make it centered and Italic

**19.** Close the text editor

**20.** Erase the rectangle

**21.** Go to ISO A1 – Architectural Details

**22.** Using the rectangle, start Multiline text, select two opposite corners, and import the same file: Notes.txt

**23.** Do the same thing you did above for the numbered and bulleted lists

**24.** Start the Column command and set two static columns, then using the Column Settings dialog box, set Column Width = 125 and Gutter = 22.5

**25.** Using the two arrows at the lower left corner, make sure that all six points are in the first column

**26.** Erase the rectangle

**27.** This what you should receive:

General Notes:

These notes apply for villas type 01, 02, and 03:

1. If you have any architectural question, please ask Eng. John
2. If you have any structural question, please ask Eng. Rajeev
3. If you have any electro-mechanical question, please ask Eng. Mo
4. AIA layering system is our standard for naming layers
5. For mechanical drawings use always ANSI standards
6. For Structural detailing use ACI code

- As for drawing numbering use the following numbering guidlines:
- All architectural drawings will be: XYZ-001-NNNN
- All electro-mechanical drawings will be: XYZ-002-NNNN
- All structural drawings will be: XYZ-003-NNNN

*XYZ Management*

**28.** Save and close the file

## 7.3    TEXT EDITING

There are several ways to edit the text, such as using double-click, using Quick Properties, using Properties, and using Grips.

### 7.3.1    Double-Click Text

To edit the text, whether it is Single Line Text or Multiline text, simply double-click it. If it is Single Line Text you will see the text selected, so you will be able to add to it or modify it. If it is Multiline text, you will see the Text Editor reopened, and the Text Edit context tab appears at the top. Make all of your desired changes.

### 7.3.2    Quick Properties and Properties

To show the Quick Properties for either Single Line Text or Multiline Text, simply click the desired text. You will see the following (Text in the below pictures means Single Line Text and MText is Multiline Text).

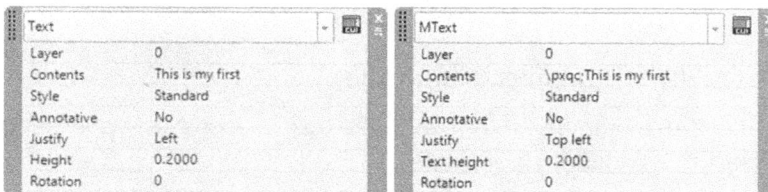

| Text | |
|---|---|
| Layer | 0 |
| Contents | This is my first |
| Style | Standard |
| Annotative | No |
| Justify | Left |
| Height | 0.2000 |
| Rotation | 0 |

| MText | |
|---|---|
| Layer | 0 |
| Contents | \pxqc;This is my first |
| Style | Standard |
| Annotative | No |
| Justify | Top left |
| Text height | 0.2000 |
| Rotation | 0 |

You can change the following things: Layer, Contents, the Style, Annotative (Yes or No), Justification, Height, and Rotation angle.

If you select Single Line Text or Multiline Text, right-click, and then select the Properties option, you will see the following (Text in the below pictures means Single Line Text and MText is Multiline Text):

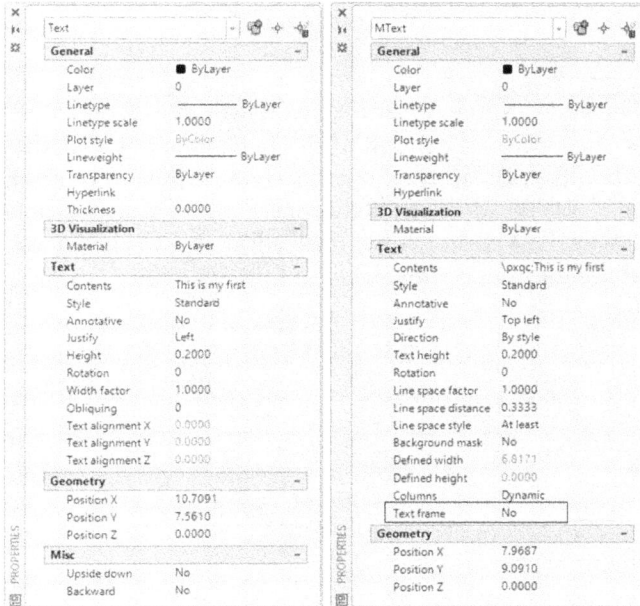

As you can see from the above pictures, you can change anything related to the selected text. Properties for MTEXT include an option to add a Text frame around the text. Your text will appear similar to the following:

For any structural
question, please ask
Eng. Gubta

### 7.3.3 Editing Using Grips

You will see a grip showing at the start point of the baseline, and another one at the justification point (you may see one only, in the case of both points coinciding); if you click a Single Line text you will see:

.This is my first text

Whereas you will see a single grip appears at the Justification point when Multiline text is selected (in the lower example, the Justification point is TL), and you will see two triangles, one at the lower part and one at the right side. The lower part of the triangle will allow you to cut your Multiline Text to columns; simply stretch it up and see the text cut into two columns. The triangle at the right will allow you to increase/decrease the horizontal distance of the text, hence, increasing/ decreasing the height of the text.

These notes apply for villas type
01, 02, and 03:

If you click any multicolumn text, you will see the same thing for the first column at the left, except the arrow at the bottom is pointing downward instead of upward. For the other columns you will see an arrow at the right side. For the last column at the right, in addition to the normal arrow at the right, you will see an arrow at the lower right corner, which will allow the whole text to increase/decrease the width and height in one shot. See the following illustration:

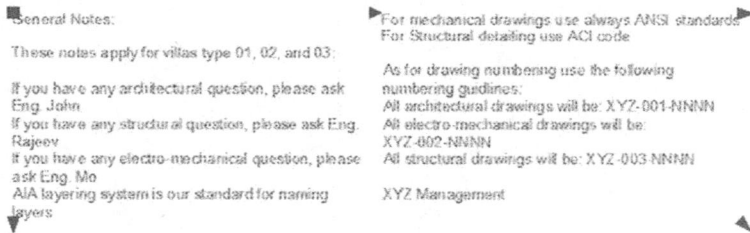

General Notes:

These notes apply for villas type 01, 02, and 03:

If you have any architectural question, please ask Eng. John
If you have any structural question, please ask Eng. Rajeev
If you have any electro-mechanical question, please ask Eng. Mo
AIA layering system is our standard for naming layers

For mechanical drawings use always ANSI standards
For Structural detailing use ACI code

As for drawing numbering use the following numbering guidlines:
All architectural drawings will be: XYZ-001-NNNN
All electro-mechanical drawings will be: XYZ-002-NNNN
All structural drawings will be: XYZ-003-NNNN

XYZ Management

## 7.4 SPELLING CHECK AND FIND AND REPLACE

While you are at the Text Editor, you can Spell Check and Find and Replace, but what about if you are not in the Text Editor: can you Spell Check and Find and Replace? The answer to that question is yes, AutoCAD can spell check not only the whole drawing and the current text but can also check the current space and/or layout (layout will be discussed in Chapter 9) or any desired selected text. To issue the Check Spelling command, go to the **Annotate** tab, locate the **Text** panel, and click the **Check Spelling** button:

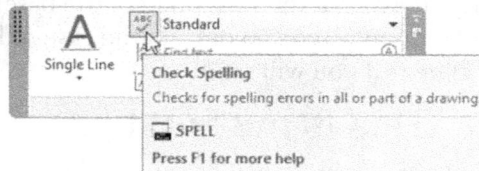

A
Single Line

Standard

Find text

**Check Spelling**
Checks for spelling errors in all or part of a drawing

SPELL
Press F1 for more help

You will see the following dialog box:

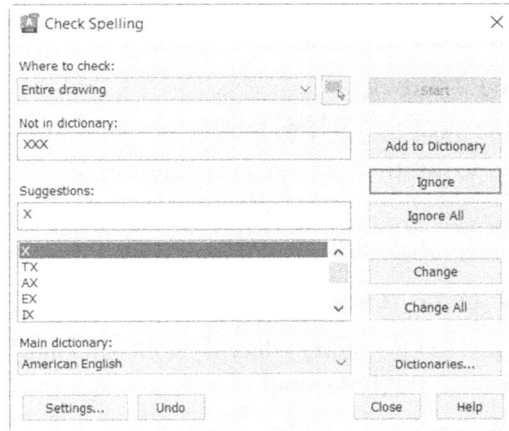

If AutoCAD finds any misspelled word, it will provide you with suggestions for a proper replacement or you can simply ignore the AutoCAD findings. Also, AutoCAD can find any word or part of a word and replace it in the entire drawing file.

To issue the **Find and Replace** command, go to the **Annotate** tab, locate the **Text** panel, type the desired word you want to replace in the **Find text** field as shown below, then click the small button at the right:

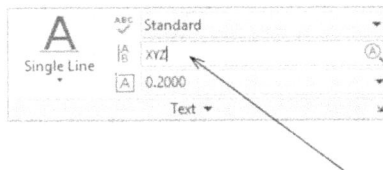

You will see the following dialog box:

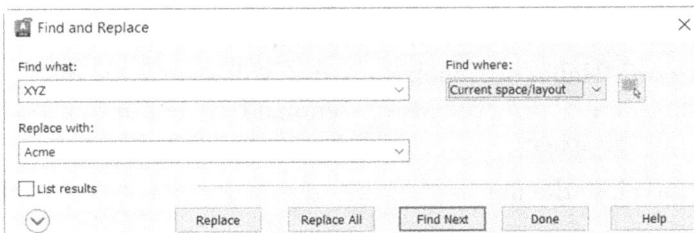

Under **Replace with**, type the new word(s) you want to replace. Select one of the following choices: **Find**, **Replace**, and **Replace All**. When done, click **Done**.

## PRACTICE 7-3 EDITING TEXT

**1.** Start AutoCAD 2025

**2.** Open **Practice 7-3.dwg**

**3.** You are now in Cover layout. Click the text, and using the arrow at the bottom, drag it upward to cut the text into two columns, each holding two lines

**4.** Using the arrow at the right of the second column, drag it until the two columns touch each other. Press [Esc] to end the editing process

**5.** Go to layout ISO A1 – Overall

**6.** Zoom to the text at the right, select it, right-click and choose the Properties option, and change the following:

    **a.** Change the style from Notes to Standard

    **b.** Change Justify to be Middle Center

    **c.** Change Text Height to be 4.5

**7.** Using the arrow at the right, stretch the text to the right by 10 units (if OSNAP is frustrating you, switch it off)

**8.** Press [Esc] to end the editing process

**9.** Go to the Annotate tab, locate the Text panel, and at the Find and Replace field type XYZ, then click the small button at the right. When the dialog box comes up, type ACME in the Replace field, then click Replace All

**10.** Go to layout ISO A1 Architectural Details, double-click the multicolumn text, and convert it to single column text.

**11.** Press [Esc] to end the editing process

**12.** Go to Model space

**13.** Click word "Hall" and Quick Properties will appear; set the Justification to be Middle Center. Using the Move command, try to put this word at the middle center of the space

**14.** Save and close the file

# NOTES

## CHAPTER REVIEW

1. There are two types of text in AutoCAD; single line and multiline:

   **a.** True

   **b.** False

2. One of the following is not a panel in the Text Editor context tab:

   **a.** Insert

   **b.** Options

   **c.** Text

   **d.** Paragraph

3. When you start the Single line text command, AutoCAD will inform you the current text style:

   **a.** True

   **b.** False

4. While you are in multiline command, select _____ option to make the rest of the column go to the next column

5. In the multiline text command, you can set the background for a text:

   **a.** True

   **b.** False

6. Degrees, Centerline, and Not Equal are _____ you can insert in the multiline text command

## CHAPTER REVIEW ANSWERS

**1.** a

**3.** a

**5.** a

## CHAPTER 8

# DIMENSIONS

## In This Chapter

- What is dimensioning and what are dimension types?
- How to insert different types of dimensions
- How to edit dimension block

## 8.1 WHAT IS DIMENSIONING IN AUTOCAD?

Dimensioning in AutoCAD is similar to Text; the user should prepare dimension style as the first step, and then use it to insert dimensions. Dimension Style will control the overall outcome of the dimension block generated by the different types of dimension commands.

To insert a dimension, depending on the type of the dimension, the user should specify points or select objects, and then a dimension block will be added to the drawing. For example, in order to add a linear dimension, the user will select two points representing the distance to be measured, and a third point will be the location of the dimension block. See the illustration below:

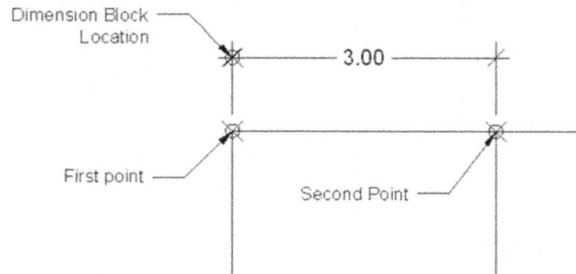

The generated block consists of three portions, which are:

- Dimension line
- Extension lines
- Dimension Text

See the following illustration:

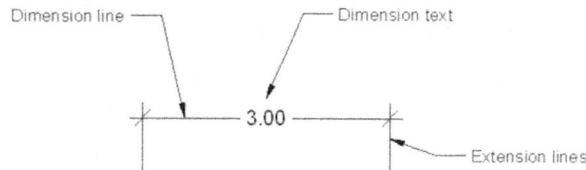

## 8.2   DIMENSION TYPES

The following are the dimension types in AutoCAD:

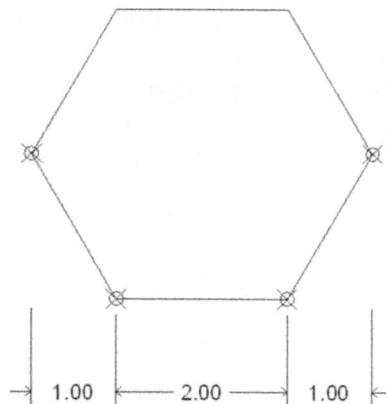

Linear

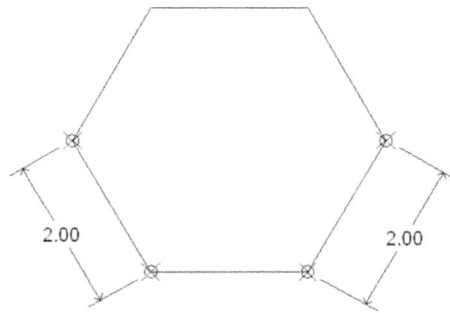

Aligned

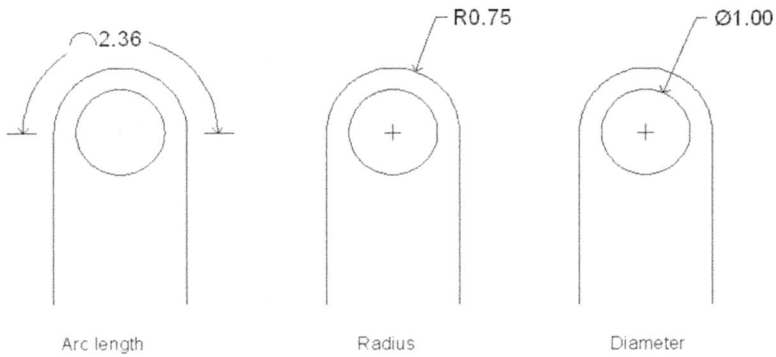

Arc length          Radius          Diameter

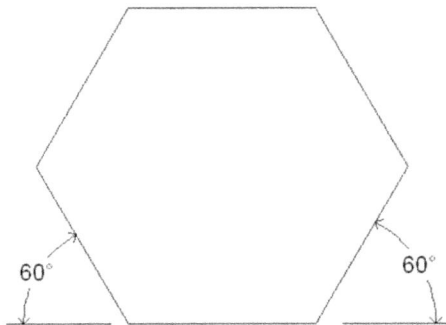

Angular

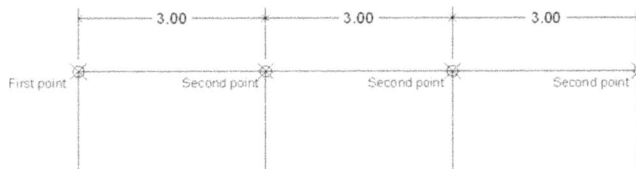

Continuous Dimension

Baseline Dimension

Ordinate

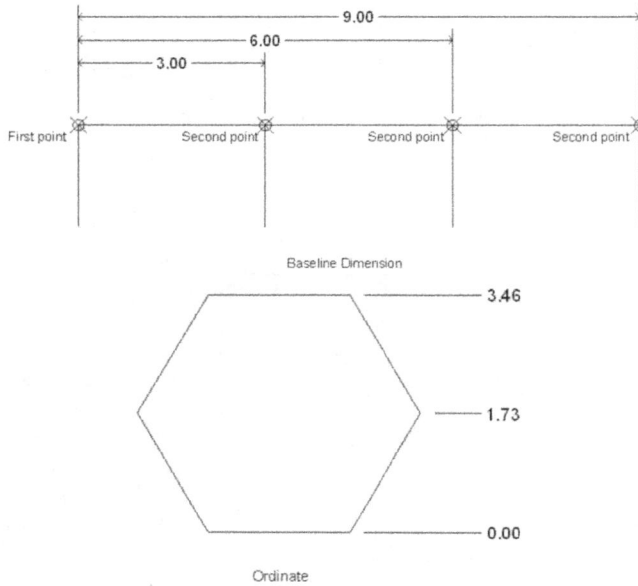

## 8.3    HOW TO INSERT A LINEAR DIMENSION

This command will create a horizontal or vertical dimension. To start the Linear command, go to the **Annotate** tab, locate the **Dimensions** panel, then select the **Linear** button:

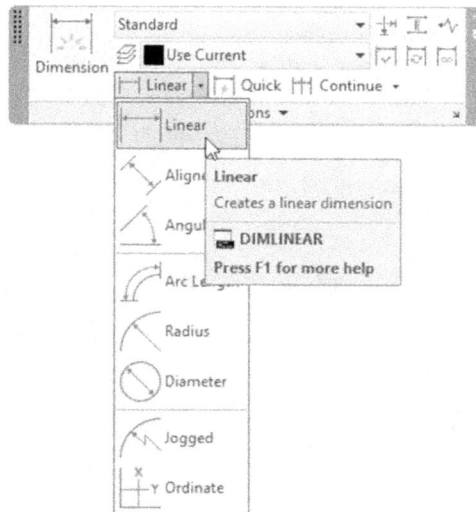

You will see the following prompts:

```
Specify first extension line origin or <select object>:
Specify second extension line origin:
Specify dimension line location or
[Mtext/Text/Angle/Horizontal/Vertical/Rotated]:
```

Specify the first point and second point of the dimension distance to be measured, and then specify the location of the dimension block by specifying the location of the dimension line. The following shows the result:

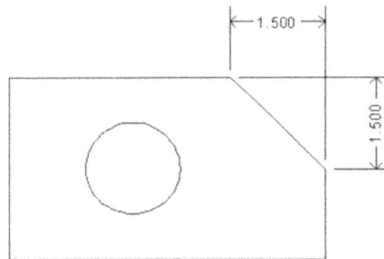

Prompts contain other options, which are:

- Mtext
- Text
- Angle
- Horizontal
- Vertical
- Rotated

Mtext will edit the measured distance in **MTEXT** mode, whereas Text will edit the measured distance in **DTEXT** (Single line) mode. Angle changes the angle of the text (default is 0 (zero)). Horizontal and Vertical will force the dimension to be either horizontal or vertical (default will specify either by the movement of the mouse). Finally, Rotated creates a dimension line parallel to another angle given by you.

## 8.4 HOW TO INSERT AN ALIGNED DIMENSION

This command will create a dimension parallel to the two points specified. To start this command, go to the **Annotate** tab, locate the **Dimensions** panel, and then select the **Aligned** button:

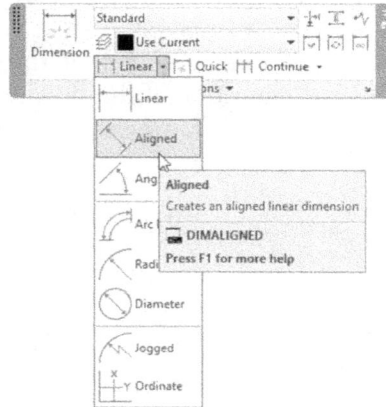

The following prompt will appear:

```
Specify first extension line origin or <select object>:
Specify second extension line origin:Specify dimension
line location or[Mtext/Text/Angle]:
```

Specify the first point and second point of the dimension distance to be measured, then specify the location of the dimension block by specifying the location of the dimension line, as seen in the following illustration:

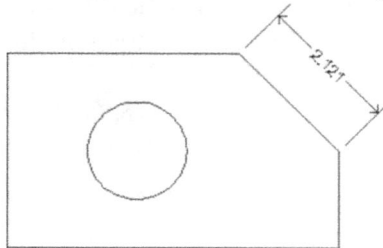

The rest of the options are identical to the **Linear** command prompts.

## 8.5 HOW TO INSERT AN ANGULAR DIMENSION

This command will insert an angular dimension between two lines, the included angle of a circular arc, two points and the center of a circle, or three points. To start this command, go to the **Annotate** tab, locate the **Dimensions** panel, then select the **Angular** button:

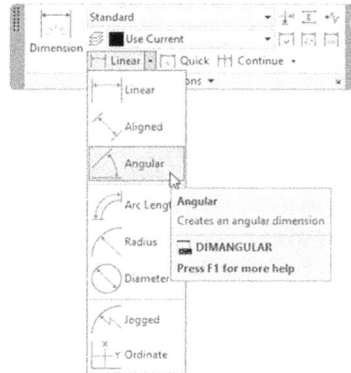

AutoCAD may use one of the following methods based on the selected objects:

- If you select a circular arc, AutoCAD will measure the included angle
- If you select a circle, your selecting points will be the first point, the center of the circle will be the second point, and then the user will select the third point
- If you select a line, it will ask you to select a second line
- If you select a point, it will be considered as a center point, and AutoCAD will ask the user to specify two more points

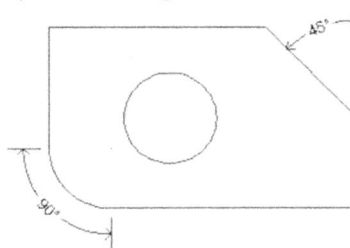

Based on the above discussion, when you start the command, you will see the following prompts (the below example is when we select an arc):

```
Select arc, circle, line, or <specify vertex>:
Specify dimension arc line location or
[Mtext/Text/Angle]:
```

## PRACTICE 8-1   INSERTING LINEAR, ALIGNED, AND ANGULAR DIMENSIONS

**1.** Start AutoCAD 2025

**2.** Open **Practice 8-1.dwg**

**3.** Make layer Dimension the current layer

**4.** Make sure that Dimension Style = Part

**5.** Insert the dimension as shown below:

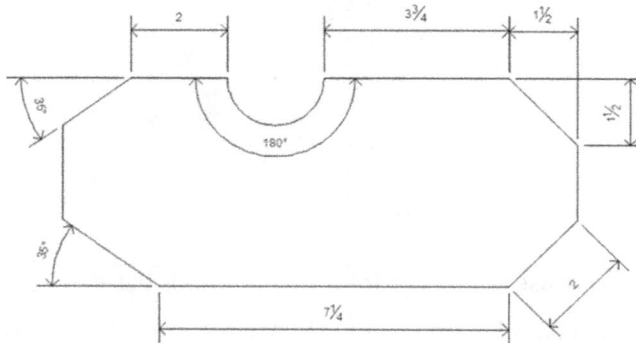

**6.** Save and close the file

## 8.6 HOW TO INSERT AN ARC-LENGTH DIMENSION

This command will create a dimension measuring the length of an arc. To start this command, go to the **Annotate** tab, locate the **Dimensions** panel, then select the **Arc Length** button:

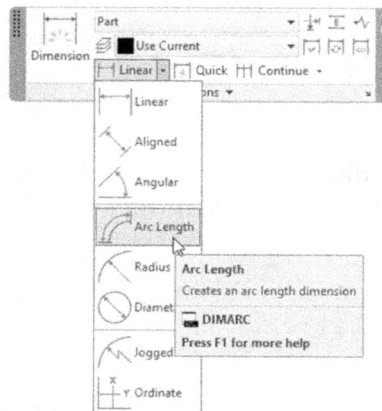

You will see the following prompts:

```
Select arc or polyline arc segment:
Specify arc length dimension location, or
[Mtext/Text/Angle/Partial/Leader]:
```

Select the desired arc and then locate the dimension block, either inside or outside the arc. You will receive the following:

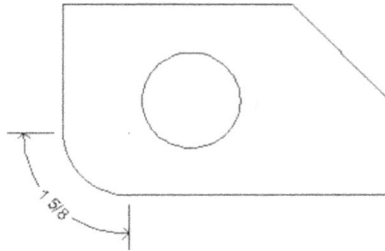

As for the options: Mtext, Text, and Angle were already tackled previously. Partial option means you want to insert an arc-length dimension on part of the arc. The procedure is to select the arc, then select two internal points on the arc to obtain the following:

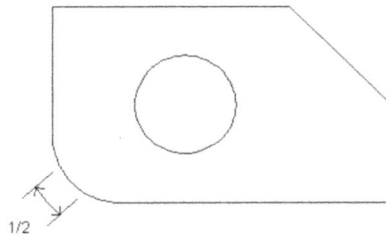

## 8.7   HOW TO INSERT A RADIUS DIMENSION

This command will insert a radius dimension on an arc or circle. To start this command, go to the **Annotate** tab, locate the **Dimensions** panel, then select the **Radius** button:

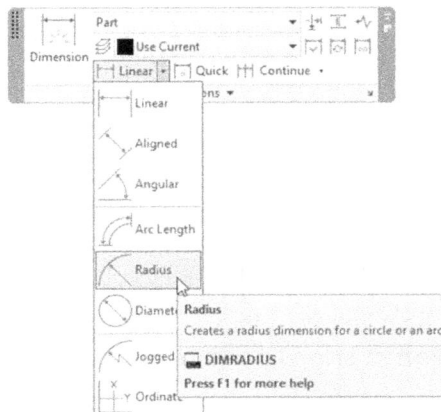

You will see the following prompts:

```
Select arc or circle:
Specify dimension line location or [Mtext/Text/ Angle]:
```

Select the desired arc or circle and then locate the dimension block. You will receive the following:

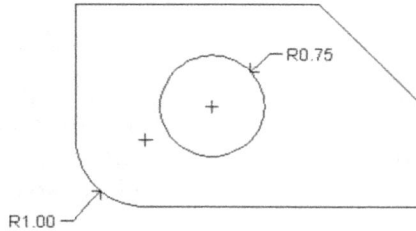

## 8.8   HOW TO INSERT A DIAMETER DIMENSION

This command will insert a diameter dimension on an arc or circle. To start this command, go to the **Annotate** tab, locate the **Dimensions** panel, then select the **Diameter** button:

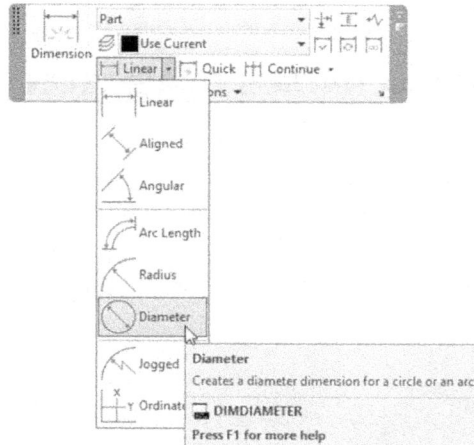

You will see the following prompts:

```
Select arc or circle:
Specify dimension line location or [Mtext/Text/Angle]:
```

Select the desired arc or circle, and then locate the dimension block. You will receive the following:

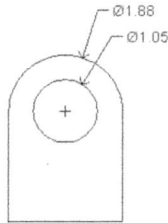

## PRACTICE 8-2   INSERTING ARC LENGTH, RADIUS, AND
## DIAMETER DIMENSION

**1.** Start AutoCAD 2025

**2.** Open **Practice 8-2.dwg**

**3.** Make layer Dimension the current layer

**4.** Make sure that the current Dimension Style is Part

**5.** Insert the dimension as shown below:

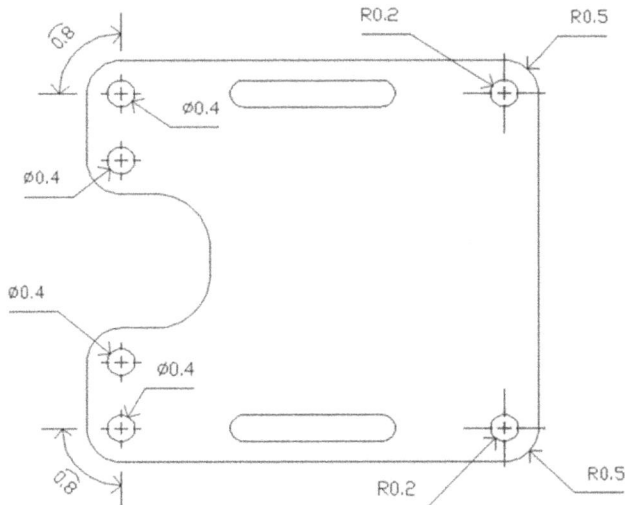

**6.** Save and close the file

## 8.9 HOW TO INSERT A JOGGED DIMENSION

This command will insert a jogged arc dimension for a big arc, simulating a new center point. To start this command, go to the **Annotate** tab, locate the **Dimensions** panel, and select the **Jogged** button:

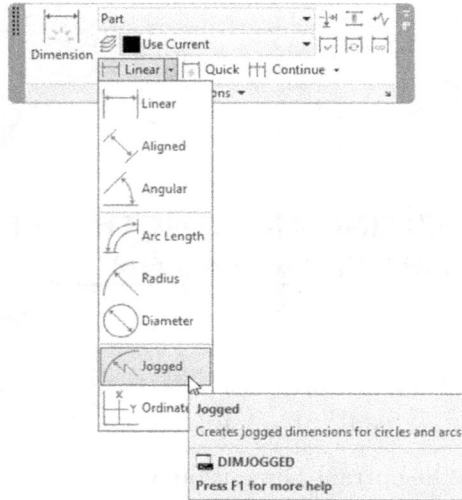

The following prompts will appear:

```
Select arc or circle:
Specify center location override:
Dimension text = 1.5 Specify dimension line location
or [Mtext/Text/ Angle]:  Specify jog location:
```

As a first step, select an arc or circle, specify the point that will be the new center point (here AutoCAD calls it override), locate the dimension line and, finally, specify the jog's location. You will receive the following:

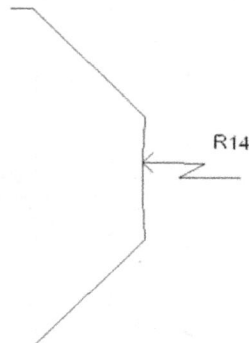

## 8.10 HOW TO INSERT AN ORDINATE DIMENSION

This command will insert dimensions relative to a datum, either in X or in Y. See the following illustration:

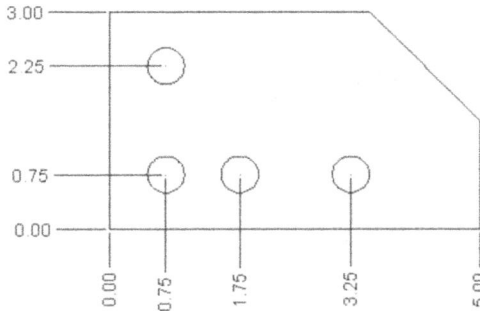

**NOTE** *Use the UCS command, Origin option to relocate the origin to one side of the shape so the values in both X and Y will be correct; if you leave the origin to the current UCS origin, the values inserted may be wrong.*

To start this command, go to the **Annotate** tab, locate the **Dimensions** panel, then select the **Ordinate** button:

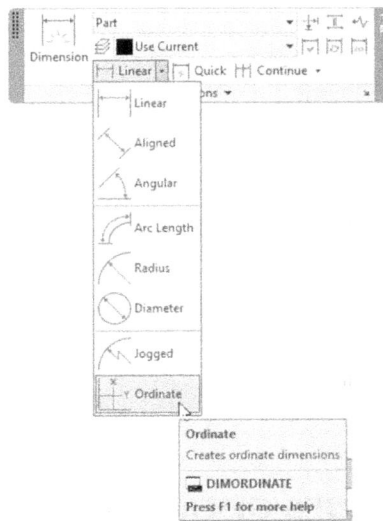

You will see the following prompts:

```
Specify feature location:
Specify leader endpoint or [Xdatum/Ydatum/Mtext/Text/Angle]:
```

First select the desired point. By default, AutoCAD will give you the freedom to go in the direction of X or Y. If you want to force the mouse to measure points related to the X axis, then select the **Xdatum** option; the same applies for the Y axis. The rest of options have already been discussed.

## PRACTICE 8-3   INSERTING DIMENSIONS

**1.** Start AutoCAD 2025

**2.** Open **Practice 8-3.dwg**

**3.** Make layer Dimension current

**4.** Make sure that Dimension Style is Part

**5.** Insert the dimensions as shown below:

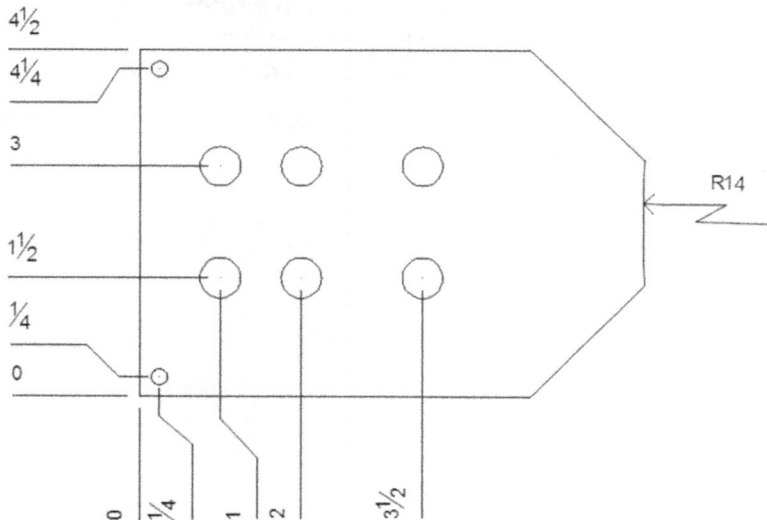

**6.** Save and close the file

## 8.11   INSERTING A SERIES OF DIMENSIONS USING THE CONTINUE COMMAND

AutoCAD allows the user to input a series of dimensions using the Continue command. The Continue command will follow the last dimension command by asking to input the second point, assuming that the last point of the last dimension will be considered the first point for the coming one. To start this command, go to the **Annotate** tab, locate the **Dimensions** panel, then select the **Continue** button:

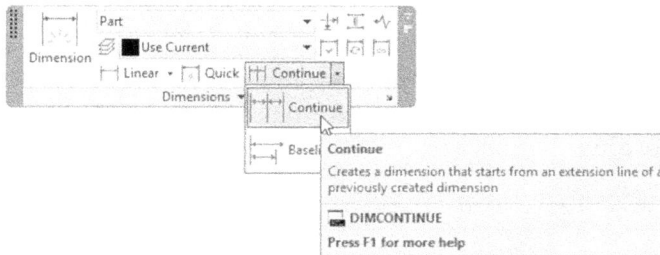

There are two scenarios in which to use Continue:

- If there was not any dimension command issued in this AutoCAD session, AutoCAD will ask you to select an existing dimension (linear, ordinate, or angular). You will see the following prompt:

```
Select continued dimension:
```

- If there was a dimension command issued in this AutoCAD session, AutoCAD then will ask you to continue this command by asking you to specify the second point. Also, you can select an existing dimension block to continue it, or you can undo the last continue command. You will see the following prompts:

```
Specify a second extension line origin or [Undo/
Select] <Select>:
```

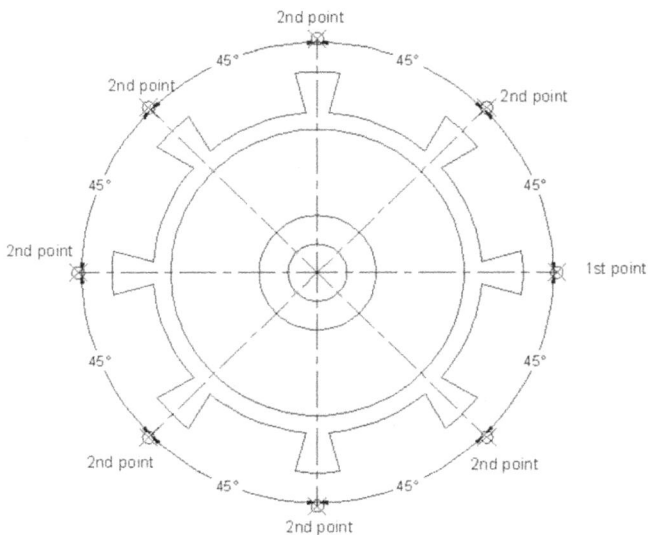

## 8.12 INSERTING A SERIES OF DIMENSIONS USING THE BASELINE COMMAND

This command is identical to the Continue command, except all dimensions will be measured referenced to the first point specified by you as the baseline. To start this command, go to the **Annotate** tab, locate the **Dimensions** panel, then select the **Baseline** button:

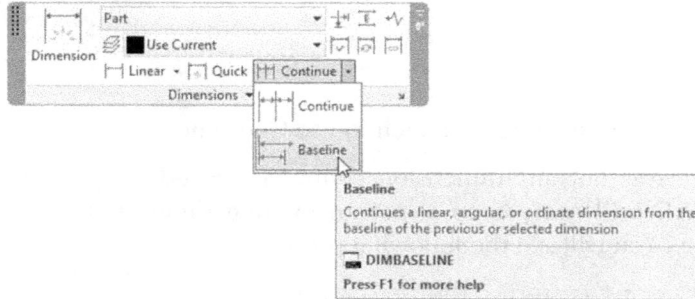

We do not need to discuss the prompts of this command, because it resembles the Continue command prompts. You will receive the following image:

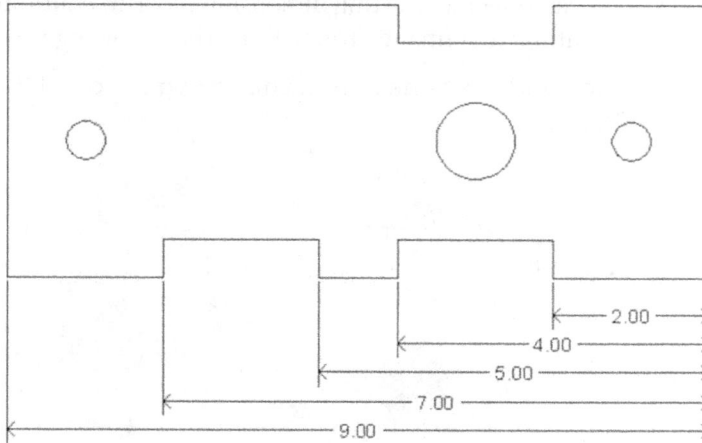

## PRACTICE 8-4    CONTINUE COMMAND

**1.** Start AutoCAD 2025

**2.** Open **Practice 8-4.dwg**

**3.** Create an angular dimension, then using the Continue command, complete the shape as follows:

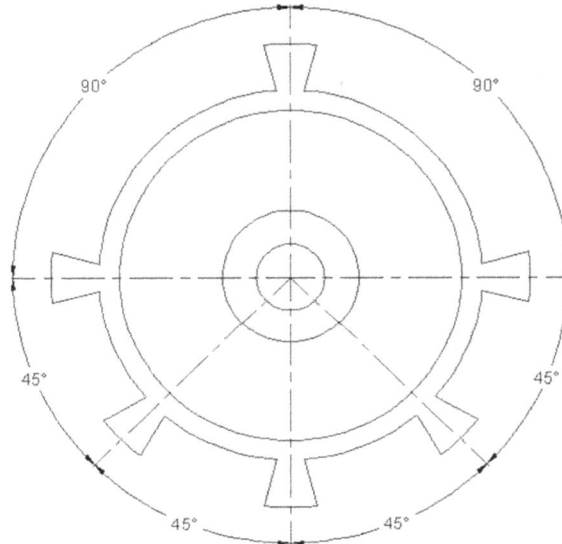

**4.** Save and close the file

## PRACTICE 8-5   BASELINE COMMAND

**1.** Start AutoCAD 2025

**2.** Open **Practice 8-5.dwg**

**3.** Create a linear dimension, then using the Baseline command, complete the shape as follows:

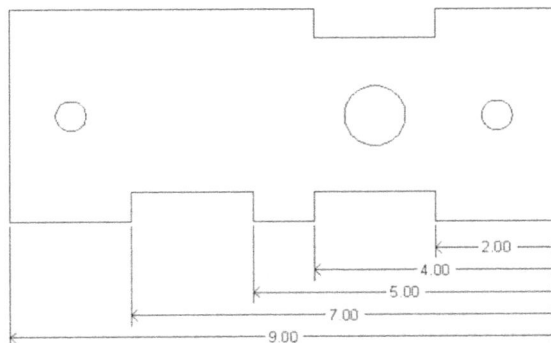

**4.** Save and close the file

## 8.13   USING THE QUICK DIMENSION COMMAND

This command will insert a group of dimensions in one shot. To start this command, go to the **Annotate** tab, locate the **Dimensions** panel, and select the **Quick Dimension** button:

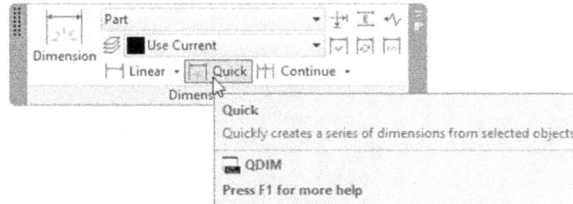

You will see the following prompt:

```
Associative dimension priority = Endpoint
Select geometry to dimension:  Specify dimension line
position, or [Continuous/ Staggered/Baseline/Ordinate/
Radius/ Diameter/ datumPoint/Edit/seTtings] <Continuous>:
```

As a first step, select the desired geometry you want to dimension, using clicking, window mode, crossing mode, or any other mode you know. If this the first time you are using the command in this AutoCAD session, then AutoCAD will use Continuous as the default option. But, if you right click, you will see the following menu:

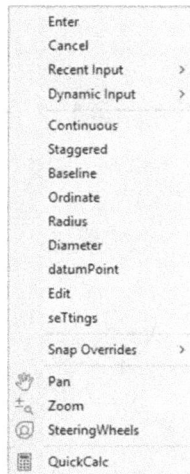

Using this shortcut menu, select the desired dimension type and then specify the dimension line location; consequently, a set of dimensions will be inserted.

Refer to the following examples:

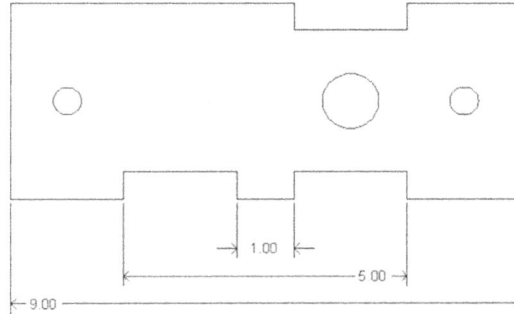

Staggered

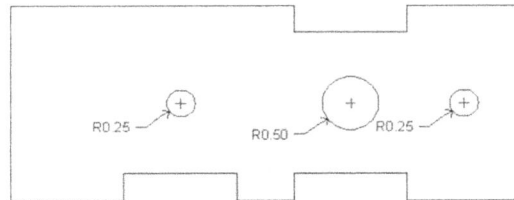

Radius

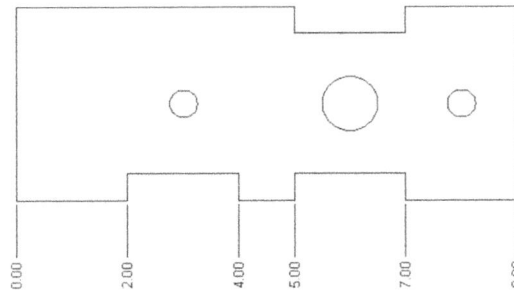

Ordinate

If you select the Settings option, you will see the following menu:

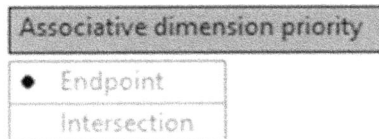

This option will enable you to set the default object snap for specifying extension line origins.

## 8.14 USING THE DIMENSION COMMAND

This command is the summation of the eight basic dimensioning commands. The essence of it is so simple: move your mouse pointer over the object and it will predict the proper dimension type, and then you will locate it in the drawing. To issue this command go to the **Annotate** tab, locate the **Dimension** panel, and click the **Dimension** command:

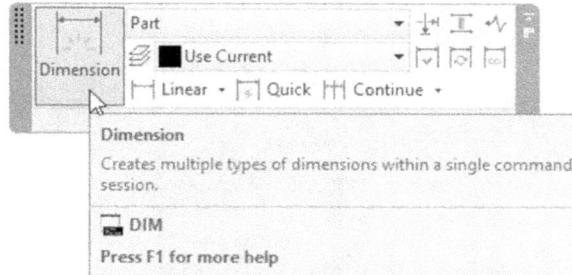

Roll your mouse pointer over one of the objects, and depending on the following, you will see a dimension type generated:

- If you roll your mouse cursor over a horizontal or vertical line (or line segment in case of a polyline), the Dimension command will produce a Linear dimension
- If you roll your mouse cursor over an inclined line (or line segment in case of a polyline), the Dimension command will produce an Aligned dimension
- If you roll your mouse cursor over an arc (or an arc segment in case of a polyline), the Dimension command will produce a Radius dimension
- If you roll your mouse cursor over a circle, the Dimension command will produce a Diameter dimension
- To produce an Angular dimension, first hover your mouse cursor over a line (linear or aligned), and when the dimension block is previewed, click, hover your mouse over the other line, and the Angular dimension will be created

When you start the command, you will see the following prompt:

```
Select objects or specify first extension line origin
or [Angular/Baseline/Continue /Ordinate /aliGn
/Distribute/Layer/Undo]:
```

Notice that Continue and Baseline are listed in the above prompt. If you want to use the Dimension command for a continuous or baseline dimension, you should start with these two options as a first step.

## 8.15   DIMENSION SPECIAL LAYER

You can specify the name of the layer that dimension objects produced from the Dimension command should reside in. This can be accomplished through doing one of two things:

- Type at the command prompt the system variable: DIMLAYER, then type the name of the desired layer
- Using the **Annotate** tab, locate the **Dimension** panel and you will see the following:

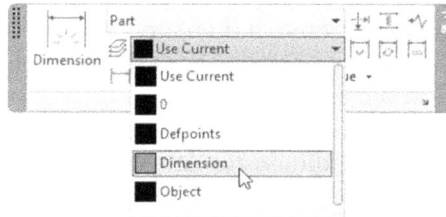

- Using the drop-down list, select the name of the desired layer

Using either method, and regardless of the current layer shown in the Layers panel, dimension objects will go directly to the selected layer.

## 8.16   EDITING A DIMENSION BLOCK USING GRIPS

After inserting a dimension block, it is easy to edit it using grips or using right-clicking. Depending on the type of dimension, if you click any dimension block you will see grips in certain places. Below are some examples:

- With linear and aligned dimensions, grips will appear in five places: at the two points measured, the two ends of the dimension line, and finally at the dimension text. If you hover over the text grip, you will see the following:

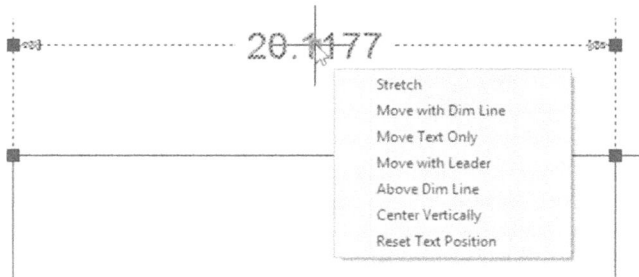

- AutoCAD will Stretch, Move with Dim Line, Move Text Only, as well as the rest of the list of actions you can make while holding this grip. If you hover over the two grips at the ends, however, you will see:

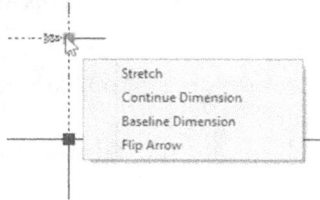

- This list of actions will Stretch and create a continuous or baseline dimension based on the selected type of dimension. The final thing the user can do is flip the arrow nearest to the selected grip.
- In Angular dimension, grips will appear at five places: at the endpoints of the two lines involved, at the dimension line, and finally at the dimension text. If you hover over the text grip, you will see the following:

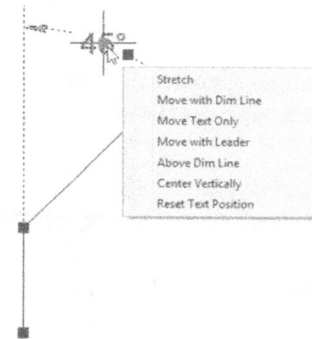

- The commands are identical to the linear and aligned. If you hover over the two end grips, you will see the following:

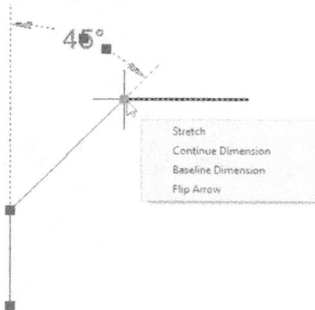

- These are the same commands mentioned for linear and aligned
- In the Ordinate dimension, grips will appear at four places: the origin point and the measured point, the dimension line, and finally at the dimension text. If you hover over the text grip, you will see the following:

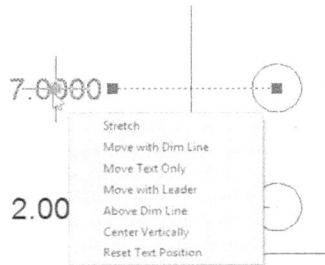

7.0000

Stretch
Move with Dim Line
Move Text Only
Move with Leader
2.00    Above Dim Line
Center Vertically
Reset Text Position

- This is the same set of commands as mentioned above. Hovering over the end of the line grip will show the following:

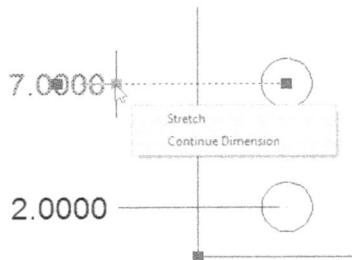

7.0000

Stretch
Continue Dimension

2.0000

- Ordinate will never work with Baseline and does not have arrows; hence, these two options are not mentioned for this type of dimension
- Radius and diameter dimension grips will appear at three places: the selected point, at the center, and finally at the dimension text. If you hover over the text grip, you will see the following:

R1.0

Stretch
Move with Dim Line
Move Text Only
Above Dim Line
Center Vertically
Reset Text Position

- This is the same list of commands mentioned earlier. If you hover over the grip at the end of the arrow you will see the following:

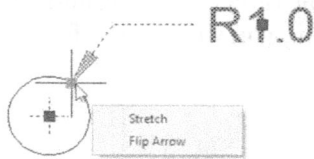

- Since Radius and Diameter do not have the ability to use Continue or Baselines, these two commands are not mentioned. You can use only Stretch and Flip Arrow.
- Arc length will show four grips, one at the two ends of the arc, one at the text, and finally one near the text. If you hover over the text grip, you will see the following:

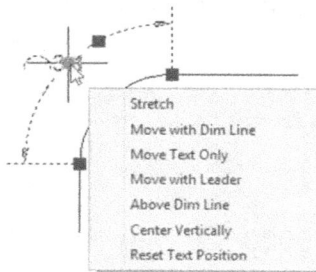

- This is the same list of commands mentioned above. If you hover over the endpoint grip, you will see the following:

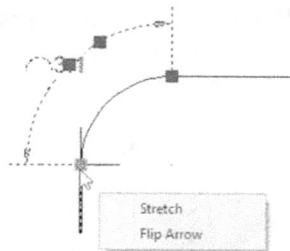

- Since Arc Length will not work with Continue and Baseline, only Stretch and Flip Arrow will appear

## 8.17 EDITING A DIMENSION BLOCK USING THE RIGHT-CLICK MENU

On the other hand, if you select a dimension block and right-click, you will see the following shortcut menu:

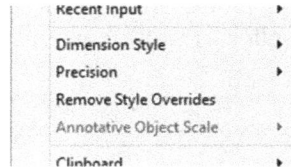

You can change the dimension style used to insert the selected dimension block. Or better, the changes you made can be saved as a new dimension style, as shown below:

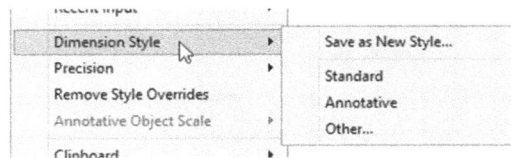

You can change the Precision of the dimension text:

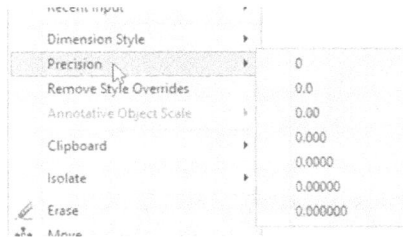

## 8.18 EDITING A DIMENSION BLOCK USING QUICK PROPERTIES AND PROPERTIES

If you select a dimension block, the Quick Properties will appear automatically. You will see the following:

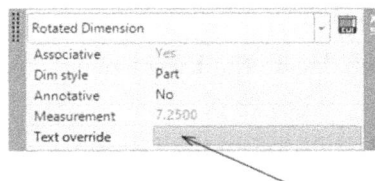

You have the ability to modify the dimension style and the annotative feature (which will be discussed in the later chapters). Also, you will find the exact measurement of the selected dimension, and you can change the number by utilizing the Text override field.

Properties, however, will make global changes to the selected dimension blocks. Simply select the dimension blocks, right-click, and then select the **Properties** option (you can use double-click as well). You will see the following:

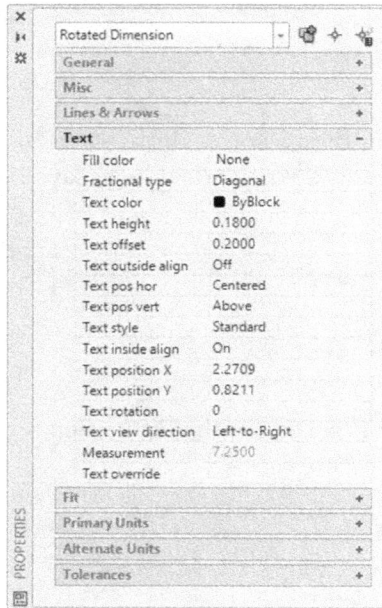

Using the Properties palette, you can change everything related to the dimension block or the dimension style of this block (in the previous example we showed the variables you can control under the Fit category).

## PRACTICE 8-6    QUICK DIMENSION AND EDITING

**1.** Start AutoCAD 2025

**2.** Open **Practice 8-6.dwg**

**3.** Go to the **Home** tab, locate the **Layer** panel, and check the name of the current layer

**4.** Go to the **Annotate** tab, locate the **Dimension** panel, and using the Dimension Layer drop-down list, pick the Layer Dimension (color is RED). Now regardless of the current layer, AutoCAD will input all dimension objects in the Dimension layer

**5.** Make Part the current Dimension Style

**6.** Using the Dimension command ONLY, try to create the following dimensions:

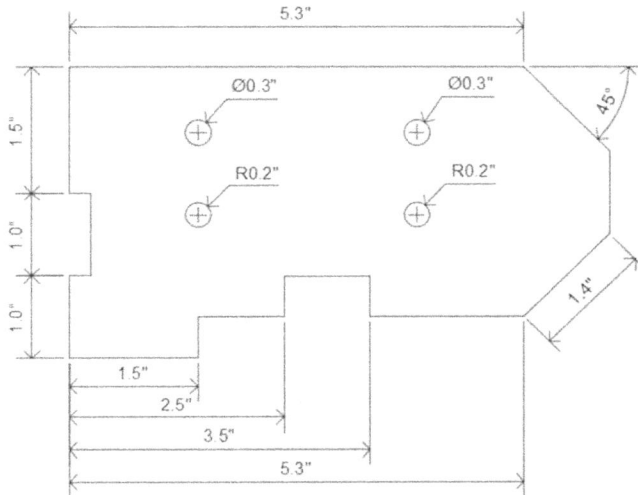

**7.** Pan to left to find another shape

**8.** Make the current layer Dimension

**9.** Using Quick Dimension create continuous for the top part of the shape

**10.** Using Quick Dimension create baseline for the bottom part of the shape

**11.** Using Quick Dimension create Radius for all circles inside, making the radius dimension pointing up and right

**12.** Using grips make sure to get all Radius dimensions to be inside the shape

**13.** Select the only dimension shown at the left to show its grips. Hover over the grip below, and when the menu comes up, select Continue Dimension and add two more dimensions

**14.** Zoom to the dimension reading 1.5". Click it to show the grips, right-click and select Dimension Style, then select Standard. What happened to this specific dimension block?

_____

_____

_____

**15.** Using grips make the dimension reading 2.5" to be Center Vertically

**16.** Make the Precision of the dimension reading 3.5 to be 0.000

**17.** Using Quick Properties make the dimension reading 6.3" to 6.5"

**18.** You should have an image to the following:

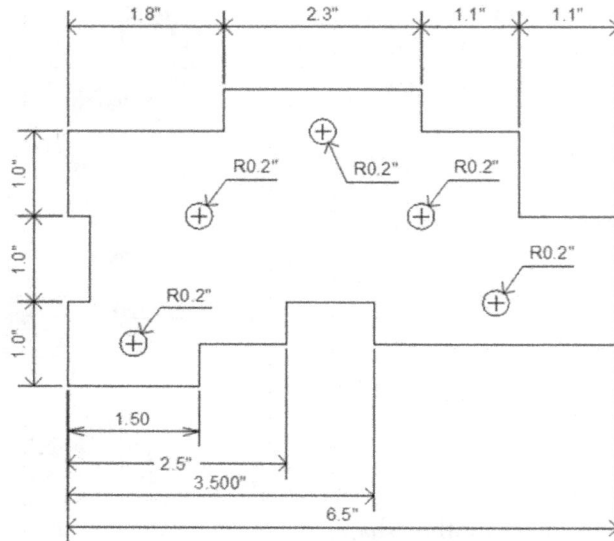

**19.**      Save and close the file

# NOTES

## CHAPTER REVIEW

**1.** Arc length command should be used only with arcs and polylines, not with circles:

    **a.** True

    **b.** False

**2.** Continue and Baseline cannot work with:

    **a.** Linear

    **b.** Radius

    **c.** Ordinate

    **d.** Angular

**3.** _____ will insert dimensions relative to a datum, either in X or in Y

**4.** Inputs to use this dimensioning command may be three points or two lines:

    **a.** Linear

    **b.** Radius

    **c.** Ordinate

    **d.** Angular

**5.** One of the following is not among the commands of Quick Dimension:

    **a.** Continuous

    **b.** Radius

    **c.** Staggered

    **d.** Angular

## CHAPTER REVIEW ANSWERS

**1.** a

**3.** Ordinate

**5.** d

## CHAPTER 9

# PLOTTING

## In This Chapter

- What is the difference between Model Space and Paper Space?
- How to create a new layout using different methods
- How to create and control viewports

## 9.1   WHAT IS MODEL SPACE AND WHAT IS PAPER SPACE?

AutoCAD provides two spaces, one for creating your drawing, called Model Space, and the other for plotting your drawing, called Paper Space. There is only one Model space in each drawing; in contrast, there are an infinite number of Paper spaces per file, and each one is called a layout.

Each layout is linked to a Page Setup (which is similar to other computer software). The user should specify everything related to plotting in Page Setup: for example, the user should specify the plotter, the paper size, and the paper orientation (Portrait or Landscape).

After you create the layout and link it to page setup, you will insert a title block and then add viewports (which represent a portion of the Model space). Afterward, the user will set up the scale of the viewports.

## 9.2 INTRODUCTION TO LAYOUTS

Layout is where you will plot your drawing. Each layout will be linked to a Page Setup, Objects (like a title block), text, dimensions, and finally **Viewports**, which will be covered separately in the coming discussion. View the roadmap illustrated below:

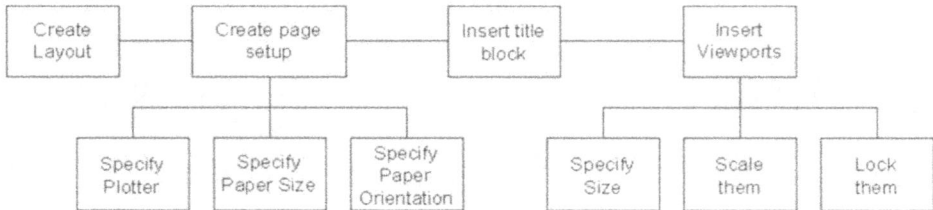

Each Layout should have a name; by default, when you create a new drawing using the *acad.dwt* template, two layouts will come with it: Layout1 and Layout2. These two preset layouts are empty of anything, so if you want to use them, you must undergo the following steps:

- Rename them to the desired name
- Create the page setup and link the layout to it, or you can simply link an existing page setup
- Insert the title block
- Insert Viewports, scale them, and lock them

## 9.3 STEPS TO CREATE A NEW LAYOUT FROM SCRATCH

This method will create a new layout from scratch. You can use one of the following methods:

- At the right of the names of the existing layouts there is a (+) sign; click it to add a new layout. Check the illustration below:

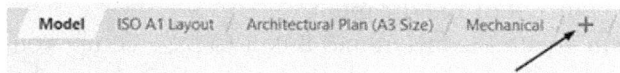

- Right-click on any existing layout name (the tab at the lower left corner of the screen and above the command window), and a shortcut menu will appear; select the **New Layout** option:

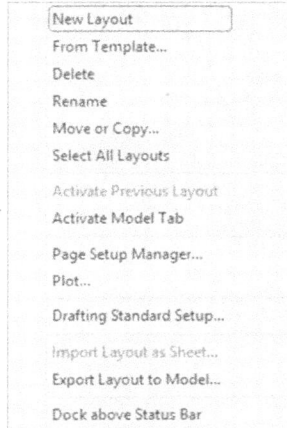

New Layout

From Template...

Delete

Rename

Move or Copy...

Select All Layouts

Activate Previous Layout

Activate Model Tab

Page Setup Manager...

Plot...

Drafting Standard Setup...

Import Layout as Sheet...

Export Layout to Model...

Dock above Status Bar

- Click Layout Tab Menu at the left of the Model tab, to show the following menu, and select New Layout option:

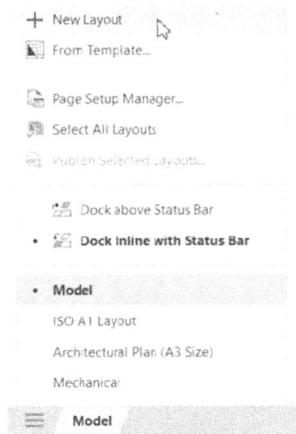

+ New Layout

From Template...

Page Setup Manager...

Select All Layouts

Publish Selected Layouts...

Dock above Status Bar

• Dock Inline with Status Bar

• Model

ISO A1 Layout

Architectural Plan (A3 Size)

Mechanical

Model

After the creation of the new layout, do the following steps:

- A new layout will be added with a temporary name. Rename it by right-clicking and selecting the **Rename** option (you can double-click the temporary name; the name will be editable, and you can input the new name)

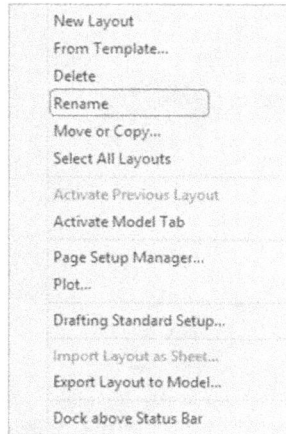

```
New Layout
From Template...
Delete
Rename
Move or Copy...
Select All Layouts

Activate Previous Layout
Activate Model Tab

Page Setup Manager...
Plot...

Drafting Standard Setup...

Import Layout as Sheet...
Export Layout to Model...

Dock above Status Bar
```

- Link the new layout with a page setup. To do that, right-click the name of the layout, and then select the **Page Setup Manager**.

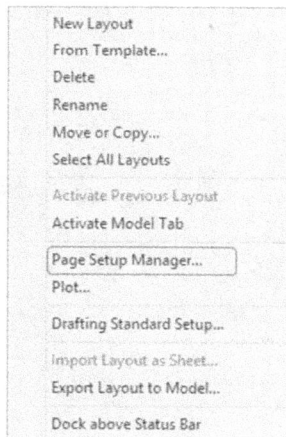

```
New Layout
From Template...
Delete
Rename
Move or Copy...
Select All Layouts

Activate Previous Layout
Activate Model Tab

Page Setup Manager...
Plot...

Drafting Standard Setup...

Import Layout as Sheet...
Export Layout to Model...

Dock above Status Bar
```

- The following dialog box will be shown:

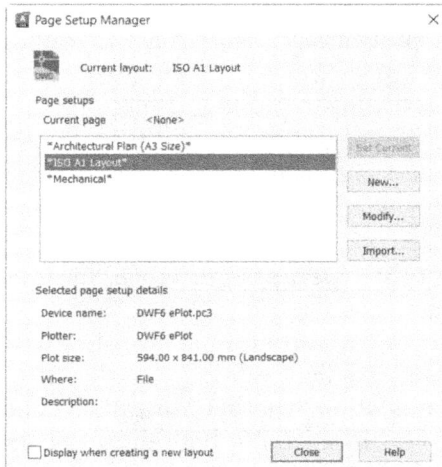

At the top of the dialog box, you can see the name of the **Current layout**; on the other hand, at the bottom you will see **Selected page setup details**, which is a summary of the current page setup. Finally, you will see a checkbox that says **Display when creating a new layout**; this checkbox will force this dialog box to appear every time you go to a newly created layout.

AutoCAD by default will create a page setup holding the same name of the layout. So, click the **Modify** button to modify it and you will see the following dialog box:

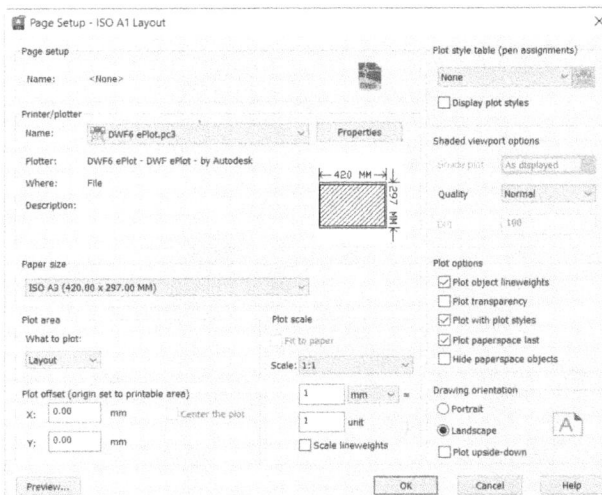

- Select the desired plotter (this plotter should be connected and configured)
- Select the desired **Paper Size**
- Specify **What to plot**: always leave it as **Layout** (the other options are for Model space printing)
- Input **Plot Offset:** if you are printing from layout, leave the values as zeros
- Input **Plot Scale**: If you are printing from layout, then you will use viewports (this is our next topic), and each viewport will hold its own scale. Hence, you will set the layout plot scale to 1=1. Specify as well if you want to **scale lineweights** or not
- Select **Plot style table (pen assignment)**: this topic will be discussed at the end of this chapter
- **Shaded viewport options** are for 3D modeling in AutoCAD
- Leave **Plot options** to default values
- Select Drawing orientation, whether Portrait or Landscape
- The Plotter will print from top to bottom; select the checkbox if you want it otherwise
- Click **OK**. The Page Setup you create will be available for all layouts in the current drawing file
- You will be back to the first dialog box; select the page setup and click **Set Current** (also, you can double-click the name of the Page Setup). Now the current layout is linked to the page setup you select
- To modify the settings of an existing page setup click **Modify**. To bring a saved Page Setup from an existing file click **Import**

## 9.4   STEPS TO CREATE A NEW LAYOUT USING A TEMPLATE

This procedure will import any layout from a template file, including the page setup and any other contents like title block, viewports, text, etc. To do this, you must undergo the following steps:

- Right-click on any existing layout, and you will see the following menu: select the **From template** option:

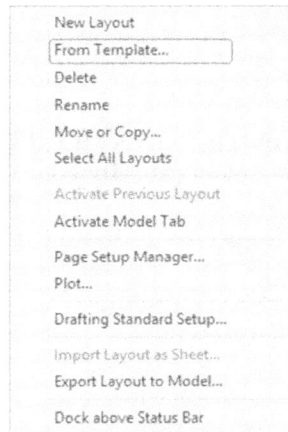

```
New Layout
From Template...
Delete
Rename
Move or Copy...
Select All Layouts

Activate Previous Layout
Activate Model Tab

Page Setup Manager...
Plot...

Drafting Standard Setup...

Import Layout as Sheet...
Export Layout to Model...

Dock above Status Bar
```

- The normal Open file dialog box will appear to select the desired template file. Select the desired template and click **Open**, and you will see the following dialog box:

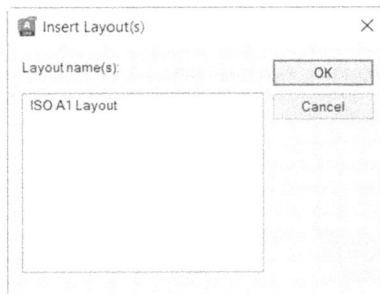

```
Insert Layout(s)                    ×

Layout name(s):              [   OK   ]

ISO A1 Layout                [ Cancel ]
```

- Click one of the listed layouts and click **OK**. You will see that the newly imported layout resides in your drawing

## 9.5 CREATING LAYOUTS USING COPYING

These two methods will create copies of an existing layout. The first method is similar to something we do in Microsoft Excel. Do the following steps:

- Select the desired layout
- Hold the [Ctrl] key at the keyboard, then hold and drag the mouse to the new position of the newly copied layout

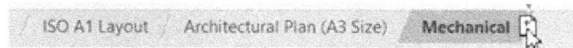

ISO A1 Layout    Architectural Plan (A3 Size)    Mechanical

- Rename the new layout

Another way to copy layouts is to select the desired layout and then right-click. A shortcut menu will appear; select the **Move or Copy** option:

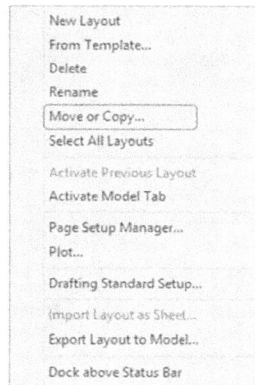

New Layout
From Template...
Delete
Rename
Move or Copy...
Select All Layouts

Activate Previous Layout
Activate Model Tab

Page Setup Manager...
Plot...

Drafting Standard Setup...

Import Layout as Sheet...
Export Layout to Model...

Dock above Status Bar

You will see the following dialog box:

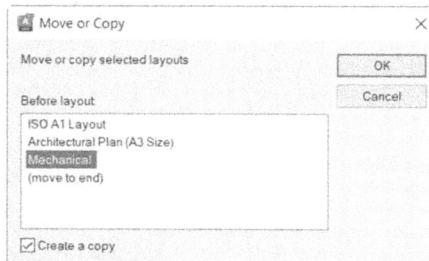

Move or Copy                                    ×

Move or copy selected layouts                    OK

Before layout                                    Cancel

ISO A1 Layout
Architectural Plan (A3 Size)
Mechanical
(move to end)

☑ Create a copy

A list of the current layouts is displayed; select one of them and click on the checkbox **Create a copy**. A copy will be created; you can rename it and make the necessary changes.

Using the same option, the user can move a layout from its current position to the left or to the right. Alternatively, you can move a layout without this command by clicking the layout name and holding and dragging it to the desired location.

**NOTE** *To rearrange the position of the layouts, simply hold the name in the tab and click and drag.*

By now, you have accomplished the following steps:

- Creating a new layout
- Creating a Page Setup
- Linking a Page Setup to the layout

Accordingly, when you select the newly created layout, you will see the following:

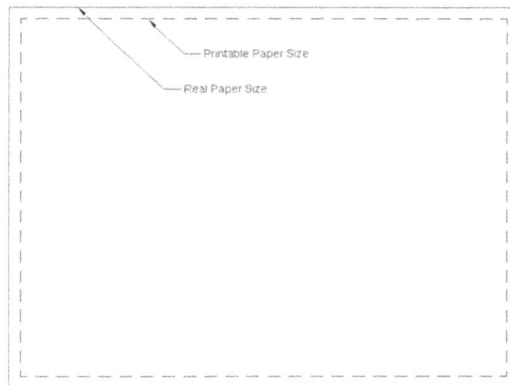

The outer frame is the real paper size, and the inside frame (the dashed line) is the printable paper size, which is the paper size minus the printer's margins. Based on this layout, you will see exactly what will be printed and what will not, because any object outside the dashed line will not be printed. This proves that printing from layouts is WYSIWYG (What You See Is What You Get).

You can reach some of the commands mentioned above if you go to the **Layout** tab (this tab will not be visible unless you are at one of the layouts). Locate the **Layout** panel and you will see the following buttons:

In this panel, the user can create a new layout from scratch or from a template. The second button will access the page setup dialog box.

## PRACTICE 9-1    CREATING NEW LAYOUTS

1. Start AutoCAD 2025

2. Open **Practice 9-1.dwg**

3. Select Layout 1 and delete the existing viewport (select it from its border lines)

4. Rename the layout to be Architectural Plan (A3 Size)

5. Start Page Setup Manager for this layout

6. Create a new page setup using the following information:

   **a.** Plotter = DWF6 ePlot.pc3

   **b.** Paper Size = ISO A3 (420.00 × 297.00 MM)

   **c.** What to plot = Layout

   **d.** Plot scale = 1:1

   Drawing orientation = Landscape

7. Click OK to end the creation process

8. Click Close to close the dialog box

9. Make layer Title Block the current layer

10. Insert the file called A3 Size Title Block.dwg to be your title block, using an insertion point of 0,0

11. Create a copy of the newly created layout, and name it Mechanical

12. Delete Layout 2

13. Right-click any existing layout, and select the From template option

14. Import ISO A1 Layout from Tutorial-march.dwt

15. Move ISO A1 Layout to be the first layout after the Model tab

16. Save and close the file

## 9.6    CREATING VIEWPORTS

When you visit a layout for the first time (just after creation), you will see that a single viewport appears at the center of the paper size. **Viewport** iis a window containing the view of your Model Space, initially scaled to the size of the window. Viewports inserted in layouts can be tiled, will be scaled, and can be printed.

The user can add viewports to layouts using different methods:

- Adding single rectangular viewport
- Adding multiple rectangular viewports
- Adding single polygonal viewport
- Converting object to be a viewport
- Clipping an existing viewport

**NOTE** *The Layout tab appears only when you are in a layout.*

### 9.6.1 Adding Single Rectangular Viewports

This command will add single rectangular viewports in a layout. The user should specify two opposite corners to specify the area of the viewport. To issue this command, go to the **Layout** tab, locate the **Layout Viewports** panel, then select the **Rectangular** button:

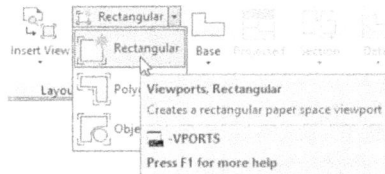

The following prompts will be shown:

```
Specify corner of viewport or [ON/OFF/Fit/Shadeplot /
Lock/Object/Polygonal/Restore/LAyer/2/3/4] <Fit>:
Specify opposite corner:
```

Specify the two opposite corners to create a single rectangular viewport. This is what you will receive:

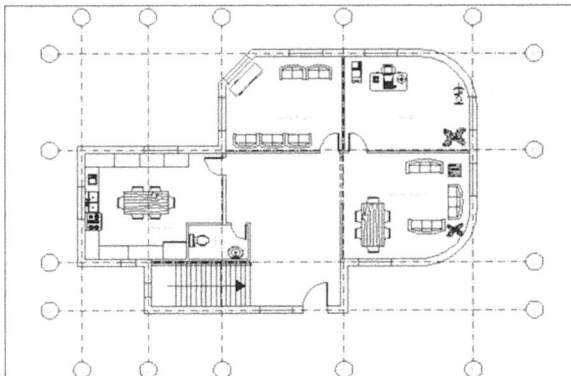

### 9.6.2 Adding Multiple Rectangular Viewports

This command will add multiple rectangular viewports in a layout. The user should specify two opposite corners to specify the area of the viewports. To issue this command, go to the **Layout** tab, locate the **Layout Viewports** panel, then click **New Viewports**, as shown below:

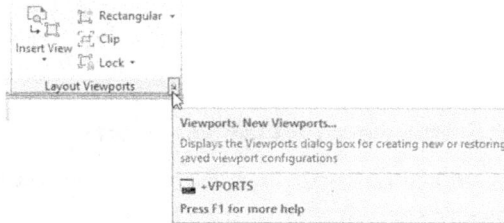

You will see the following dialog box:

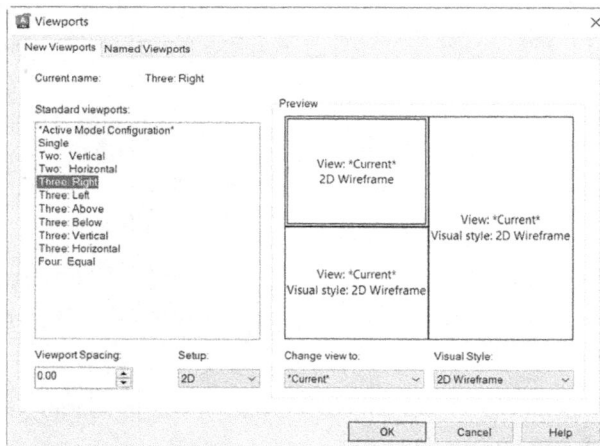

Select the desired display from the list. By default, the **Viewport Spacing** value = 0 (zero), which means viewports will be tiled. If you want them separated, input a value greater than 0 (zero). Click **OK**, and you will see the following prompt:

```
Specify first corner or [Fit] <Fit>:
Specify opposite corner:
```

Check the following illustration:

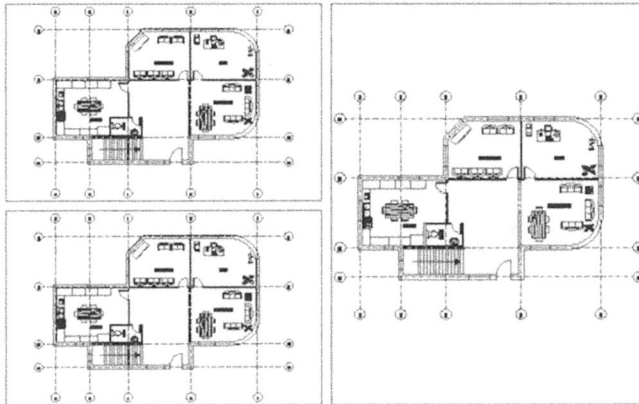

### 9.6.3 Adding Polygonal Viewports

This command will add a polygonal viewport that consists of both straight lines and arcs. To start this command, go to the **Layout** tab, locate the **Layout Viewports** panel, then select the **Polygonal** button:

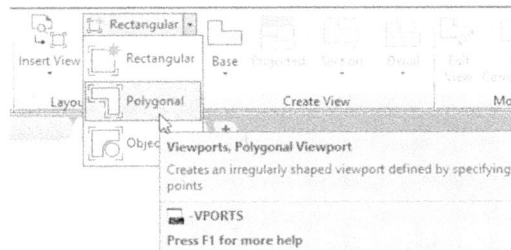

You will see the following prompts:

```
Specify start point:
Specify next point or [Arc/Length/Undo]:
Specify next point or [Arc/Close/Length/Undo]:
```

They look like the Polyline command prompts.

See the below image:

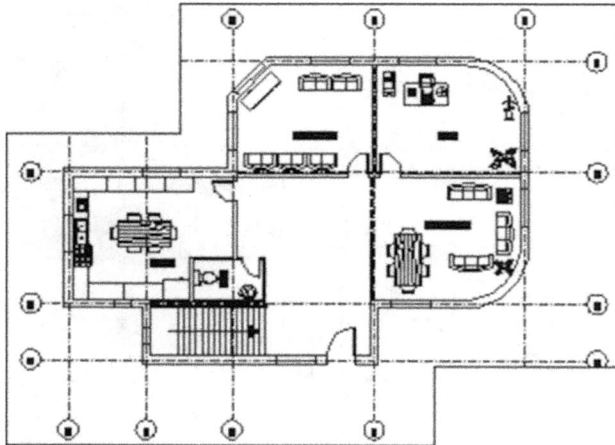

### 9.6.4   Creating Viewports by Converting Existing Objects

This command will convert any existing object (a single object, like a polyline or circle—lines and arcs are not allowed) to a viewport. To start this command, go to the **Layout** tab, locate the **Layout Viewports** panel, then select the **From Object** button:

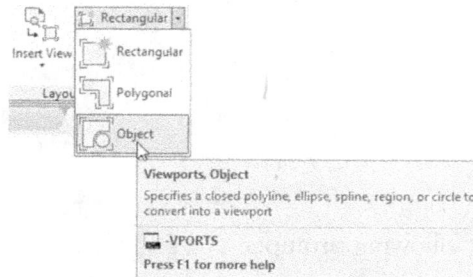

You will see the following prompt:

```
Select object to clip viewport:
```

See the illustration below:

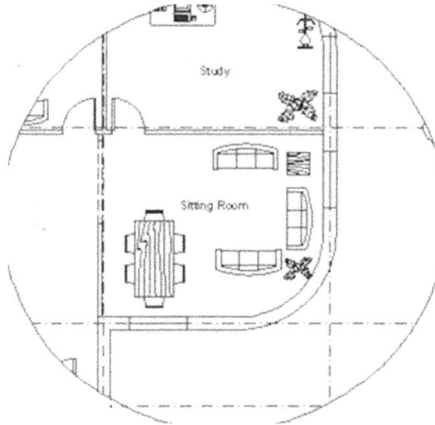

### 9.6.5   Creating Viewports by Clipping Existing Viewports

This command will clip an existing viewport, and hence create a new shape. To start this command, go to the **Layout** tab, locate the **Layout Viewports** panel, then select the **Clip** button:

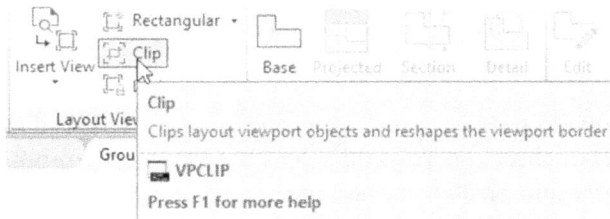

You will see the following prompt appear:

```
Select viewport to clip:
Select clipping object or [Polygonal] <Polygonal>:
Specify start point: Specify next point or
[Arc/Length/Undo]: Specify next point or
[Arc/Close/Length/Undo]:
```

First select an existing viewport. You can select a polyline that was drawn previously, or you can draw any irregular shape using the **Polygonal** option (which is identical to the Polygonal viewport command discussed above). See the illustration below:

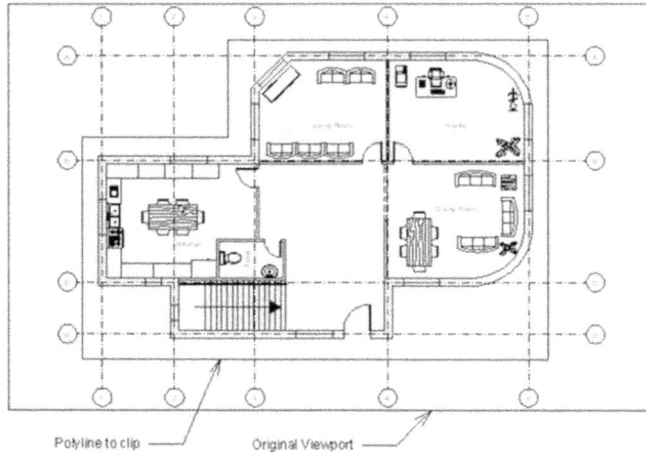

Polyline to clip ———        Original Viewport ———

### 9.6.6 Dealing with Viewports After Creation

There are two modes to work with viewports:

- You are outside the viewport, which means you will deal with it as any other object in your drawing. You can select the viewport from its frame and erase, copy, move, scale, stretch, rotate, etc.

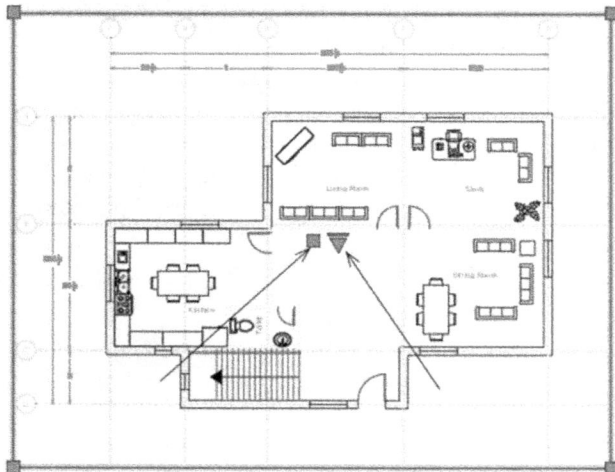

- In the second mode, you are inside the viewport; you achieve this by double-clicking inside the viewport. This mode will allow you to zoom, pan, scale, etc. the objects inside the viewport. To return to the first mode, double-click outside the viewport.

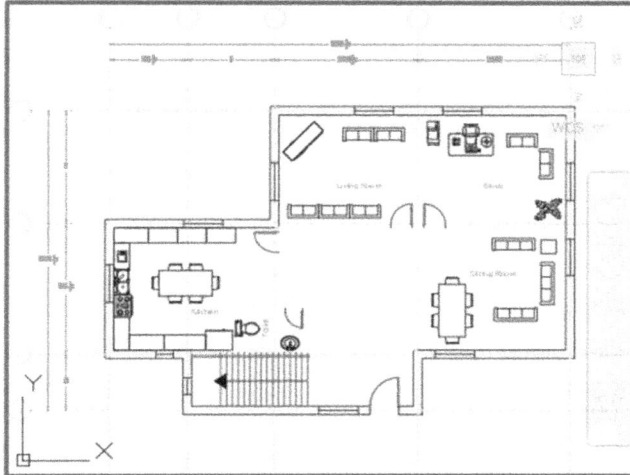

## 9.7   SCALING AND MAXIMIZING VIEWPORTS

By default, when you insert a viewport, AutoCAD will zoom the whole drawing into the area you specified as the viewport size. This viewport is not to scale, and the user should set the scale relative to the Model Space units. There are multiple ways to do that:

**Method one**

- Click the frame of the viewport
- Triangle will appear at the center of the viewport
- Click the triangle. A list of scales will appear. Select the desired scale
- Viewport size will increase or decrease depending on the scale

**Method two**

- Double-click inside the desired viewport (or you can select only the viewport's frame)

- Look at the right side of the Status bar and you will see Viewport Scale list as shown below:

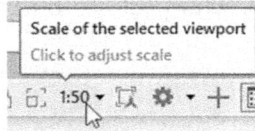

Click the list which contains all scales, similar to the following:

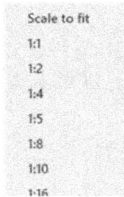

Select the desired scale for your viewport. If you did not find your desired scale, select the **Custom** option, and the following dialog box will be displayed:

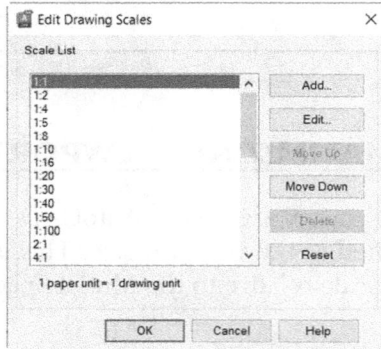

To add a new scale, select the **Add** button, and the following dialog box will be displayed:

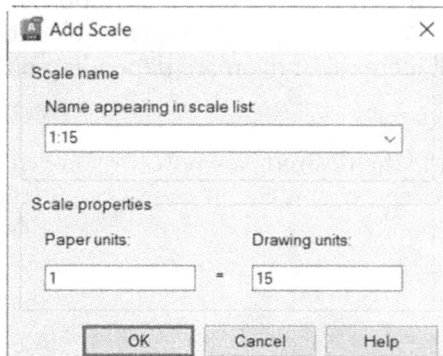

Input the desired scale, and then click **OK** twice.

If you are inside the viewport, and after setting the scale, it is permissible to use the Pan command but not the Zoom command, because this will ruin the scale you set. To avoid this problem, you can lock the display of the viewport by selecting the viewport and then clicking the golden opened lock in the status bar (you have to be inside the viewport or border of the viewport has to be selected); the golden lock will change to a blue lock, and the viewports will be locked accordingly.

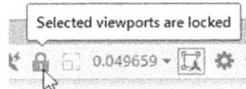

The Maximize function will maximize the viewport to fit the size of the screen temporarily. This will give you the needed space to edit objects as you wish without leaving the layout and going to the Model space. Go to status bar and click the **Maximize Viewport** button as shown:

The same button will restore the original size of the viewport. Or the user can go to the **Layout** tab, the **Layout Viewports** panel, and use the following two buttons:

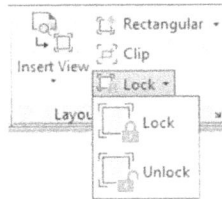

## 9.8   FREEZING LAYERS IN VIEWPORTS

The Freezing function will freeze layers in Model space and all viewports in all layouts. If you want to freeze a layer in a certain viewport, you have to do the following steps:

- Double-click inside the desired viewport

- Go to the **Home** tab, locate the **Layers** panel, select the layer list, and click the icon **Freeze or thaw in current viewport** for the desired layer. See the illustration below:

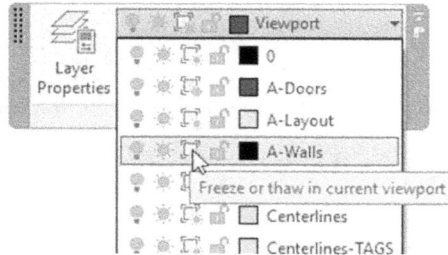

Also, the user has the ability to freeze/thaw layers in all viewports except the current viewport. To do this start the Layer Properties Manager, select the desired layers, right-click and select **VP Freeze layer**, then select the **In All Viewports Except Current** option, as shown below:

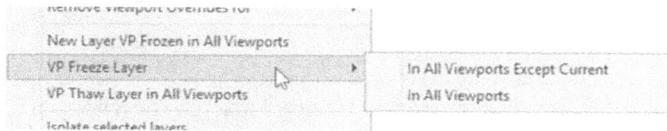

## 9.9   LAYER OVERRIDE IN VIEWPORTS

In all viewports, a layer will be displayed with the same color, linetype, lineweight, and plot style. You can change these settings in one viewport to display it differently. This is called layer override. Do the following steps:

- Double-click inside the desired viewport
- Issue the Layer Properties Manager command
- Under VP Color, VP Linetype, VP Lineweight, or VP Plot Style, make the desired changes
- These changes will take place only in the current viewport

See the following illustration:

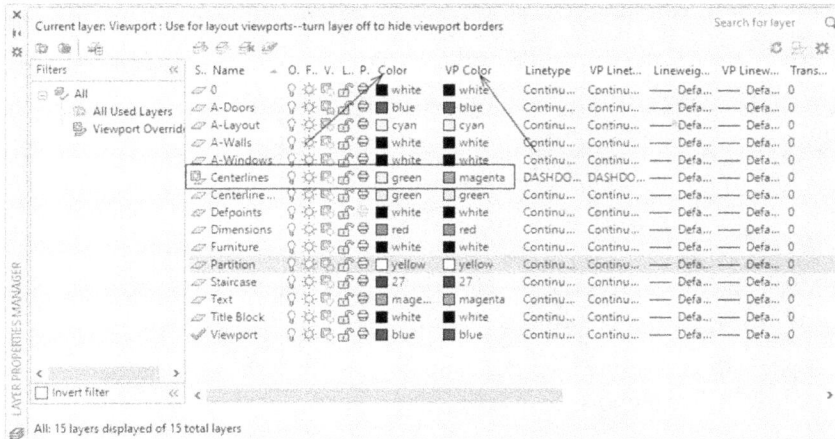

- In the above example layer, **Centerlines** has a color (applies in Model Space and all other viewports) equal to Green, and an override color equals Magenta in the current viewport
- Also, note that symbol at the left of layer name

## PRACTICE 9-2   CREATING AND CONTROLLING VIEWPORTS

1. Start AutoCAD 2025

2. Open **Practice 9-2.dwg**

3. Make layer Viewport the current layer

4. Switch to D-Size Architectural Plan layout

5. Insert a single viewport to fill the entire area

6. Set the scale to ½" = 1', and lock the viewport

7. Switch to D-Size Arch Details layout

8. Insert viewports using Three: Right, with Viewport Spacing = 0.25, and select an area to fill the paper size

9. Select the borders of the big viewport at the right, scale it to be ¼" = 1'

10. Select the two viewports at the left, and set the scale to ¾" = 1'

11. For the top viewport and using the Pan command, set the view to show the Kitchen and Toilet

**12.** For the bottom viewport and using the Pan command, set the view to show the Living Room

**13.** Select the three viewports and lock them

**14.** Switch to ANSI B Size Architectural Plan layout

**15.** Convert the big circle to be a viewport

**16.** Double-click inside the new viewport, and set the scale to ¾" = 1', then pan to show the Toilet

**17.** Click inside the rectangle viewport, and freeze in the current viewport the following layers: Centerline, Centerline-TAGS, and Dimensions

**18.** Switch to the other layouts to make sure that these three layers were frozen only in this viewport

**19.** Change the color of layer A-Walls in this viewport to red

**20.** Save and close the file

## 9.10    PLOT COMMAND

This will be our final step. This command mission is to send whatever we set in the layout to the specified plotter. To issue this command, go to the Output tab, locate the **Plot** panel, then select the **Plot** icon:

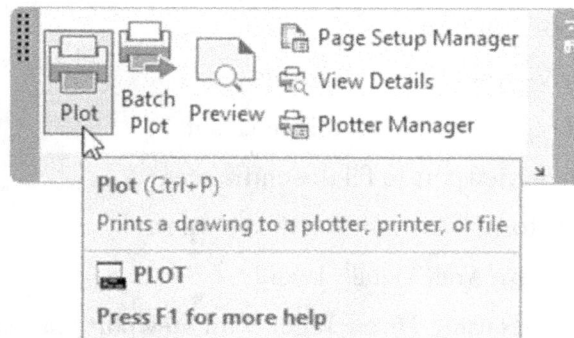

If you have multiple layouts in your drawing, and you chose Plot command, and not Batch Plot command, AutoCAD will show you the following message:

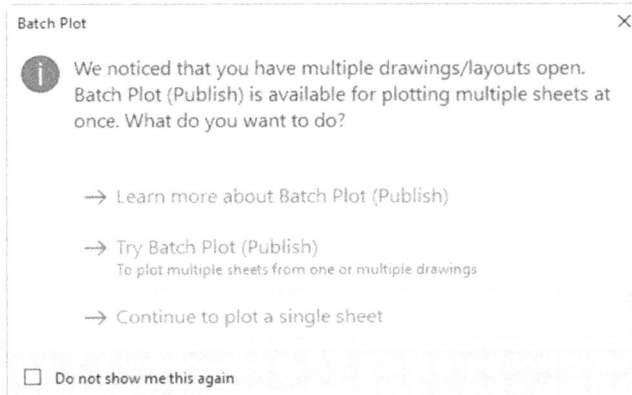

You can learn more about Batch Plot command, try it, or Continue to plot a single sheet.

If you choose to carry on with printing a single sheet you will see the following dialog box:

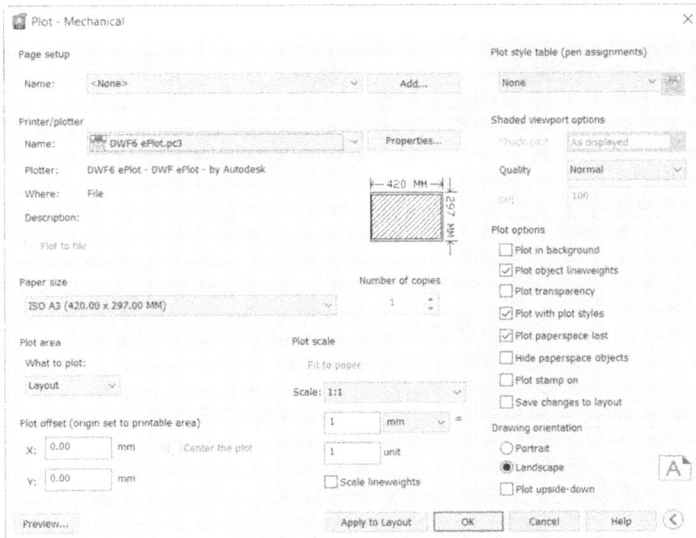

This dialog box is identical to the Page setup settings. If you modify any of these settings, AutoCAD will separate the Page setup from the current layout. To send the layout to the plotter, click **OK**. To save the settings of this dialog box with the layout, select the **Apply to Layout** button. Click the **Preview** button to see the final printed drawing on the screen before the real printout.

You can preview your drawing from outside this dialog box by going to the **Output** tab, locating the **Plot** panel, then selecting the **Preview** button:

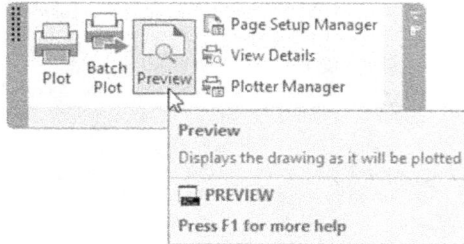

## NOTES

## CHAPTER REVIEW

**1.** What is true about creating a new layout?

    **a.** You can bring a layout from an existing template

    **b.** You can create a new layout using the right-click menu at the name of an existing layout

    **c.** You can click the (+) sign beside the last layout name at the right

    **d.** All of the above

**2.** Objects to be converted to a viewport should be _____ objects like Circles and Polylines

**3.** Using the Pan command after setting the scale of a viewport will ruin the scaling process

    **a.** True

    **b.** Flase

**4.** The user can freeze a layer in a certain viewport. The user can change the color of a layer in a certain viewport:

    **a.** The first statement is correct, yet the second is wrong

    **b.** The two statements are correct

    **c.** The two statements are wrong

    **d.** The first statement is wrong, yet the second is correct

**5.** _____ involves specifying plotter, page size, and orientation

## CHAPTER REVIEW ANSWERS

**1.** d

**3.** b

**5.** Page Setup

## CHAPTER 10

# PROJECTS

## In This Chapter

- How to draw a new project

## 10.1 HOW TO PREPARE YOUR DRAWING FOR A NEW PROJECT

In order to prepare your drawing for a new project, do the following steps:

- Start a new drawing based on acad.dwt or acadiso.dwt

- Set up the Drawing Units. From the Application Menu select **Drawing Utilities/Units**

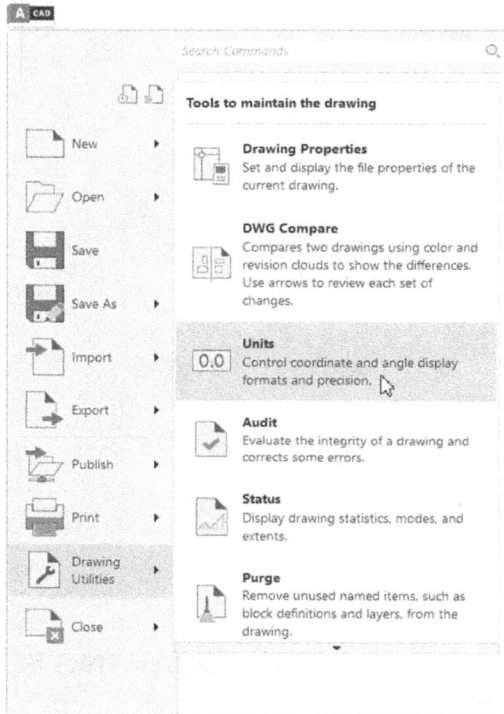

You will see the following dialog box:

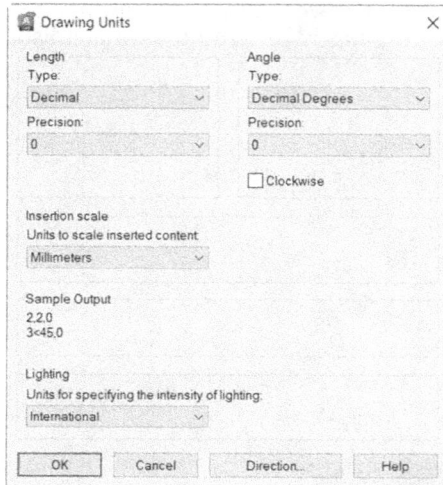

- Pick the desired **Length Type** by selecting one of the following: Architectural (example: 2'-4 8/16"), Decimal (example: 25.5697), Engineering (example: 3'-5.6688"), Fractional (example: 16 3/16), or Scientific (example: 8.9643E+03)
- Pick the desired **Angle Type** by selecting one of the following: Decimal Degrees (example: 36.7), Deg/Min/Sec (example: 47d25'31"), Grads (example: 60.7g), Radians (example: 0.6r), or Surveyor's Units (example: N 51d25'31" E)
- Set up the **Precision** for both length and angle units. For instance, pick the number of decimals for Decimal units, from no decimal places to eight decimal places
- The default AutoCAD setting for angles is counterclockwise, but you can switch it to be **clockwise**
- Under the **Insertion scale**, specify **Units to scale inserted content**, which was discussed in Chapter 6
- Select the **Direction** button and you will see the following dialog box:

- Change the East to be 0 (zero) angle, hence, changing the other angles as well (we recommend this setting to be left as is)
- This step will end the Units command
- Set up Drawing Limits. Drawing limits is your working area. You will specify it using two opposite corners, lower left corner, and upper right corner. To set up the drawing limits correctly, answer these two questions: what is the longest dimension in my drawing in both X and Y? And what does AutoCAD unit mean to me (is it meters, centimeters, inches, or feet)?
- If you are showing the menu bar, then choose **Format/Drawing Limits**, or you can type **limits** at the command window. You will see the following prompts:

```
Specify lower left corner or [ON/OFF] <0,0>:
Specify upper right corner <12,9>:
```

- Specify the lower left corner and right upper corner by typing or by clicking. To forbid yourself from using any area outside these limits, AutoCAD gives you the ability to turn it on and off
- Create layers
- Start drafting

## 10.2 ARCHITECTURAL PROJECT (IMPERIAL)

Do the following steps:

**1.** From the designated folder, open file Ground Floor_Starter.dwg

**2.** Switch off the grid

**3.** Set the units to the following:

    **a.** Architectural

    **b.** Precision 0'- ½"

    **c.** Units to scale inserted contents = inch

**4.** Set up drawing limits to be:

    **a.** Lower left corner = 0,0

    **b.** Upper right corner = 50',50'

**5.** Double-click the mouse wheel in order to see the new limits

**6.** Create the following layers:

| Layer Name | Layer Color |
|------------|-------------|
| A-Door | Blue |
| A-Wall | White |
| A-Window | White |
| Dimension | Red |
| Furniture | White |
| Staircase | Magenta |
| Text | Green |
| Title Block | White |
| Viewport | 9 |
| Hatch | 8 |

**7.** Save your file in the Chapter 10/Imperial folder and name it Ground Floor.dwg

**8.** Make layer A-Wall current

**9.** Draw the below architectural plan and the partitions inside using the following guidelines:

    **a.** Draw the outer shape using polylines

    **b.** Offset it to the inside using 6" as distance

    **c.** Explode the two polylines

    **d.** Use the outer wall to draw the inner walls, using all the commands you learned in this book. The inner wall is 4"

**10.** This is the architectural plan:

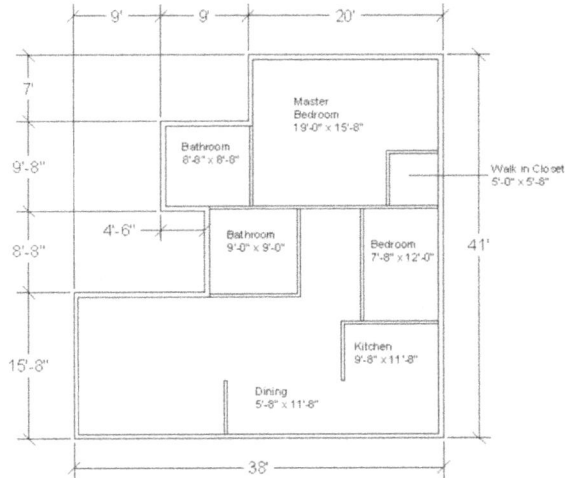

**11.** Create a 36" door opening as follows (you can always take 4" clearance from the wall). The main entrance, master bedroom, and walk-in closet door are at the middle of the wall

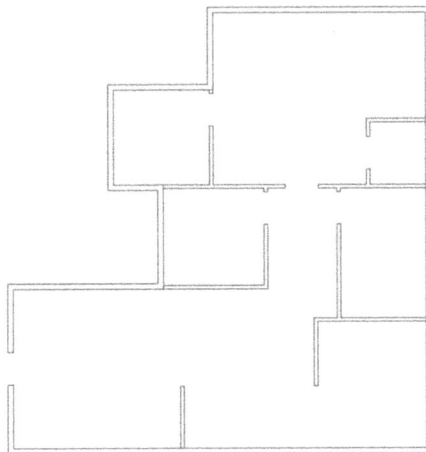

**12.** Make layer 0 (zero) current

**13.** Create the following door blocks using the name beneath (the base point is the lower left point of the jamb):

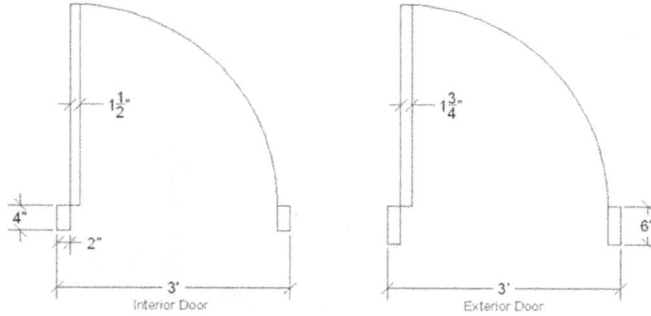

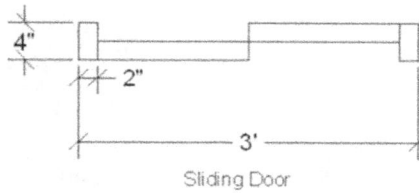

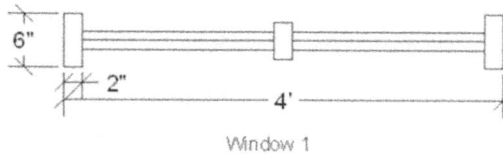

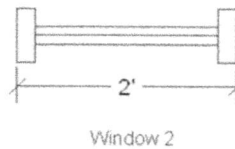

Interior Door

Exterior Door

**14.** Create the following door block using the name beneath (base point is the lower left point of the jamb):

Sliding Door

**15.** Create the following window blocks using the name beneath (base point is the lower left point of the jamb):

Window 1

Window 2

**16.** Insert the doors and windows in their respective layers to obtain the following result:

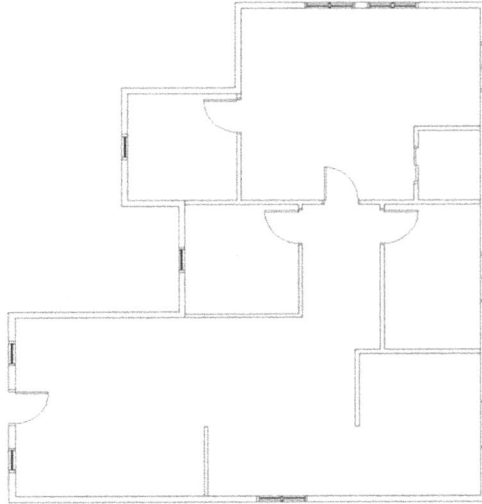

**17.** Make layer Furniture current

**18.** Using the Insert command insert the following blocks:

**19.** Make layer Hatch current

**20.** Using Solid hatching, hatch both the outside and the inside wall

**21.** Using ANSI37 and scale = 100, hatch the kitchen (hint: draw a line to separate the kitchen from the adjacent room)

**22.** Using a user-defined hatch (switch the Double checkbox on) and scale = 20, hatch both bathrooms

**23.** You should have the following:

**24.** Make layer Text current and freeze layer Hatch

**25.** Make Room Titles the current text style, then write text using Multiline text to add the room titles as shown in the following, making sure the Justify is Middle Center:

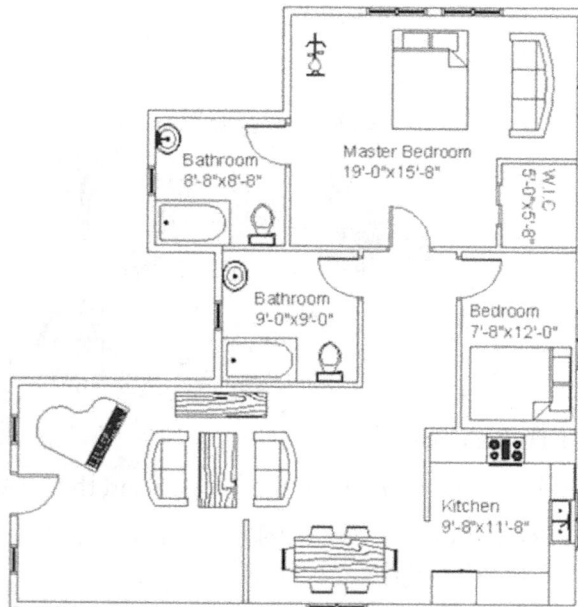

**26.** Thaw layer Hatch

**27.** Select the hatch of one of the two bathrooms. When the context tab appears, locate the **Boundaries** panel and click the **Select** button, then choose the text, press [Enter], and then press [Esc]. Do the same for the other bathroom and the kitchen

**28.** Make Outside Walls the current dimension style

**29.** Make layer Dimension current

**30.** Insert the dimensions as shown below (use the Continue command whenever possible):

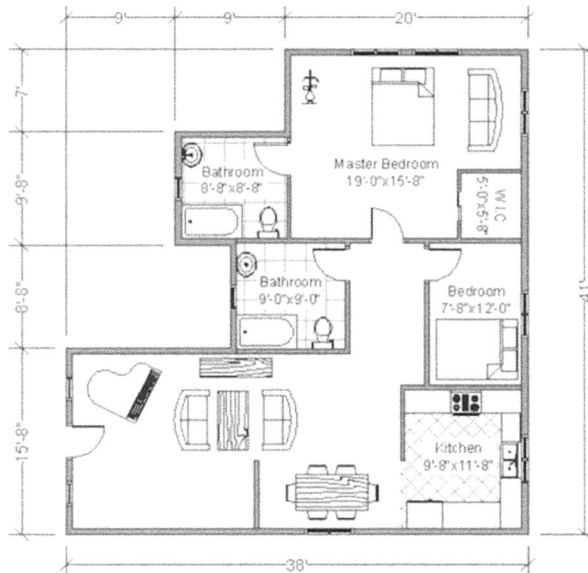

**31.** Go to Layout1 and rename it to Full Plan

**32.** Using the Page Setup Manager, modify the existing page setup to be as follows:

    **a.** Printer = DWF6 ePlot.pc3

    **b.** Paper = ANSI B (17x11 inch)

    **c.** Drawing orientation = Landscape

**33.** Erase the existing viewport

**34.** Make layer Title Block current

**35.** Insert the file ANSI B Landscape Title Block.dwg in the layout, using 0,0,0 insertion point.

**36.** Create a copy of the layout and name it Details

**37.** Erase Layout 2

**38.** Go to layout1 named Full Plan

**39.** Make layer Viewport current

**40.** Insert a single viewport to fill the space, and set the viewport scale to be 3/16" = 1', then lock the viewport

**41.** Go to layout named Details

**42.** Create a single viewport to occupy half of the space of the paper

**43.** Set the scale to be ½" = 1" and lock the viewport

**44.** Pan to the entrance, making sure to show the dimension, the two windows, and the door

**45.** You are still in the Details layout. Create another viewport to occupy half of the remaining area of the paper

**46.** Set the scale to ¼" = 1', then lock the viewport

**47.** Double-click inside the viewport, and pan to the two windows of the master bedroom

**48.** Save and close the file

## 10.3   ARCHITECTURAL PROJECT (METRIC)

Do the following steps:

**1.** From the designated folder, open file Ground Floor_Starter.dwg

**2.** Switch off the grid

**3.** Set the units to the following:

   **a.** Decimal

   **b.** Precision = 0

   **c.** Units to scale inserted contents = Millimeters

**4.** Set up drawing limits to be:

    **a.** Lower left corner = 0,0

    **b.** Upper right corner = 15000,15000

**5.** Double-click the mouse wheel in order to see the new limits

**6.** Create the following layers:

| Layer Name | Layer Color |
|---|---|
| A-Door | Blue |
| A-Wall | White |
| A-Window | White |
| Dimension | Red |
| Furniture | White |
| Staircase | Magenta |
| Text | Green |
| Title Block | White |
| Viewport | 9 |
| Hatch | 8 |

**7.** Save as your file in the Chapter 10\Metric folder and name it Ground Floor. dwg

**8.** Make layer A-Wall current

**9.** Draw the below architectural plan and the partitions inside using the following guidelines:

    **a.** Draw the outer shape using polylines

    **b.** Offset it to the inside using 150 as distance

    **c.** Explode the two polylines

    **d.** Use the outer wall to draw the inner walls using all the commands you learned in this book. The inner wall is 100

**10.** This is the architectural plan:

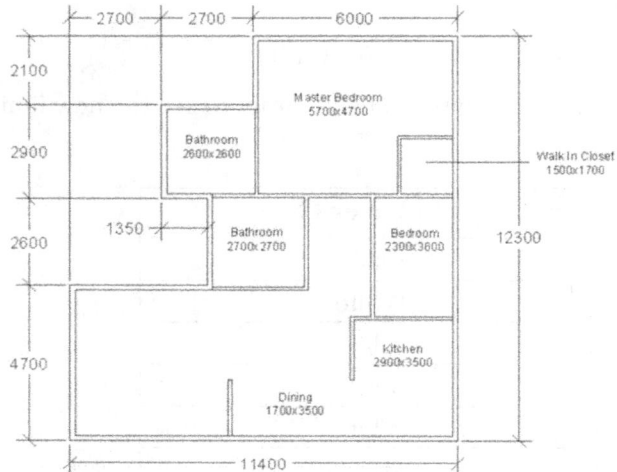

**11.** Create a 900 door opening as follows (you can always take 100 clearance from the wall). The main entrance, master bedroom, and walk-in closet door are at the middle of the wall:

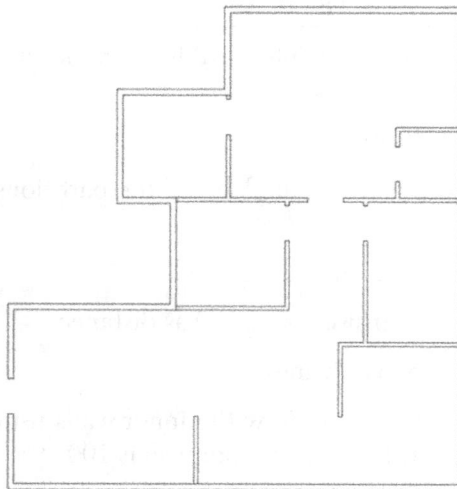

**12.** Make layer 0 (zero) current

**13.** Create the following door blocks using the name beneath (base point is the lower left point of the jamb):

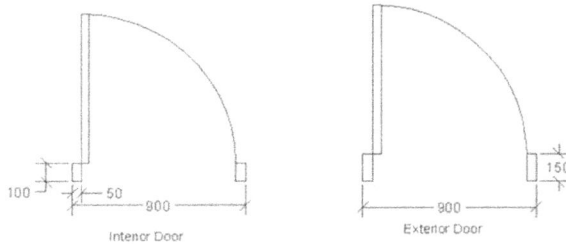

Interior Door

Exterior Door

**14.** Create the following door block using the name beneath (the base point is the lower left point of the jamb):

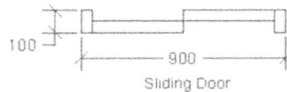

Sliding Door

**15.** Create the following window blocks using the name beneath (base point is the lower left point of the jamb):

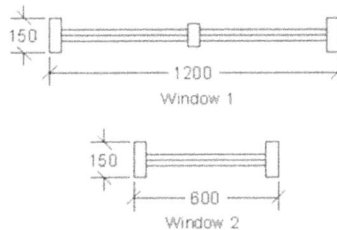

Window 1

Window 2

**16.** Insert the doors and windows in their respective layers to get the following result:

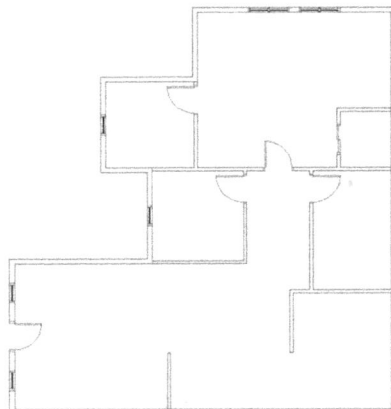

**17.** Make layer Furniture current

**18.** Use the Insert command to insert the following blocks

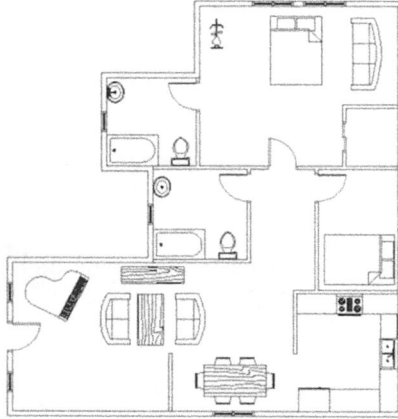

**19.** Make layer Hatch current

**20.** Using Solid hatching, hatch both the outside and the inside wall

**21.** Using ANSI37 and scale = 2000, hatch the kitchen (hint: draw a line to separate the kitchen from the adjacent room)

**22.** Using a user-defined hatch (switch Double checkbox on) and scale = 500, hatch both bathrooms

**23.** You should have the following:

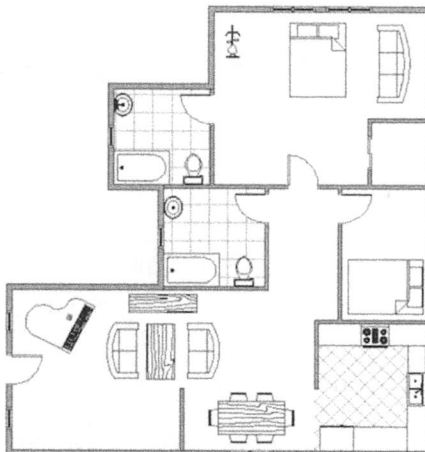

**24.** Make layer Text current and freeze layer Hatch

**25.** Make Room Titles the current text style

**26.** Write text using Multiline text to add the room titles similar to the following, making sure the Justify is Middle Center:

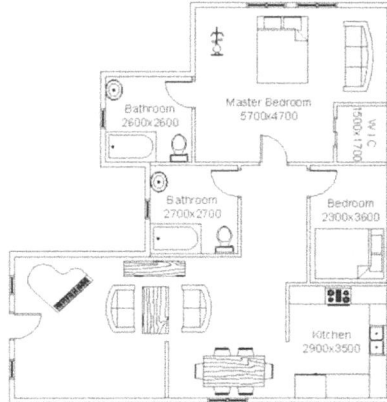

**27.** Thaw layer Hatch

**28.** Select the hatch of one of the two bathrooms. When the context tab appears, locate the **Boundaries** panel and click the **Select** button; then choose the text, press [Enter], and press [Esc]. Do the same for the other bathroom and the kitchen

**29.** Make Outside Walls the current dimension style

**30.** Make layer Dimension current

**31.** Insert the dimensions as shown below (use the Continue command whenever possible):

**32.** Go to Layout1 and rename it to Full Plan

**33.** Using the Page Setup Manager, modify the existing page setup to be as follows:

    **a.** Printer = DWF6 ePlot.pc3

    **b.** Paper = ISO A3 (420×297 MM)

    **c.** Drawing orientation = Landscape

**34.** Erase the existing viewport

**35.** Make layer Title Block current

**36.** Insert the file ISO A3 Landscape Title Block.dwg in the layout, using 0,0,0 insertion point.

**37.** Create a copy of the layout and name it Details

**38.** Erase Layout 2

**39.** Go to the layout named Full Plan

**40.** Make layer Viewport current

**41.** Insert a single viewport to fill the space and set the viewport scale to be 1:100, then lock the viewport

**42.** Go to the layout named Details

**43.** Create a single viewport to occupy half of the space of the paper

**44.** Set the scale to be 1:20 and lock the viewport

**45.** Pan to the entrance, making sure to show the dimension, the two windows, and the door.

**46.** Make layer Viewport current

**47.** You are still in the Details layout. Create another viewport to occupy half of the remaining area of the paper

**48.** Set the scale to 1:40, then lock the viewport

**49.** Double-click inside the viewport and pan to the two windows of the master bedroom

**50.** Save and close the file

## 10.4    MECHANICAL PROJECT – I (METRIC)

Do the following steps:

**1.** From the designated folder, open file Mechanical-1_Starter.dwg

**2.** Switch off the grid

**3.** Set the units to the following:

   **a.** Decimal

   **b.** Precision 0

   **c.** Units to scale inserted contents = Millimeters

**4.** Set up drawing limits to be:

   **a.** Lower left corner = 0,0

   **b.** Upper right corner = 350,250

**5.** Double-click the mouse wheel in order to see the new limits

**6.** Create the following layers:

| Layer Name | Layer Color | Layer Linetype |
|------------|-------------|----------------|
| Centerline | Green | Center × 2 |
| Dimension | Red | Continuous |
| Hatch | 8 | Continuous |
| Hidden | Cyan | Hidden × 2 |
| Part | White | Continuous |
| Text | Magenta | Continuous |
| Title Block | White | Continuous |
| Viewport | 9 | Continuous |

**7.** Save as your file in Chapter 10\Metric folder and name it: Mechanical–1.dwg

**8.** Make layer Part current

**9.** Draw the below plan, section, and elevation of the mechanical part using the following guidelines:

   **a.** All lines of the shape in layer Part

   **b.** All centerlines in layer Centerline

   **c.** All hidden lines in layer Hidden

    **d.** Change the **Linetype scale** using the **Properties** palette for both centerlines and hidden lines to be 5; but for the two holes at the right and left, make the Linetype scale to be 2.

**10.** Draw the shape without dimensioning for now:

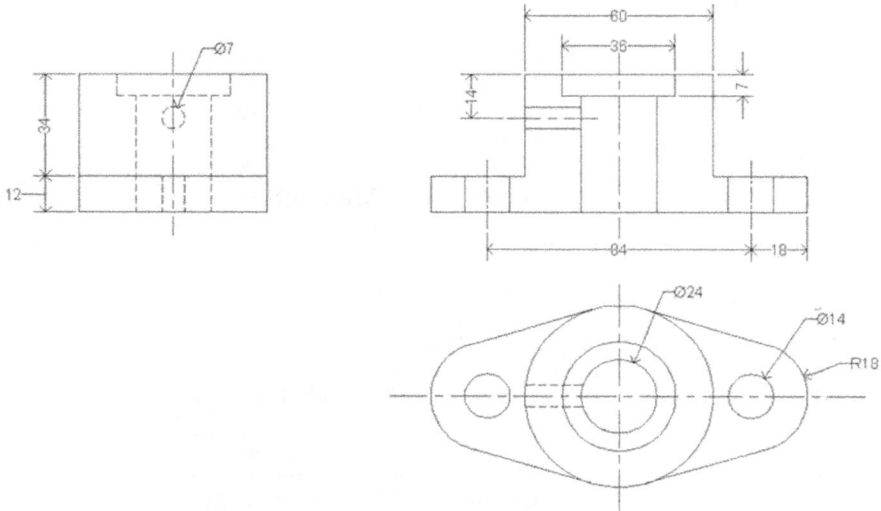

**11.** Make layer Hatch current

**12.** Using ANSI31 with scale = 20, hatch the shape as shown below:

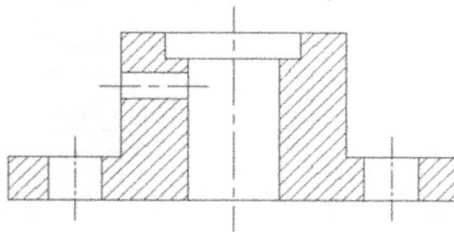

**13.** Make Part Dim the current dimension style

**14.** Make layer Dimension current then insert the dimensions as shown below:

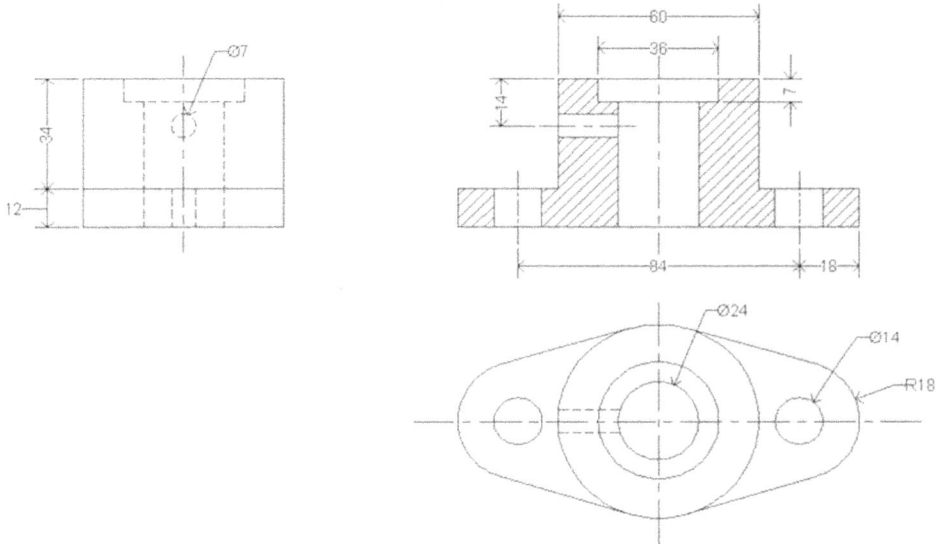

**15.** Go to Layout1 and rename it to Plan

**16.** Using the Page Setup Manager, modify the existing page setup to be as follows:

    **a.** Printer = DWF6 ePlot.pc3

    **b.** Paper = ISO A3 (420×297 MM)

    **c.** Drawing orientation = Landscape

    **d.** Make sure Scale = 1:1

**17.** Erase the existing viewport

**18.** Make layer Title Block current

**19.** Insert the file ISO A3 Landscape Title Block.dwg in the layout, using 0,0,0 insertion point.

**20.** Make two copies of Plan layout and rename it Section and Elevation

**21.** Delete Layout 2

**22.** Make layer Viewport current

**23.** Insert a single viewport like the following, doing the following steps:

    **a.** Set the scale to 2:1

    **b.** Lock the viewport

    **c.** Pan to the shape as shown below

    **d.** You will notice that hidden lines and centerlines look like continuous lines. To solve this problem, type at the command window **psltscale** command, and set this variable to 0. Then type **regenall** command to regenerate all viewports and you will receive the result shown below:

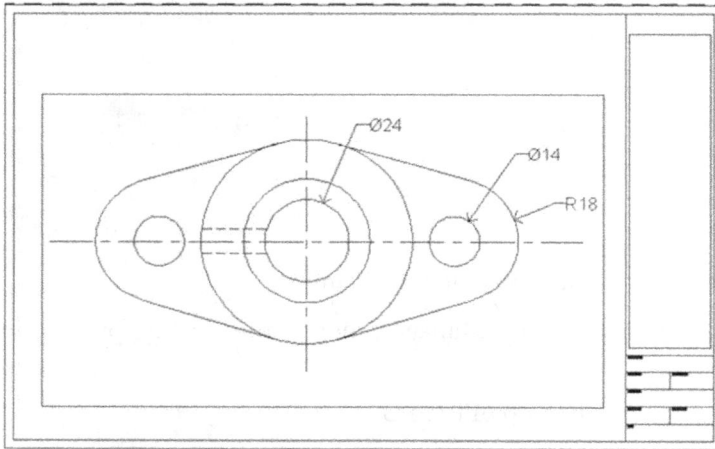

**24.** Repeat the same procedure to create section viewport in Section layout and elevation viewport in Elevation layout

**25.** Save and close the file

# 10.5   MECHANICAL PROJECT – I (IMPERIAL)

Do the following steps:

**1.** From the designated folder, open file Mechanical-1_Starter.dwg

**2.** Switch off the grid

**3.** Set the units to the following:

    **a.** Fractional

    **b.** Precision 0 – 1/16

    **c.** Units to scale inserted contents = Inches

**4.** Set up drawing limits to be:

    **a.** Lower left corner = 0,0

    **b.** Upper right corner = 18",9"

**5.** Double-click the mouse wheel in order to see the new limits

**6.** Create the following layers:

| Layer Name | Layer Color | Layer Linetype |
|------------|-------------|----------------|
| Centerline | Green | Center2 |
| Dimension | Red | Continuous |
| Hatch | 8 | Continuous |
| Hidden | Cyan | Hidden2 |
| Part | White | Continuous |
| Text | Magenta | Continuous |
| Title Block | White | Continuous |
| Viewport | 9 | Continuous |

**7.** Save your file in Chapter 10\Imperial folder and name it: Mechanical–1.dwg

**8.** Make layer Part current

**9.** Draw the below plan, section, and elevation of the mechanical part using the following guidelines:

    **a.** All lines of the shape in layer Part

    **b.** All centerlines in layer Centerline

    **c.** All hidden lines in layer Hidden

    **d.** Change the **Linetype scale** using the **Properties** palette to be 0.5 for any line you like

**10.** Draw the shape without dimensions for now:

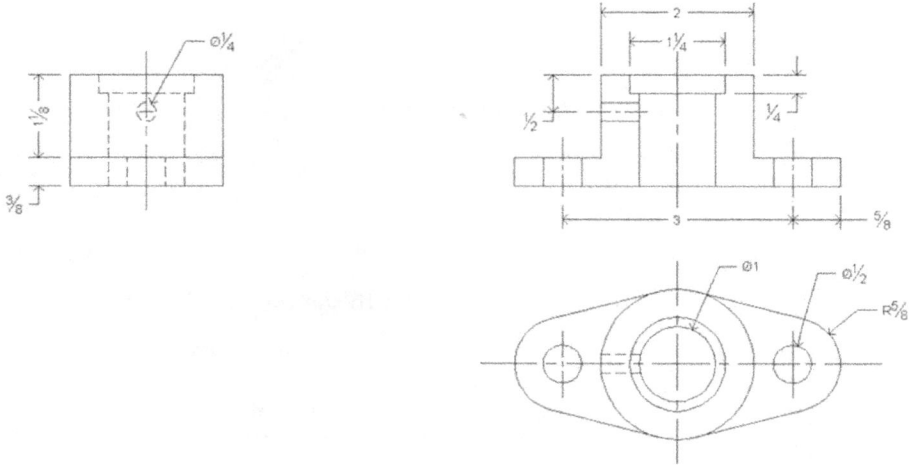

**11.** Make layer Hatch current

**12.** Using ANSI31 with scale = 1, hatch the shape as shown below:

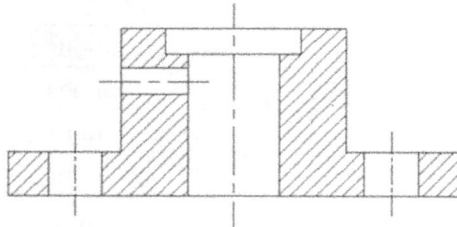

**13.** Make Part Dim the current dimension style

**14.** Make layer Dimension current

**15.** Insert the dimensions as shown below:

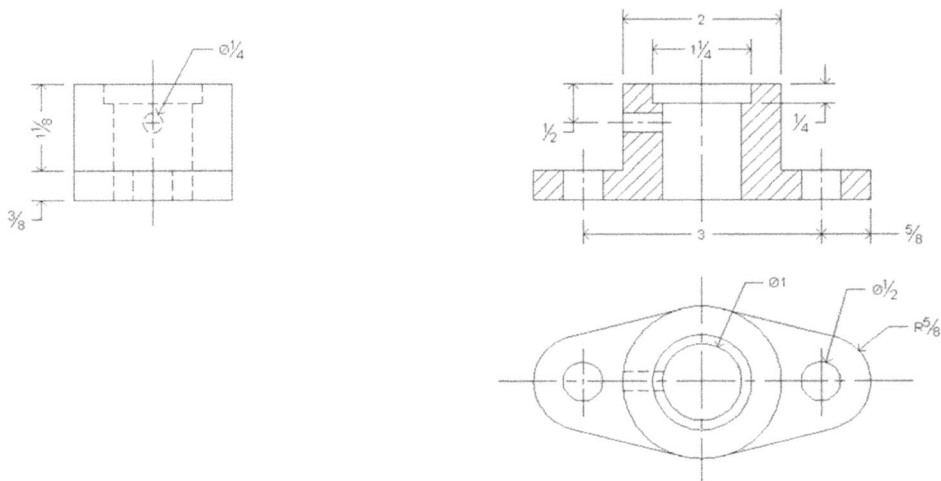

**16.** Go to Layout1 and rename it Details

**17.** Using the Page Setup Manager, modify the existing page setup to be as follows:

  **a.** Printer = DWF6 ePlot.pc3

  **b.** Paper = ANSI B (17×11 in)

  **c.** Drawing orientation = Landscape

  **d.** Make sure Scale = 1:1

**18.** Erase the existing viewport

**19.** Make layer Title Block current

**20.** Insert the file ANSI B Landscape Title Block.dwg in the layout, using 0,0,0 insertion point.

**21.** Delete Layout 2

**22.** Make layer Viewports current

**23.** Insert three single viewports like the following, doing the following steps:

  **a.** Set the scale for the three viewports to be 1' = 1'

  **b.** Lock the viewports

  **c.** Pan to the shape as shown below

**d.** If you noticed that hidden lines and centerlines look like continuous lines, type at the command window the **psltscale** command, and set this variable to 0. Then type the **regenall** command to regenerate all viewports, and you will receive the result shown below:

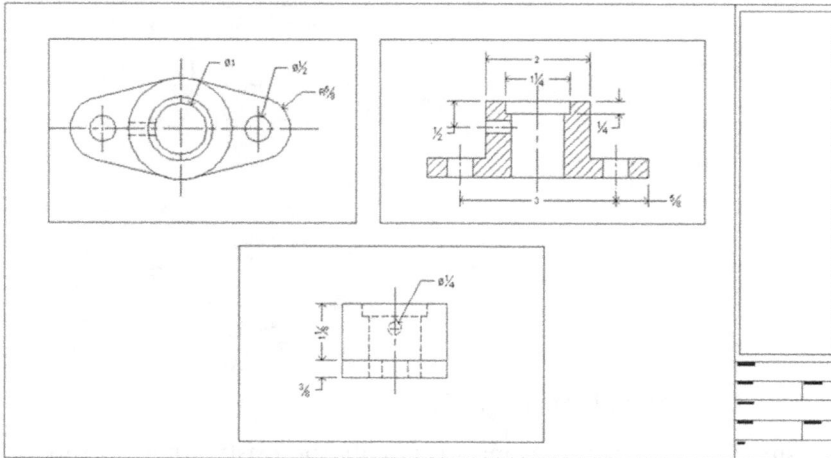

**24.** Save and close the file

## 10.6   MECHANICAL PROJECT – II (METRIC)

Using the same methodology we used in Mechanical Project – I (Metric), draw the following project using Mechanical-1_Starter.dwg:

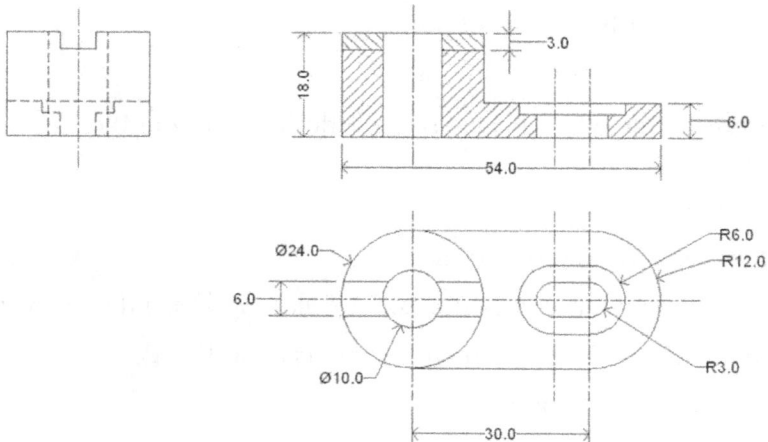

## 10.7   MECHANICAL PROJECT – II (IMPERIAL)

Using the same methodology we used in Mechanical Project – I (Imperial), draw the following project using Mechanical-1_Starter.dwg:

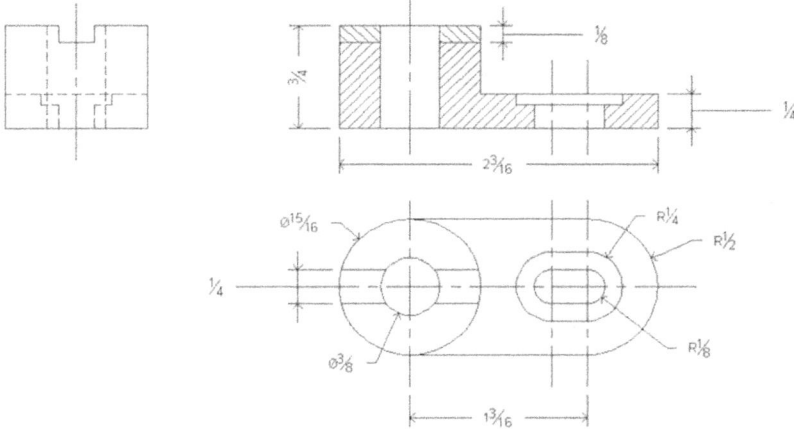

# NOTES

## CHAPTER 11

# MORE ON 2D OBJECTS

## In This Chapter

- The Polyline command and other drafting and editing commands
- How to use both Constructions Line & Ray
- How to use a point with different styles, along with Divide and Measure
- Using Spline and Ellipse
- Using Boundary and Region commands along with Boolean operations

## 11.1 INTRODUCTION

This chapter is dedicated to discuss the rest of the 2D objects not discussed in Chapter 2, as we will cover all the other 2D objects based on polyline, covering special features of polylines in some of the editing commands. Then we will cover other 2D commands like Spline and Ellipse. These two commands have some unique features, which will draw exact 2D curves. Then we will delve into other commands like Point (Divide and Measure) and Revision Cloud. We will close the chapter discussing Boundary and Region commands.

## 11.2   DRAWING USING THE RECTANGLE COMMAND

This command will draw a rectangle or square shape. The Rectangle command will use a polyline as an object. To issue this command go to the **Home** tab, locate the **Draw** panel, then select the **Rectangle** button:

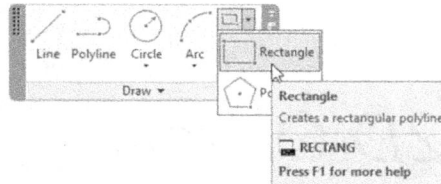

You will see the following prompts:

```
Specify first corner point or [Chamfer/Elevation/
Fillet /Thickness/Width]: Specify other corner point
or [Area/Dimensions/ Rotation]:
```

By default, you can draw a rectangle by specifying two opposite corners. The other options are:

### 11.2.1   Chamfer Option

This option will draw a rectangle with chamfered edges. The user will see the following prompts:

```
Specify first chamfer distance for rectangles <0.00>:
Specify second chamfer distance for rectangles <0.2>:
```

Specify the first and second distance, then the Rectangle command will continue with normal prompts. See the following example:

### 11.2.2   Elevation Option

This option is used for 3D only.

### 11.2.3 Fillet Option

This option is identical to the Chamfer option, except the user has to input the Radius instead of Distances. The user will see the following prompt:

```
Specify fillet radius for rectangles <0.0000>:
```

Specify the fillet radius, then the Rectangle command will continue with normal prompts. See the following example:

### 11.2.4 Thickness Option

This option is available for 3D only.

### 11.2.5 Width Option

The user will be able to draw a rectangle with width by using this option. The user will see the following prompt:

```
Specify line width for rectangles <0.0000>:
```

Specify the width value, and then the rectangle command will continue normally. See the following example:

### 11.2.6 Area Option

This option will specify the total area of the rectangle prior to specifying the second corner. The user will see the following prompts:

```
Enter area of rectangle in current units <25.0000>:
Calculate rectangle dimensions based on [Length/Width]
<Length>: Enter rectangle length <10.0000>:
```

AutoCAD asks the user to input the total area and then asks the user to input either the length (in X-axis) or width (in Y-axis). AutoCAD will then draw a rectangle above and to the right of the first corner.

### 11.2.7 Dimensions Option

This option will draw a rectangle by specifying length (in X-axis) and width (in Y-axis). The user will see the following prompts:

```
Specify length for rectangles <10.0000>:
Specify width for rectangles <10.0000>:
Specify other corner point or [Area/Dimensions/ Rotation]:
```

AutoCAD asks you to input the length and the width; in the final prompt, the user is invited to input the position of the second point.

### 11.2.8 Rotation Option

This option will draw a rectangle with a rotation angle. The user will see the following prompt:

```
Specify rotation angle or [Pick points] <0>:
```

Specify the rotation angle either by typing the value or by specifying points.

## 11.3 DRAWING USING THE POLYGON COMMAND

This command will draw an equilateral polygon with 3 sides to 1,024 sides. Polygon uses the polyline as an object. To issue this command, go to the **Home** tab, locate the **Draw** panel, then select the **Polygon** button:

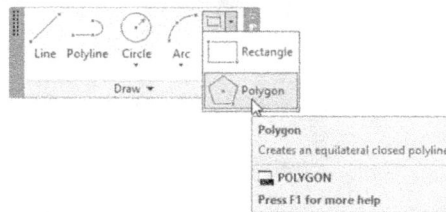

The user will see the following prompt:

```
Enter number of sides <6>:
```

Input the number of sides for your polygon. Next, you will see the following prompt:

```
Specify center of polygon or [Edge]:
```

AutoCAD offers two methods to draw a polygon, either by using an imaginary circle or by specifying the length and angle of one of the sides.

### 11.3.1   Using an Imaginary Circle

This method depends on an imaginary circle. The polygon is either inscribed inside it or circumscribed around it. The center of the circle and the polygon coincide, hence the radius of the circle will decide the size of the polygon. The question is: when should you use this or that method? The answer to that is the available information. If you know the distance between one of the edges and the polygon's center, then use the Inscribed option. However, if you know the distance between the midpoint of one of the edges and the center, then the Circumscribed option is the solution. Refer to the following picture:

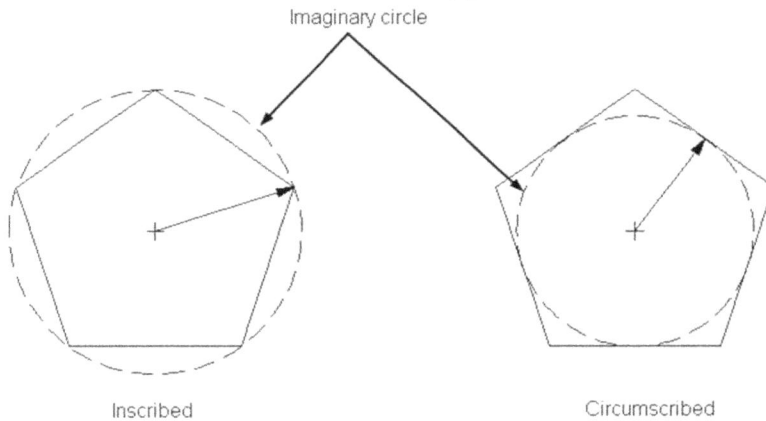

Inscribed                                        Circumscribed

The user will see the following prompts:

```
Enter an option [Inscribed in circle/Circumscribed
about circle] <I>: Specify radius of circle:
```

### 11.3.2   Using the Length and Angle of One of the Edges

If you do not know the center of the polygon, then you cannot use the above method. Alternatively, if you specify the length of one side, the other side's length will be known automatically. While you are specifying the two points as a length of one of the sides, you are also specifying the angle of this side; accordingly, the angles of the other sides will be defined. You will see the following prompts:

```
Enter number of sides <4>:
Specify center of polygon or [Edge]:
Specify first endpoint of edge:
Specify second endpoint of edge:
```

## PRACTICE 11-1   DRAWING RECTANGLES AND POLYGONS

1. Start AutoCAD 2025

2. Open **Practice 11-1.dwg** file

3. Use the Rectangle and Polygon commands to complete the practice to become as follows. Use OSNAP = Node to select the two points:

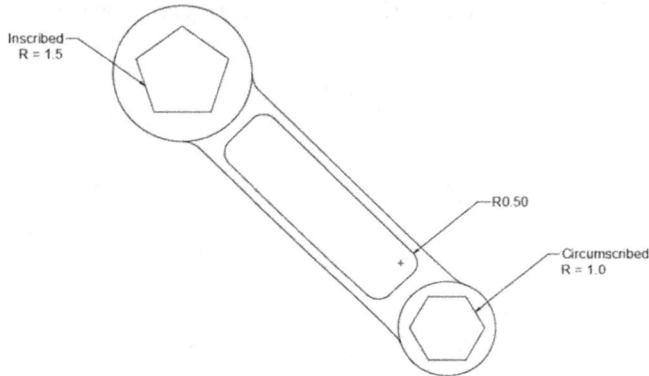

4. Freeze layer Points

5. Save and close

## 11.4   DRAWING USING THE DONUT COMMAND

This command will draw either a circle with width or a filled circle. Donut uses the polyline as an object. To issue this command, go to the **Home** tab, locate the **Draw** panel, then select the **Donut** button:

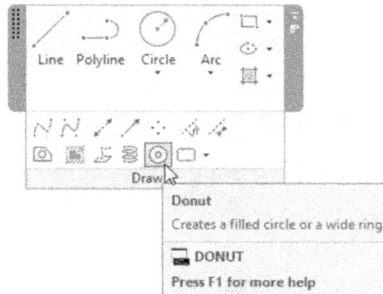

The following prompt will appear:

```
Specify inside diameter of donut <0.5000>:
Specify outside diameter of donut <1.0000>:
Specify center of donut or <exit>:
```

AutoCAD will ask you to input the inside and outside diameter, then to specify the center of the donut. The user can insert as many donuts as needed.

## 11.5 DRAWING USING THE REVISION CLOUD COMMAND

This command will draw a revision cloud using polyline arcs. To issue this command, go to the **Home** tab, locate the **Draw** panel, then select the **Revision Cloud** button:

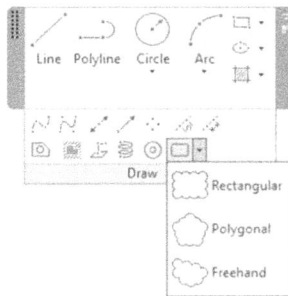

There are three types of Revision Clouds:

- Rectangular
- Polygonal
- Freehand

Using one of the three types, you will see the following prompts:

```
Minimum arc length: 0.5000 Maximum arc length: 0.5000
Style: Normal Type: Rectangular
Specify first corner point or [Arc length/Object/
Rectangular/Polygonal/Freehand/Style/Modify] <Object>:
```

### Arc Length

You should input the minimum and maximum arc length. Select Arc length and you will see the following prompt:

```
Specify minimum length of arc <15>:
Specify maximum length of arc <30>:
```

### Object

Instead of drawing a revision cloud, you can create a revision cloud by converting a closed 2D object (circle, polyline, ellipse, etc.); you will see the following prompts:

```
Select object:
Reverse direction [Yes/No] <No>:
```

### Rectangular

You can draw a rectangular revision cloud by specifying two opposite corners (just like the Rectangular command). You will see the following prompts:

```
Specify first corner point or [Arc length/ Object
/Rectangular /Polygonal/ Freehand/Style/Modify] <Object>:
Specify opposite corner:
```

You will receive the following:

### Polygonal

You can draw a polygonal revision cloud by specifying multiple corners. You will see the following prompts:

```
Specify first corner point or [Arc length/ Object
/Rectangular /Polygonal/ Freehand/Style/Modify] <Object>:
Specify next point:
Specify next point or [Undo]:
```

You will receive the following:

**Freehand**

You can draw a freehand revision cloud by specifying multiple points. You will see the following prompts:

```
Specify first corner point or [Arc length/ Object
/Rectangular /Polygonal/ Freehand/Style/Modify] <Object>:

Guide crosshairs along cloud path...
```

You will a shape similar to the following:

**Style**

If you select Style option, you will see the following prompt:

```
Select arc style [Normal/Calligraphy] <Normal>:
```

To know the difference between the two styles, refer to the following picture:

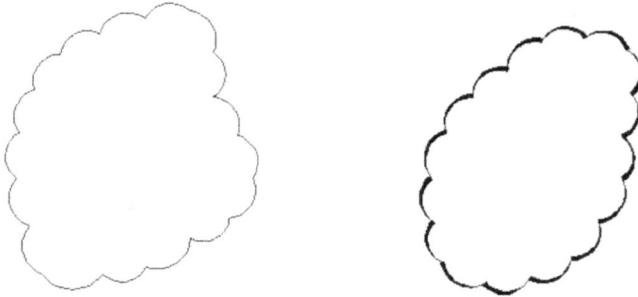

**Modify**

You can modify the shape of any existing revision cloud by adding new segments on one of the sides, then erasing any part desired. You will see the following prompts:

```
Select polyline to modify:
```

The position of the selection is important, as the new revision will start from there. You will see the following prompts:

```
Specify next point or [First point]:
Specify next point or [Undo]:
Specify next point or [Undo]:
```

You will receive the following:

You will see the following prompt:

```
Pick a side to erase:
Reverse direction [Yes/No] <No>:
```

Click on the side you want to remove, then answer whether you want to reverse the revision cloud or not. This is the final result:

## PRACTICE 11-2    DRAWING A DONUT AND A REVISION CLOUD

1. Start AutoCAD 2025

2. Open **Practice 11-2.dwg** file

**3.** Using the horizontal lines and the vertical lines, insert donuts with inside diameter = 0 and outside diameter = 0.25

**4.** Freeze Scratch layer

**5.** Make Redline the current layer

**6.** Using the Revision Cloud command, change the min arc length = 0.75 and max arc length t, and draw a rectangular revision cloud as seen in the following:

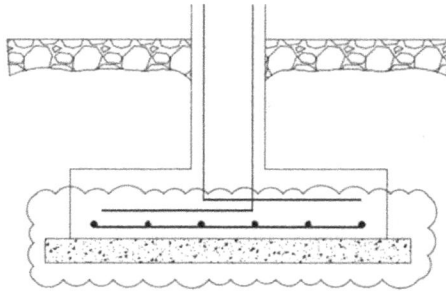

**7.** Using the Modify Option, add and remove segments of the revision and convert the circle to a revision cloud to look similar to the following:

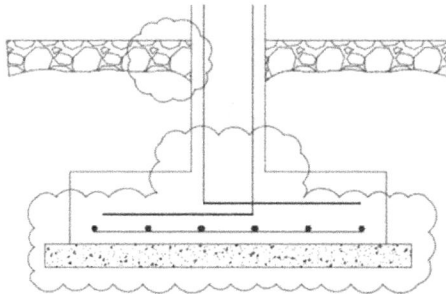

**8.** Save and close

## 11.6   USING THE EDIT POLYLINE COMMAND

This is a special editing command that can deal only with polylines. It can do certain things that the normal modifying commands cannot. We saw some of its power when we touched on the subject of converting lines and arcs to polylines. To issue this command, go to the **Home** tab, locate the **Modify** panel, then select the **Edit Polyline** button:

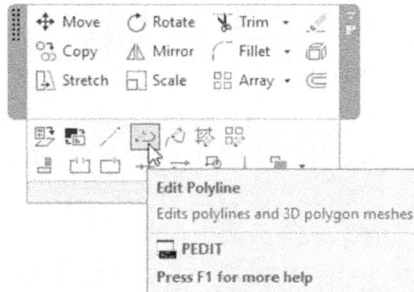

Another way is to double-click the polyline. The user will see the following prompts:

```
Select polyline or [Multiple]:
Enter an option [Close/Join/Width/Edit vertex/Fit/
Spline/Decurve/Ltype gen/Reverse/Undo]:
```

As you can see, AutoCAD is asking to select a single polyline to perform one of many editing options. AutoCAD, however, is also able to deal with **Multiple** polylines. We will first tackle editing options for a single polyline, then we will cover options for multiple polylines.

### 11.6.1   Open and Close Options

The Open option will be displayed if the polyline selected is closed, and vice versa. There are no prompts for these two options, as AutoCAD remembers the last segment drawn and will erase it to create an opened polyline.

### 11.6.2   Join Option

This option will join lines and arcs to the first selected polyline. The user will see the following prompt:

```
Select objects:
```

You are invited to select objects to join them to the selected polyline.

### 11.6.3   Width Option

This option will give a width to the selected polyline. The user will see the following prompt:

```
Specify new width for all segments:
```

### 11.6.4   Edit Vertex Option

This option will select a vertex in the polyline, then perform an editing option on this vertex. As agreed upon by many AutoCAD experts, this option is extremely lengthy, tedious, and very difficult. Instead, the user can explore the polyline, perform all/any normal modifying commands, then join the lines and arcs in a single polyline. The user will see the following prompt:

```
[Next/Previous/Break/Insert/Move/Regen/Straighten
/Tangent/Width/eXit] <N>:
```

**NOTE**   *The user can use the Grips and Properties palette to edit vertices.*

### 11.6.5   Fit, Spline, and Decurve Options

Fit and Spline are both options that will convert a straight-line segment polyline to a curved polyline using two different methods; check the following illustration:

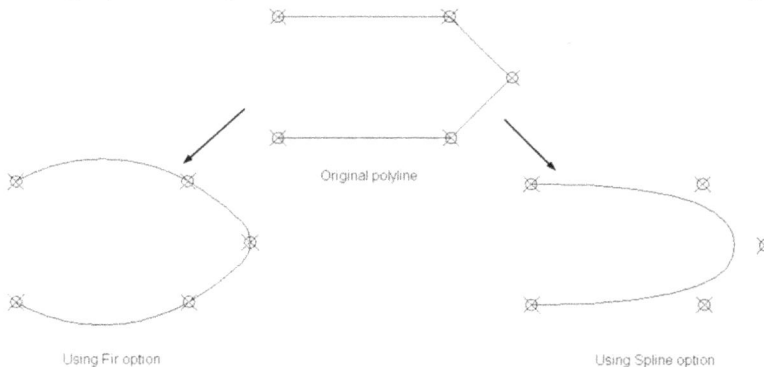

As you can see from the above illustration, each command is handling the process in a different way:

- The **Fit** option will use the same vertices and connect them using a curve, so it is considered to be an approximate method
- The **Spline** option will use the vertices as controlling points to draw the needed curve. This method will display a more accurate curve

The **Decurve** option will convert back from curved shape polyline to straight lines polyline.

### 11.6.6  Ltype Gen Option

This option will allow polylines converted from straight lines to curve to retain the original linetype. See the following illustration:

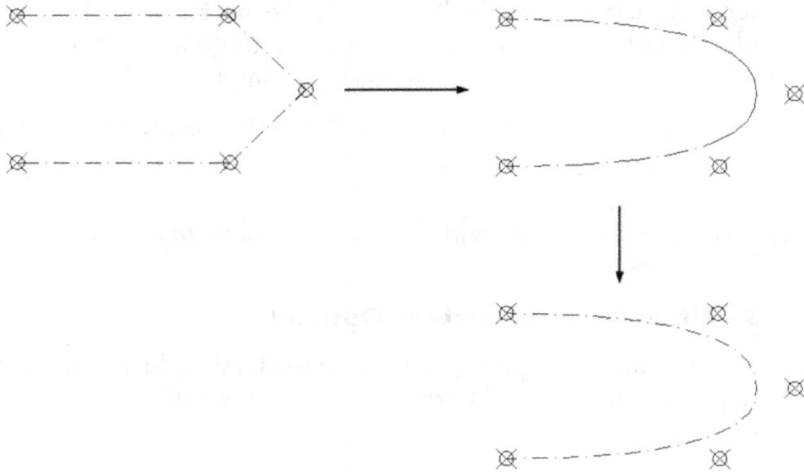

The user will see the following prompt:

```
Enter polyline linetype generation option [ON/OFF] <Off>:
```

In order to retain the linetype, input ON as an answer for this prompt.

### 11.6.7  Reverse Option

This option will reverse the order of vertices in a polyline. It will be evident when using special linetype such as the following:

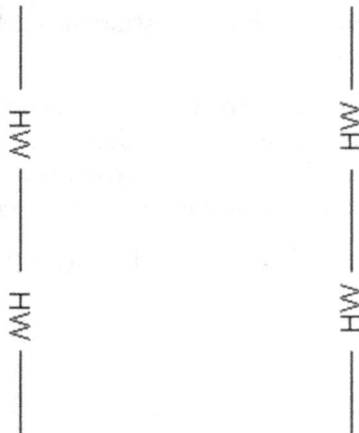

There is a system variable PLINEREVERSEWIDTHS will control whether to reverse the polyline width or not. If the value is 0 (zero) the polyline will not be reversed, but if the value is 1, the polyline width will be reversed. See the following illustration:

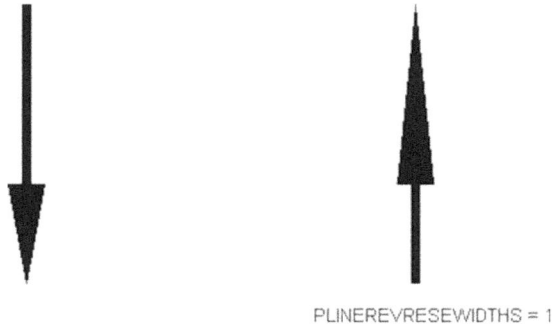

PLINEREVRESEWIDTHS = 1

### 11.6.8   Multiple Option

This command will modify multiple polylines using a single modifying command. Another mission for this command is to join polylines together. This is different from the join option discussed previously, which is for single polylines joining lines and arcs; this option is to join polylines together in a single polyline. You will see the following prompts:

```
Select objects:
Select objects:
Enter an option [Close/Open/Join/Width/Fit/Spline /
Decurve/Ltype gen/Reverse/Undo]:
```

It is identical to editing a single polyline, except for Edit Vertex option. If you select the Join option, you will see the following prompt:

```
Join Type = Extend
Enter fuzz distance or [Jointype] <0.0000>:
```

The first line is showing you the current value for Join Type which is Extend; in order to change it, invoke the **Jointype** option to see the following prompt:

```
Enter join type [Extend/Add/Both] <Extend>:
```

There are three types of joining:

- **Extend**: which means AutoCAD will extend the two ends to each other to join the multiple polyline
- **Add**: which means AutoCAD will add a line between the two ends
- **Both**: which means AutoCAD will use both methods

AutoCAD also will need to know the Fuzz distance, which is the maximum acceptable distance between the ends of the two polylines to join. With anything greater than this value, AutoCAD will decline to join the polylines. See the following illustration:

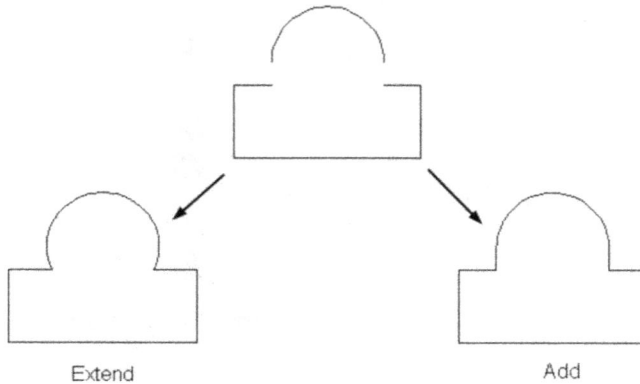

Extend          Add

One final note for the Polyline Edit command: the professional user tends to cut steps while drafting and editing and AutoCAD can assist them with that. System variable **PEDITACCEPT** can help them cut one step, by assuming that when selecting the first object in the Polyline command, the answer is always yes:

```
Object selected is not a polyline
Do you want to turn it into one? <Y>
```

PEDITACCEPT has two values:

- Either 0 (zero) which means AutoCAD will ask the question and wait for you to confirm
- Or 1 which means assume the answer is always Yes

### Special OSNAP

If you have a closed polyline, you can snap to its centroid using the Center Object Snap. This includes both regular shapes and irregular shapes.

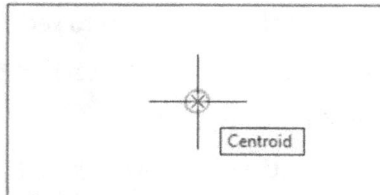

Centroid

## PRACTICE 11-3   POLYLINE EDIT COMMAND

**1.** Start AutoCAD 2025

**2.** Open **Practice 11-3.dwg** file

**3.** Check the objects drawn; you will find that all objects are polylines except the outer contour, which is lines. Using the Polyline Edit command, convert lines to polylines

**4.** Using the Polyline Edit command, convert the straight line segments to spline segments

**5.** Using the Polyline Edit command, retain the dashed line of the converted polylines

**6.** Zoom to the two small buildings inside the contours, and you will find them as polylines. But the arc is not reaching the line segments. Measure the void between the arc and the lines, and input a proper Fuzz distance. Then join the polylines using Extend at the left shape and Add at the right shape

**7.** You should receive the following picture:

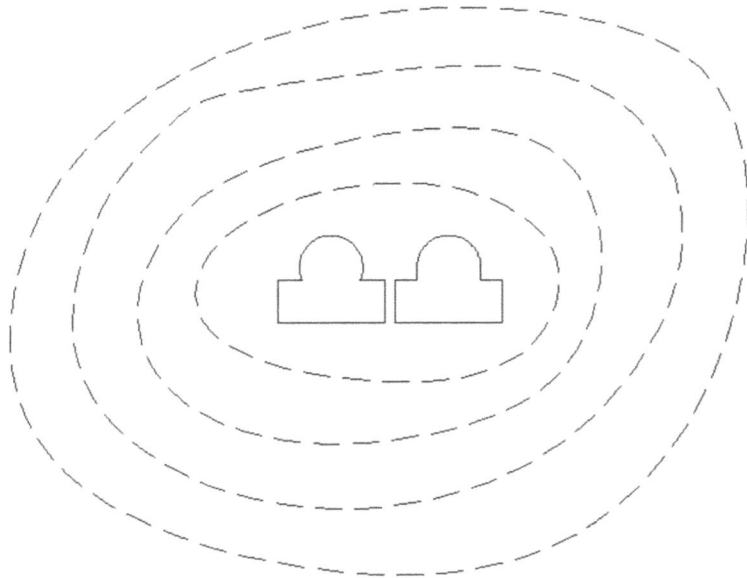

**8.** Save and close

## 11.7 USING CONSTRUCTION LINES AND RAYS

These two commands will produce objects to be used as helping tools to draw accurate drawings; they are not wanted on their own. Construction lines are objects extending beyond the screen in two directions and can be drawn using different methods. Yet Rays in AutoCAD are objects that have a known starting point, extending beyond the screen in one direction. Here is a discussion of both commands:

### 11.7.1 Using Construction Lines

This command will help you to draw a construction line which extends beyond the screen in two directions. To issue this command go to the **Home** tab, locate the **Draw** panel, then select the **Construction Line** button:

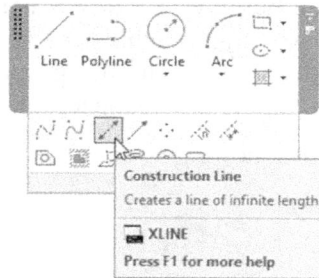

You will see the following prompt:

```
Specify a point or [Hor/Ver/Ang/Bisect/Offset]:
```

There are six methods to specify the angle of the construction line, which are:

- The first method is the default method, which is to specify two points; once you specify the first point, you will see the following prompt

```
Specify through point:
```

- The **Hor** and **Ver** options are for drawing horizontal or vertical construction lines. To complete the command, specify the through point. You will see the following prompt:

```
Specify through point:
```

- The **Ang** option is for drawing a construction line using an angle. You will see the following prompts:

```
Enter angle of xline (0) or [Reference]:
Specify through point:
```

- The **Bisect** ooption will involve specifying three points. It will pass through the first point and bisect the angle formed between the second and third points, as you will see in the following prompts:

```
Specify angle vertex point:
Specify angle start point:
Specify angle end point:
```

- The **Offset** option produces a construction line parallel to an existing line, as you will see in the following prompts:

```
Specify offset distance or [Through] <Through>:
Select a line object:
Specify side to offset:
```

### 11.7.2 Using Rays

This command will draw a ray, which has a starting point and the end extending beyond the screen. To issue this command go to the **Home** tab, locate the **Draw** panel, and select the **Ray** button:

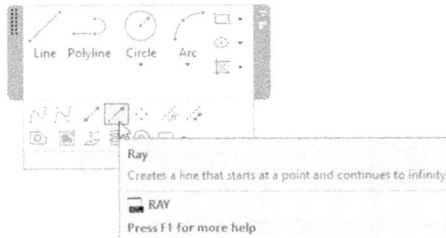

You will see the following prompts:

```
Specify start point:
Specify through point:
```

AutoCAD will ask you to specify two points: the first point is the starting point, and the second will define the angle of the ray. You can define as many rays as you prefer using the same starting point.

## PRACTICE 11-4   USING CONSTRUCTION LINES AND RAYS

1. Start AutoCAD 2025

2. Open **Practice 11-4.dwg** file

3. Make the Construction layer current

4. Insert vertical and horizontal construction lines using the center of the circle

**5.** Insert two construction lines with the offset option, using the two vertical lines with distance = 0.5, to the inside

**6.** Using the Ray command set the starting point to be the center of the circle and the second point to be with angle = 60 using Polar Tracking

**7.** Repeat the same command, using angles 120, 240, and 300

**8.** Draw a new circle with its center coinciding with the existing circle, using R=2

**9.** You should receive the following image:

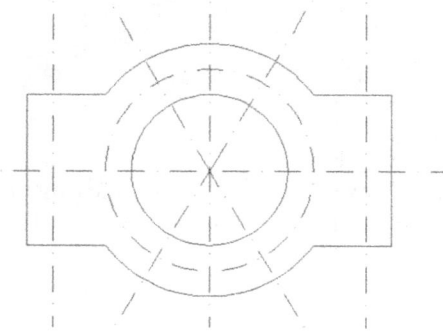

**10.** Make layer Circle current and make sure Intersection in OSNAP is turned on

**11.** Draw circles using the intersection of the four rays and the circle, along with the vertical construction line and the circle, using R=0.2

**12.** Draw two circles at the intersections of the horizontal and vertical construction lines

**13.** Freeze the Construction layer

**14.** You should receive the following image:

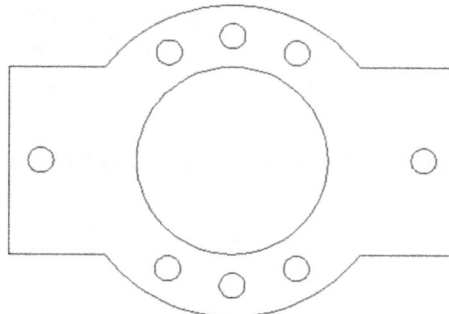

**15.** Save and close the file

## 11.8   USING POINT STYLE AND POINT COMMAND

Point style command will set the shape of the point, and Point command will insert a point in the drawing. You can change the point style as many times as you wish and points already inserted will shift to the new shape.

### 11.8.1   Using Point Style Command

This command will set the point style. To issue this command go to the **Home** tab, locate the **Utilities** panel, then select the **Point Style** button:

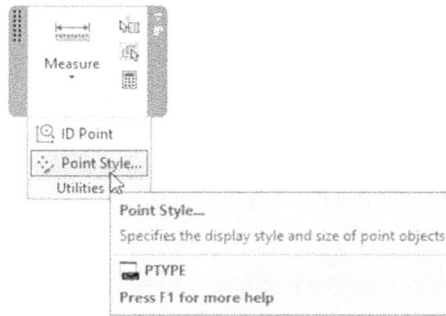

You will see the following dialog box:

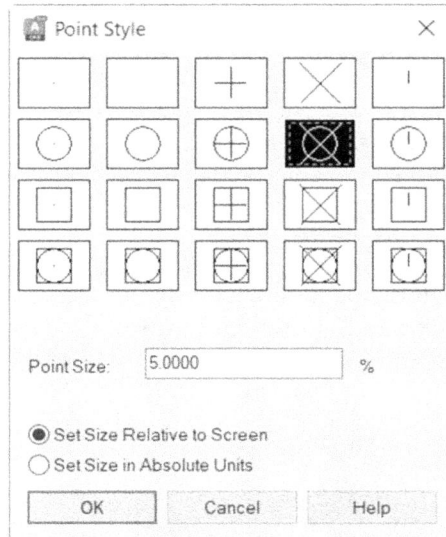

Pick one of twenty shapes available, and then set the point size, either relative to the screen or in absolute units.

### 11.8.2 Using Point Command

This command will insert as many points in the drawing as you wish.

To issue this command go to the **Home** tab, locate the **Draw** panel, then select the **Multiple Points** button:

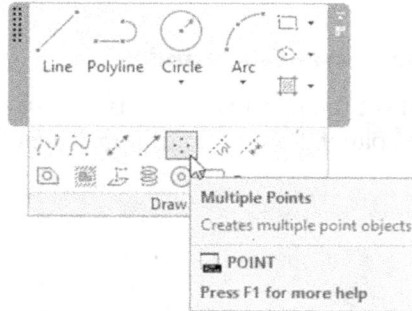

You will see the following prompts:

```
Specify a point:
```

AutoCAD will ask you to start specifying points; once done, press the [Esc] key. In order to select inserted points precisely, use **NODE** OSNAP.

## 11.9 USING DIVIDE AND MEASURE COMMANDS

The Divide command will cut an object into equal spaced intervals input by you using points, whereas the Measure command will cut an object into chunks with user-specified distance, using points.

### 11.9.1 Using the Divide Command

This command will divide an object with equally spaced intervals specified by you. To issue this command, go to the **Home** tab, locate the **Draw** panel, then select the **Divide** button:

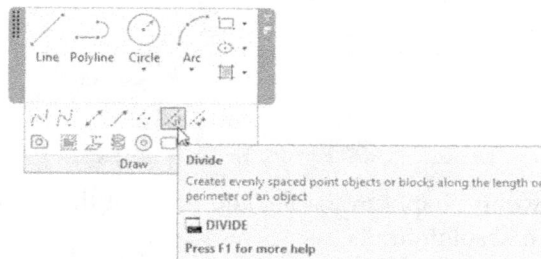

The user will see the following prompts:

```
Select object to divide:
Enter the number of segments or [Block]:
```

AutoCAD will ask you to select the desired object and then input the desired number of segments.

### 11.9.2  Using the Measure Command

This command will cut an object into segments with user-specified distance, using points. To issue this command go to the **Home** tab, locate the **Draw** panel, then select the **Measure** button:

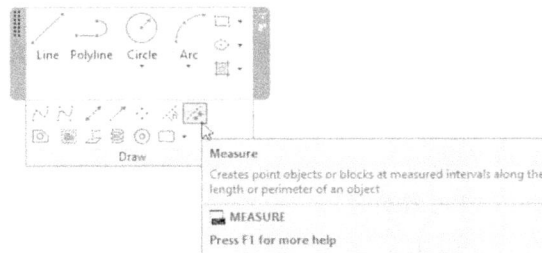

The user will see the following prompts:

```
Select object to measure:
Specify length of segment or [Block]:
```

AutoCAD will ask you to select the desired object and then specify the desired length. When you select the object, AutoCAD will start measuring from the end nearest to the selection; hence, the user should be cautious.

### 11.9.3  Using Divide and Measure Commands with Block Option

In both commands, you can use Block instead of a point and the following prompts will appear:

```
Enter the number of segments or [Block]:
Enter name of block to insert:
Align block with object? [Yes/No] <Y>:
Enter the number of segments:
```

The user should first respond with Block option to the first prompt. Then input the name of the block and whether it will be aligned or not. Finally, enter the number of segments or the distance, depending on the command used. To illustrate the Aligning concept, refer to the following image:

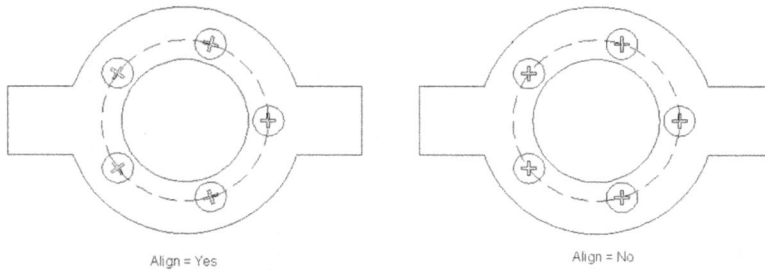

Align = Yes                    Align = No

## PRACTICE 11-5   USING POINT STYLE, POINT, DIVIDE, AND MEASURE

1. Start AutoCAD 2025

2. Open **Practice 11-5.dwg** file

3. Change the Point Style to 

4. Zoom to the upper horizontal lines. The length of the inner horizontal line is 5700. Using the Measure command, add points at 1425 starting from the left end

5. Erase the rightmost point

6. Make sure that Node is on

7. Using the Insert command, insert block = Window 1, using the three points

8. Erase the three points

9. Make layer Furniture current

10. At the middle of the room draw a circle with R = 1200. Offset the circle by 250 to the outside

11. Using the Divide command, add the block = Chair using the outside circle, using eight chairs

**12.** Erase the outside circle. You should receive the following:

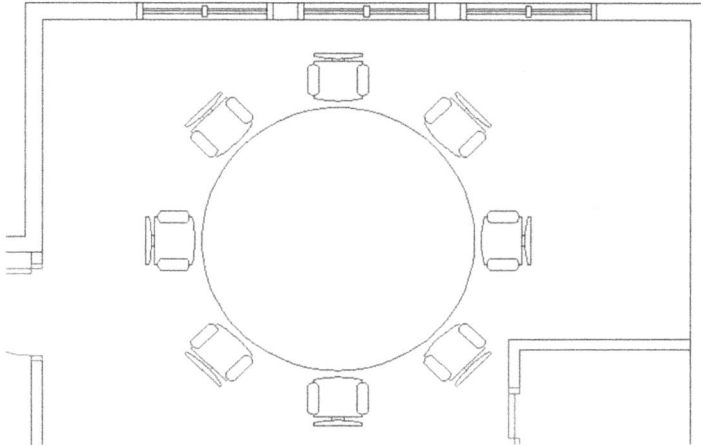

**13.** Save and close the file

## 11.10    USING THE SPLINE COMMAND

This command will draw smooth curves based on more than two points. It will draw a spline curve based on exact mathematical equations. There is one command, but two keys used to invoke two different methods, which are: Fit Points or Control Vertices. To issue these two commands, go to the **Home** tab, locate the **Draw** panel, then select one of the following two buttons:

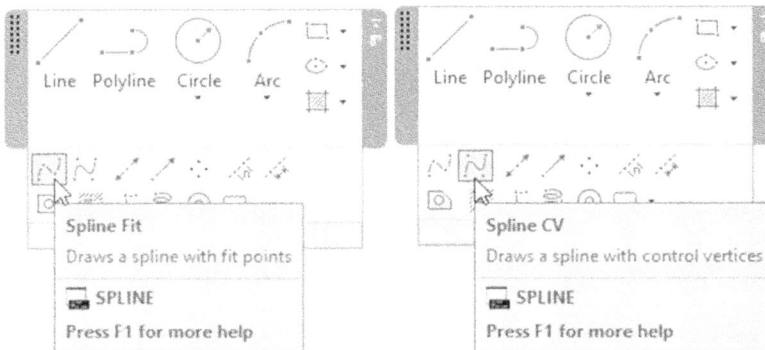

### 11.10.1  Using the Fit Points Method

This method will draw a spline with fit points coinciding with it. The following prompts will appear:

```
Specify first point or [Method/Knots/Object]:
Enter next point or [start Tangency/toLerance]:
Enter next point or [end Tangency/toLerance/Undo]:
```

AutoCAD will ask you to specify the desired points to draw the spline with an option to close the shape automatically. We used to specify start tangency and end tangency in old versions of AutoCAD, but in this version there is no need, as AutoCAD will make this based on the points specified. Though the above argument is true, AutoCAD prompts will allow you to specify a start and end tangency. AutoCAD will draw a curve connecting the points you select. See the following illustration:

The user has the ability to specify tolerance for points other than the start and end points. See the following illustration:

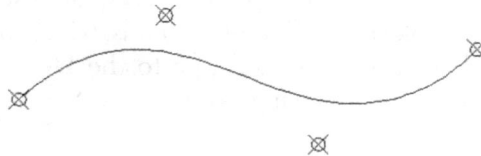

AutoCAD can also convert any polyline which was treated using the Polyline Edit command and fit in a spline to be a real spline. See the following illustration:

Polyline

Converted to Spline using Edit Polyline command

Converted to Spline using Object option

### 11.10.2 Using the Control Vertices Method

This method will draw a spline using control vertices, which will define a control frame. Control frames provide a convenient method to shape the spline. The following prompts will appear:

```
Specify first point or [Method/Degree/Object]:
Enter next point:
Enter next point or [Undo]:
Enter next point or [Close/Undo]:
```

AutoCAD will ask you to specify the desired points to draw the spline with an option to close the shape automatically. Meanwhile you can specify the degree of the spline, which sets the polynomial degree of the resulting spline. The user can input degree 1 (linear), degree 2 (quadratic), and degree 3 (cubic), and so on up to degree 10. You will receive the following:

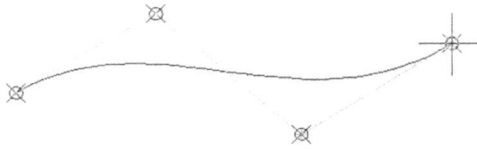

### 11.10.3 Editing a Spline

When you click a spline produced by the Fit Points method, you will see the fit points along with the triangle, which will allow you to show either the Fit Points or the Control Vertices:

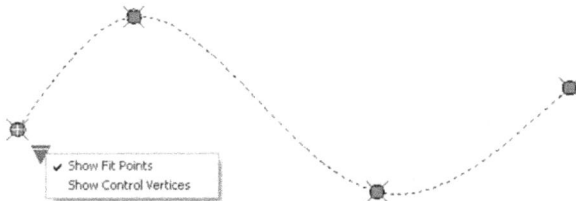

If you stay at one of the fit points, you will see the following menu:

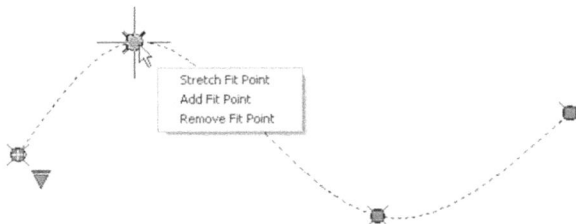

This menu will Stretch the current fit point, add a new fit point, or remove the current fit point. You will see an extra option if you stay at the start or end point of the spline, Tangent Direction, which will change the tangent direction of the spline.

If you clicked a spline drawn using Control Vertices, you will see CV, along with the triangle which will display either the fit points or CV:

If you stay at one of the CV, you will the following menu:

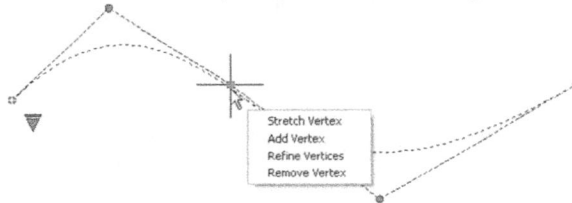

This menu will stretch the current vertex, add a vertex, or remove a vertex. The Refine Vertex option will replace the current vertex with two vertices:

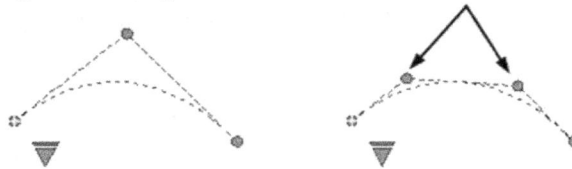

## PRACTICE 11-6    USING THE SPLINE COMMAND

**1.** Start AutoCAD 2025

**2.** Open **Practice 11-6.dwg** file

**3.** Make sure Layer Contour is current

**4.** In OSNAP, make sure that Node is on

**5.** Using the points at the left, draw an open spline using the Fit Points option

**6.** Using the points at the middle, draw a closed spline using the Control Vertices option

**7.** Thaw layer Hidden Points

**8.** Using grips and the Add Fit Points option, add the two new points at the middle

**9.** Stretch the first Fit Point (the one at the left) to the new point to its left

**10.** Now the spline does not pass through the old first point; using grips, add it up

**11.** Remove the point indicated below:

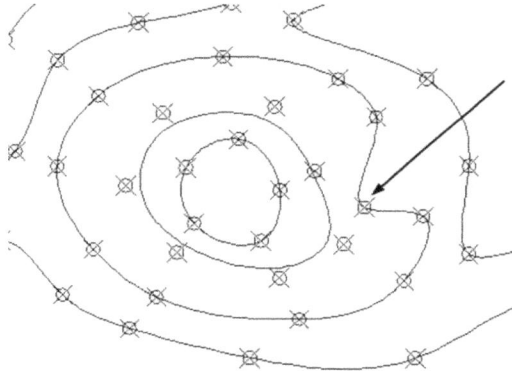

**12.** Freeze both Points and Hidden Points layers

**13.** Save and close the file

## 11.11 USING THE ELLIPSE COMMAND

This command will draw an elliptical shape or elliptical arc. To issue this command go to the **Home** tab, locate the **Draw** panel, then select the **Ellipse** button and select one of the methods:

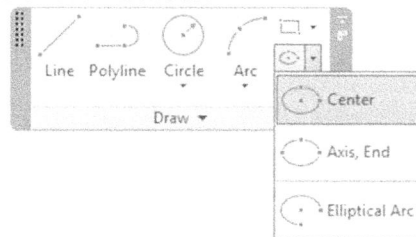

There are three options to pick from:

- Center
- Axis, End
- Elliptical Arc

Where the first two options will draw an ellipse, the third option will draw an elliptical arc. Here is a discussion for each one of them:

### 11.11.1 Drawing an Ellipse Using the Center Option

Using this method the user should specify three points, which are:

- Center point of the ellipse
- Endpoint of one of the two axes
- Endpoint of the other axis

The following image will illustrate the concept:

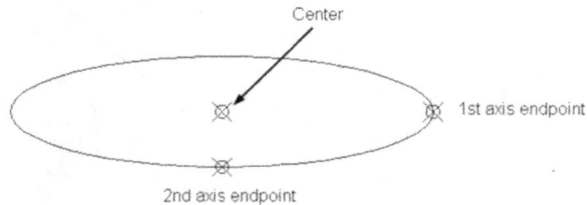

You will see the following prompts:

```
Specify axis endpoint of ellipse or [Arc/Center]: C
Specify center of ellipse:
Specify endpoint of axis:
Specify distance to other axis or [Rotation]:
```

### 11.11.2 Drawing an Ellipse Using Axis Points

The user should specify three points, which are:

- Point on one end of one of the axis
- Point on the other end of the same axis
- Point on the other axis

The following image will illustrate the concept:

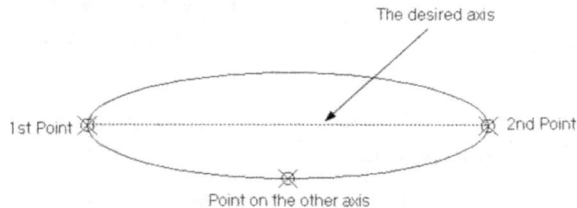

You will see the following prompts:

```
Specify axis endpoint of ellipse or [Arc/Center]:
Specify other endpoint of axis:
Specify distance to other axis or [Rotation]:
```

Using either method, the last step will include an option called Rotation. So what is rotation? After you define two points, you will draw a circle. Imagine this circle is in a plane and the plane is rotating; you will get an ellipse. See the following illustration:

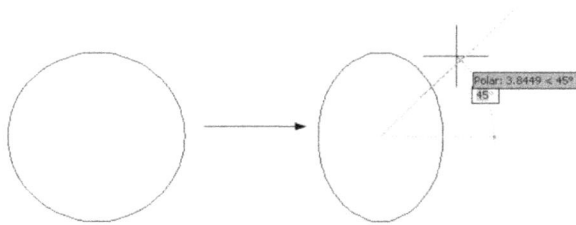

### 11.11.3 Drawing an Elliptical Arc

The first three steps for drawing an elliptical arc are identical to drawing the ellipse itself, which we discussed above. Afterward, AutoCAD will ask you to specify the starting angle and ending angle, counterclockwise. Another way is after you specify the first angle, you can input the included angle instead of the ending angle.

You will see the following prompts:

```
Specify start angle or [Parameter]:
Specify end angle or [Parameter/Included angle]:
```

## PRACTICE 11-7   USING THE ELLIPSE COMMAND

**1.** Start AutoCAD 2025

**2.** Open **Practice 11-7.dwg** file

**3.** Make layer Table current

**4.** Make sure Node is on

**5.** Using Axis, End method draw an elliptical table using the points displayed

**6.** Freeze Points layer

**7.** Make layer Window current

**8.** Using Elliptical Arc, complete the window. For the outer elliptical arc, use the endpoints of the two vertical lines and the other axis distance = 3.5, then use the offset command using distance = 0.5

**9.** You should receive the following:

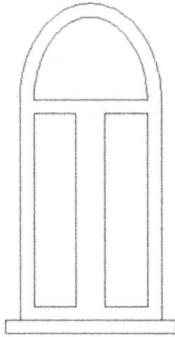

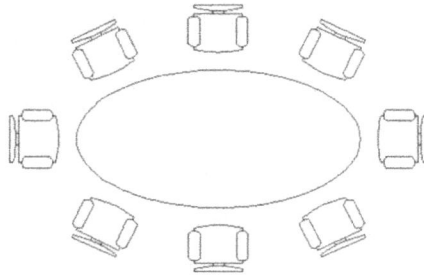

**10.** Save and close the file

## 11.12   USING THE BOUNDARY COMMAND

If you have several intersecting 2D objects (lines, arcs, circles, polylines, ellipses, etc.) and you want to calculate the net area of these objects, no command will help you more than this command. You can choose between polylines and regions as a resultant object. To issue this command go to the **Home** tab, locate the **Draw** panel, then select the **Boundary** button:

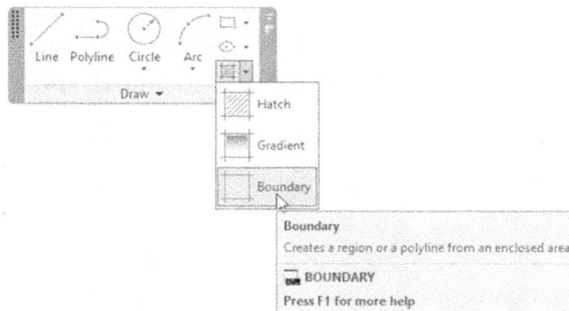

You will see the following dialog box:

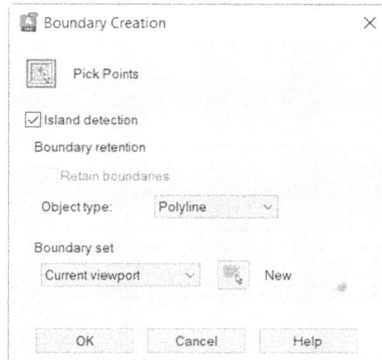

```
Boundary Creation                        ×

  [icon]   Pick Points

☑ Island detection
  Boundary retention
    ☐ Retain boundaries
  Object type:      Polyline      ∨

  Boundary set
    Current viewport   ∨   [icon]   New

        OK        Cancel        Help
```

This command depends on a simple click inside the desired area in which you want to create a polyline. Therefore, we will start with the **Pick Points** button. When you are done, click OK to end the command, and create the polyline (or region) desired. You can move it (them) outside to calculate areas or any other desired commands. Meanwhile, you can make some amendments to the command to receive different results; check the following options:

- **Island detection:** this option will control whether AutoCAD should identify an object within the area
- **Polyline or Region:** the user can pick the desired object type
- **Boundary set:** whether all objects will be involved in the creation (Current viewport option) or only selected objects (click the New button)

## PRACTICE 11-8   USING BOUNDARY COMMAND

1. Start AutoCAD 2025

2. Open **Practice 11-8.dwg** file

3. Zoom to Shape 01

4. Using the Boundary command, and without changing anything, click inside the area. To view the resultant shape, freeze layer Shap01

5. Zoom to Shape 02

6. Using Boundary command, click off the Island detection checkbox, then click inside the area. To view the resultant shape, freeze layer Shap02

7. Zoom to Shape 03

**8.** Using Boundary command, click the New button and select all of Shape 03 except the four circles at the edges, then click inside the area. To view the resultant shape, freeze layer Shap03

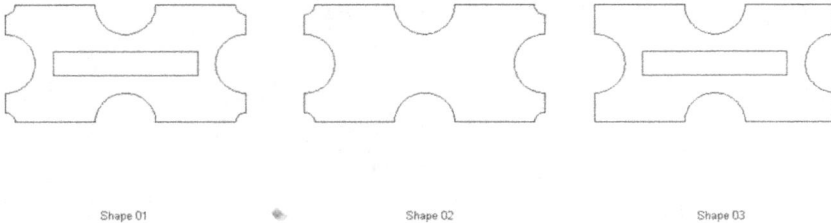

Shape 01                    Shape 02                    Shape 03

**9.** Save and close the file

## 11.13   USING THE REGION COMMAND

Assume you have some wires and you are asked to create a rectangle and circle from the wires and place the circle in the center of the rectangle. Though the circle is in the center of the rectangle, there is no relationship between them, because if you move one of the two shapes, the other will stand still.

On the other hand, if you have a piece of paper and a pair of scissors and you are asked to cut a rectangle with a hole in the shape of a circle there is a relationship between the two shapes.

This is exactly the difference between polylines and regions in AutoCAD.

To issue this command, go to the **Home** tab, locate the **Draw** panel, then select the **Region** button:

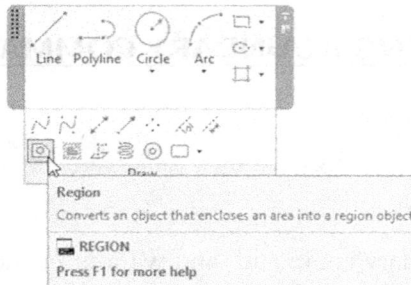

You can create a region using the previous Boundary command.

To convert wireframe 2D objects like lines, arcs, circles, polylines, etc., you have to make sure that they are formulating closed shapes only. You will see the following prompt:

```
Select objects:
```

Select the desired objects and press [Enter] when done; objects will be converted right away.

### 11.13.1 Performing Boolean Operations on Regions

Since regions are real 2D objects, AutoCAD can perform on them Boolean operations, which are: Union, Subtract, and Intersect. To issue these three commands, you have to switch to the **3D Basics** workspace, go to the **Home** tab, locate the **Edit** panel, then click one of the following buttons:

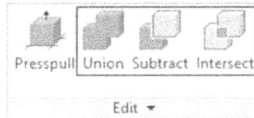

In the Union and Intersect commands, you can select objects in any order, yet in the Subtract command, first select the region(s) you want to **subtract from**, press [Enter], and then select the region(s) to be **subtracted**.

## PRACTICE 11-9  USING REGION COMMAND

1. Start AutoCAD 2025

2. Open **Practice 11-10.dwg** file

3. Using the Region command, convert all objects to regions

4. Click any object to show the Quick Properties and make sure that it has become a region

5. Using a Boolean operation, create the following shape:

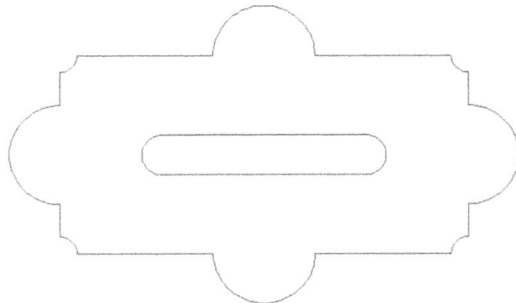

6. Using the Properties palette, what is the total area? _____ (19.2552)

7. Save and close the file

# NOTES

## CHAPTER REVIEW

1. In reality donuts, polygons, revision clouds, and rectangles are all:

   **a.** Splines

   **b.** Regions

   **c.** Polylines

   **d.** None of the above

2. In the Polyline Edit you have to select the _____ option first to join polylines to polylines

3. Construction line and Ray will produce objects as helping tools to draw accurate drawings

   **a.** True

   **b.** False

4. You can use Boolean operations with polyline objects

   **a.** True

   **b.** False

5. One of the following statements is not true:

   **a.** Divide command will cut any object to equally spaced intervals using blocks

   **b.** Measure command will cut any object to equally spaced intervals using points

   **c.** Measure command will cut any object into chunks with user-specified distance, using points

   **d.** Divide command will cut any object to equally spaced intervals using points

6. There are two methods to draw a spline in AutoCAD, Fit Points, or Control Vertices:

   **a.** True

   **b.** False

7. Boundary command can create either _____ or _____

## CHAPTER REVIEW ANSWERS

**1.** c

**3.** a

**5.** b

**7.** Polyline, Region

# *ADVANCED PRACTICES — PART I*

## In This Chapter

- The advanced features of the Offset, Trim, and Extend commands
- How to utilize Cut/Copy/Paste when you open more than one file
- How to bring in AutoCAD objects from other software
- Hyperlink
- Purge unused objects
- Views and Viewport commands

## 12.1  OFFSET COMMAND – ADVANCED OPTIONS

People who use AutoCAD on a daily basis are stuck with the default options, ignoring some powerful options that may reduce production time significantly. In this part, we will look at these options. When you start the Offset command you will see the following prompts:

```
Current settings: Erase source=No Layer=Source
OFFSETGAPTYPE=0 Specify offset distance or
[Through/Erase/Layer] <Through>:
```

The first line is a message listing the current values for the different settings of AutoCAD. The message says: Erase source = No, Layer = Source, OFFSETGAPTYPE = 0, so what are these settings and how can we change them?

### 12.1.1   Erase Source Option

By default, AutoCAD will maintain both the source and the offset object. This option will keep the offset object, but will erase the source object. You will see the following prompt:

```
Erase source object after offsetting? [Yes/No] <No>:
```

Input Yes if you want to eliminate the source object.

### 12.1.2   Layer Option

When you use any command in AutoCAD that produces a copy of the original object, the copy will always reside in the same layer of the source object. Using this option, you can request AutoCAD to send the generated object to the current layer instead. AutoCAD will show the following prompts:

```
Enter layer option for offset objects [Current /
Source] <Source>:
```

Input Current to tell AutoCAD you want the offset object in the current layer.

### 12.1.3   System Variable: Offsetgaptype

This is not an option inside Offset command, but rather a system variable that should be invoked before the command using the command window. This system variable decides the outcome of the shape to be normal (value = 0), filleted (value = 1), or chamfered (value = 2).

## 12.2   TRIM AND EXTEND – EDGE OPTION

You cannot extend an object unless there is a real intersecting point between the boundary edge and objects to be extended. In addition, you cannot trim an object unless there is a real intersecting point between the cutting edge and objects to be trimmed. See the following illustration:

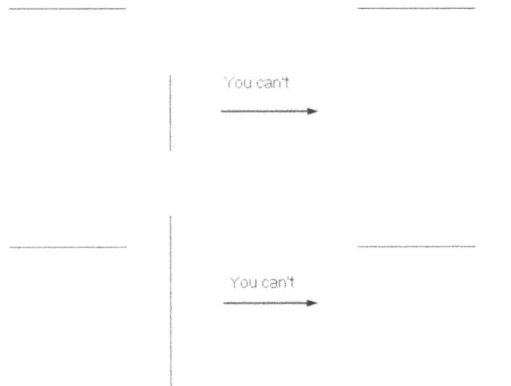

The **Edge** option will trim objects based on extended cutting edges and will extend objects based on extended boundary edges. Selecting the Edge option will invoke the following prompt (The default mode is Quick. You need to go to Standard mode first, then you can use Edge option):

```
Enter an implied edge extension mode [Extend/No
extend] <No extend>:
```

The default option is **No extend**; select the **Extend** option to allow you to trim and extend based on the extended cutting and boundary edges. The user should be careful because all the previous settings in Offset, Trim, and Extend will affect all the files from the time of change forward.

## PRACTICE 12-1   USING ADVANCED OPTIONS IN OFFSET, TRIM, AND EXTEND

1. Start AutoCAD 2025

2. Open **Practice 12-1.dwg**

3. Offset the top shape using the Offset command, making sure that the new object will reside in the current layer and the original object will be deleted, and the offset will result in a chamfered shape, using distance = 1

**4.** Explode the newly created polyline

**5.** Using the Edge option in both Trim and Extend, trim the upper horizontal line using the two vertical lines, and extend the lower horizontal lines based on the same vertical lines

**6.** Draw small vertical lines to complete the base

**7.** You should receive the following shape:

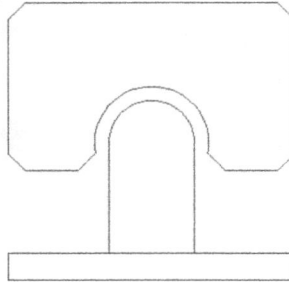

**8.** Save and close

## 12.3 USING MATCH PROPERTIES

This command will match correct properties of an object to incorrect properties of other objects. Objects here include everything in AutoCAD: lines, arcs, circles, polylines, splines, ellipses, text, hatches, dimensions, viewports, and tables. To issue this command go to the **Home** tab, locate the **Properties** panel, then select the **Match Properties** button:

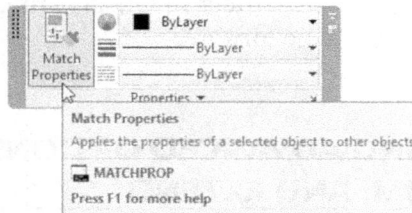

You will see the following prompt:

```
Select source object:
```

AutoCAD is requesting you to select the object that holds the correct properties; once done, you will see the following prompts:

```
Select destination object(s) or [Settings]:
```

With this prompt you will see the cursor change to:

Select objects that hold the incorrect properties, which will be matched with the source object. The user can use **Settings** option to specify the basic and advanced properties to be affected. You will see the following dialog box:

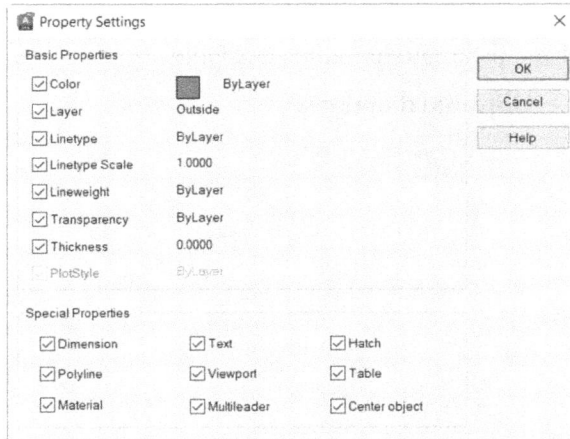

# 12.4 COPY/PASTE OBJECTS AND MATCH PROPERTIES ACROSS FILES

In AutoCAD, you can open more than one DWG file at the same time using a simple technique, which is holding the [Ctrl] key while selecting the names of the desired files in the **Open** file dialog box. However, the question is why would anybody want to open more than one file at the same time? The answer would be one or both of the following:

- To copy objects from one file to another
- To match properties across files

To tile the opened files, go to the **View** tab, locate the **Interface** panel, then use one of the following two buttons:

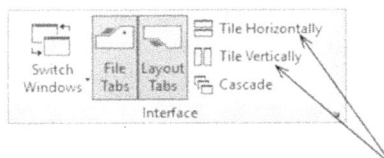

We will use the normal Copy/Paste sequence in order to copy objects from one file to another (this Copy is different compared to the Copy command in the Modify panel, as this command will copy objects from one file to another). Try the following steps:

- Without issuing any command, select the desired object(s)
- Right-click and select the Clipboard option, then select one of the two copying commands available
- Go to the other file. Right-click and select the **Clipboard** option, then select one of the three pasting commands available

These are the **Clipboard** options:

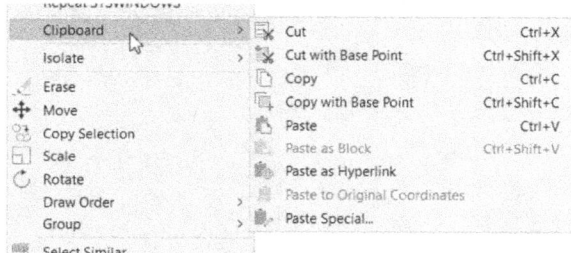

| | | |
|---|---|---|
| Clipboard ▸ | ✂ Cut | Ctrl+X |
| Isolate ▸ | ✂ Cut with Base Point | Ctrl+Shift+X |
| Erase | ▯ Copy | Ctrl+C |
| Move | ▯ Copy with Base Point | Ctrl+Shift+C |
| Copy Selection | ▯ Paste | Ctrl+V |
| Scale | Paste as Block | Ctrl+Shift+V |
| Rotate | Paste as Hyperlink | |
| Draw Order ▸ | Paste to Original Coordinates | |
| Group ▸ | Paste Special... | |
| Select Similar | | |

### 12.4.1   Copying Objects

AutoCAD allows you to copy objects from one file to another using two techniques, which are:

- Copy option which will copy objects without specifying a base point
- Copy with Base Point option which will copy objects with specifying a base point, which will produce the following prompt:

```
Specify base point:
```

### 12.4.2   Cutting Objects

AutoCAD allows you to cut objects from one file to another using two techniques, which are:

- Cut option which will cut objects without specifying a base point
- Cut with Base Point option which will cut objects with specifying a base point, which will bring up the following prompt:

```
Specify base point:
```

### 12.4.3 Pasting Objects

There are three methods to paste objects across files, which are:

- **Paste option**, which will paste the contents of the clipboard
- **Paste as Block option**, which will paste the objects as a block with an arbitrary name. The user is invited to use the Rename command to give the block the correct name
- **Paste to Original Coordinates option**, which will paste the objects to the same coordinates used in the original file

### 12.4.4 Using the Drag-and-Drop Method

If you dislike this method, you can use the Drag-and-Drop method, which includes selecting desired objects in the source file, then clicking and holding the left button going to the destination file, and dropping the objects there. You can do the same using the right button instead; with this method, when you drop the objects in the destination file, a menu such as the following will appear:

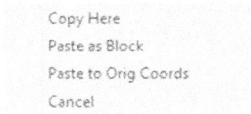

Copy Here
Paste as Block
Paste to Orig Coords
Cancel

This menu includes self-explanatory options.

While you are copying objects across files, AutoCAD will create all the necessary things to make the copying process successful, like creating layers, text styles, dimension styles, etc.

### 12.4.5 Match Properties Across Files

We will use the same command to match properties across files selecting the source object in the current file (the file you will issue the command from) and then match the objects holding the incorrect properties in the destination file. Likewise, AutoCAD will create the necessary things to make the matching process successful, like creating layers, text styles, dimension styles, etc. in the destination file.

## PRACTICE 12-2   USING MATCH PROPERTIES, COPY/PASTE ACROSS FILES

**1.** Start AutoCAD 2025

**2.** Open **Practice 12-2.dwg**, and **Ground Floor.dwg**

**3.** Tile them vertically

4. Check layers in Practice 12-2.dwg, take a note of the existing layers

5. Check the dimension styles in Practice 12-2.dwg, take a note of the existing styles

6. From Ground Floor.dwg using Copy/Paste, copy the bathroom 9′×9′ using the base point to the suitable place in Practice 12-2.dwg

7. Using Match Properties select any dimension in the Ground Floor.dwg and match it with all dimensions in Practice 12-2.dwg

8. Using Cut with Base Point, cut the couch in the Ground Floor.dwg and paste it in the entrance of Practice 12-2.dwg using the Quadrant of the outer arc. Rotate the couch to fit in the entrance

9. Close Ground Floor.dwg and maximize Practice 12-2.dwg

10. In Practice 12-2.dwg, match the properties of the copied text with all the other texts

11. Start the Match Properties command and select the hatch of the kitchen as the source object, then select the hatch of the two toilets

12. Check the layers again and see how many layers were added

13. Check the dimension styles again and see how many new styles were added

14. If you have time, put all objects in the right layer

15. Save and close the file

## 12.5   SHARING EXCEL AND WORD CONTENT IN AUTOCAD

As a Windows application, AutoCAD can give and take objects to and from any other Windows application, especially MS Office software like MS Word and MS Excel, which is the most used software nowadays. AutoCAD will use OLE (**O**bject **L**inking & **E**mbedding) to copy the contents from and to AutoCAD. The Paste command used in AutoCAD will dictate the type of object brought to AutoCAD. How to copy content from one application to another is a known and common practice among regular users of Windows.

### 12.5.1   Sharing Data Coming from MS Word

Copy the contents from Word using Copy or Cut, then go to the **Home** tab, locate the **Clipboard** panel as shown below (if this panel is not shown by default, make sure that you are at the **Home** tab, right-click any panel, select **Show Panels**, then select **Clipboard**):

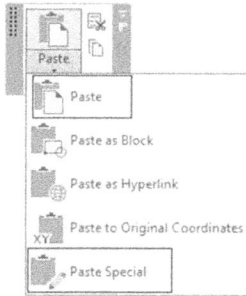

We will use either Paste or Paste Special options. See the following:

Word Content → Using Paste → OLE object embedded
Word Content → Using Paste Special / Paste / Text → MTEXT
Word Content → Using Paste Special / Paste / UniCode Text → MTEXT
Word Content → Using Paste Special / Paste / AutoCAD Entities → Text
Word Content → Using Paste Special / Paste Link → OLE object linked

Once you issue the **Paste Special** command, you will see the follow ing dialog box:

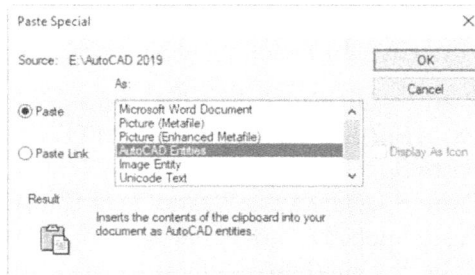

As you can see, the first option (using Paste only) and the last option using the Paste link will bring in an OLE object; the first one will not be updated when the source changes, and the last one will be. Using the first method, you will see the following dialog box:

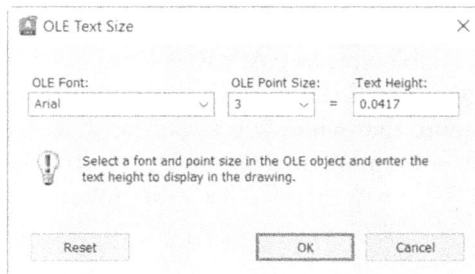

The OLE Text Size dialog box will show the font name and font size used in MS Word, and then ask you to specify the Text Height in AutoCAD units. At any time you can click the Reset button to get the original value. Once you are done, click **OK**.

After pasting an OLE object, AutoCAD allows you to edit it. To edit an OLE object, select it and right-click, and you will see a menu; select the OLE option, which will produce the following:

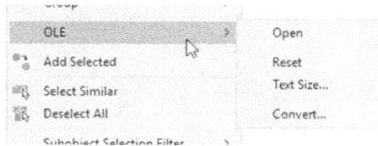

The available choices are:

- **Open option** will open the source application
- **Reset option** will retain the font and font size of the original
- **Text Size** option will change the text size
- **Convert option** will change the nature of the OLE object

### 12.5.2  Sharing Data Coming from MS Excel

Copy the contents from Excel using Copy or Cut, then go to the **Home** tab, locate the **Clipboard** panel, and choose either Paste or Paste Special options, like we did in the MS Word case. In Excel, you have the following choices:

Excel Content → Using Paste → OLE object embedded
Excel Content → Using Paste Special / Paste / Text → MTEXT
Excel Content → Using Paste Special / Paste / UniCode Text → MTEXT
Excel Content → Using Paste Special / Paste / AutoCAD Entities → Table Embedded
Excel Content → Using Paste Special / Paste Link → OLE object linked
Excel Content → Using Paste Special/Paste Link/AutoCAD objects → Table Linked

You will see the same dialog boxes discussed previously.

The interesting choice is the last choice, where you will paste a linked table in AutoCAD; this will be discussed next:

### 12.5.3  Pasting a Linked Table from Excel

This option will enable you to paste Excel content into your current drawing, linking it to the original Excel sheet. With this option the update process will be a two-way street; meaning that when you update in Excel, it will affect the table in AutoCAD, and when you update in AutoCAD, it will affect the file in Excel. When you select the table, then over it you will see the following image:

This means the table inside AutoCAD is *locked* and *linked*

The easy thing is when you change anything in the Excel sheet and save, these changes will be reflected in the AutoCAD table. At the lower right corner of the AutoCAD window you will see a chain called **Data Link** as in the following:

If you right-click this icon, you will see a menu as in the following:

Select the **Update All Data Links** option to get the latest copy of your Excel sheet. If the AutoCAD file is not opened, then the next time you open it, you will receive the newest copy of the Excel sheet.

Updating the Excel content from an AutoCAD table is slightly more difficult and it requires you to follow strict procedure:

- Go to the **Insert** tab, locate the **Linking & Extraction** panel, then select the **Data Link** button:

- The following dialog box will appear:

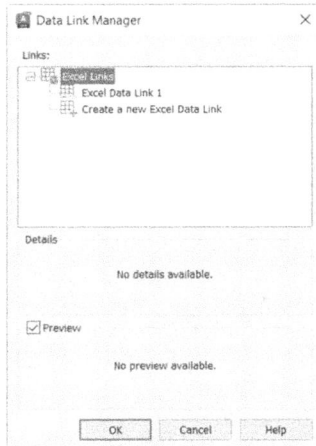

- Under **Links** you will see there is a link called **Excel Data Link1**, which was created automatically when you paste link the Excel table. This name is temporary. To rename the link, click it, and right-click then select the Rename option:

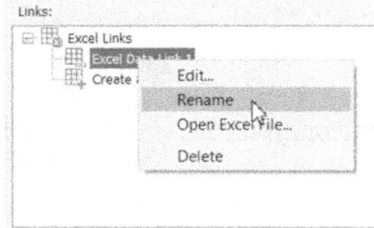

- There will be a **Preview** checkbox turned on to help you preview the link in case there is more than one link in the same file, which prevents making a mistake
- Double-clicking the link or selecting the **Edit** option from the right-click menu will lead to the following dialog box:

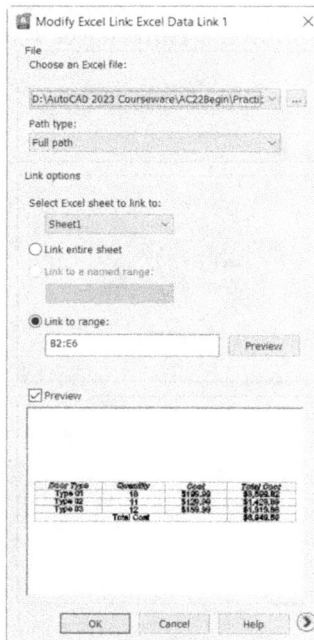

- Clicking the arrow at the lower right part of the dialog box will expand it to show more options:

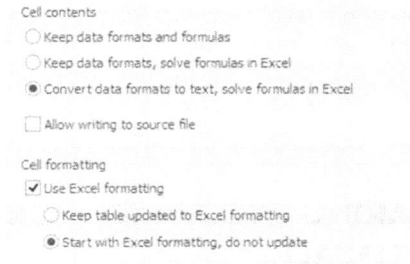

Cell contents
    ○ Keep data formats and formulas
    ○ Keep data formats, solve formulas in Excel
    ● Convert data formats to text, solve formulas in Excel

    ☐ Allow writing to source file

Cell formatting
    ☑ Use Excel formatting
      ○ Keep table updated to Excel formatting
      ● Start with Excel formatting, do not update

- Make sure that the **Allow writing to source file** checkbox is turned on, because without this we will not succeed to write back to the Excel sheet
- Click **OK** several times to close all dialog boxes
- We are ready to make the edits in the table pasted in AutoCAD
- Using crossing, select the desired cells in the table
- Right-click, the menu will appear, select the **Locking** option, then the **Unlocked** option:

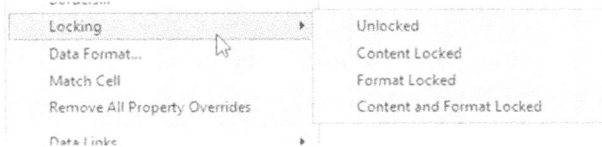

| Locking | ▸ | Unlocked |
| Data Format... | | Content Locked |
| Match Cell | | Format Locked |
| Remove All Property Overrides | | Content and Format Locked |
| Data Links | ▸ | |

- You will notice that the locking symbol disappeared, but the chain symbol is still there. Change the desired value in the unlocked cells
- To upload these changes to the original Excel sheet go to the **Insert** tab, locate the **Linking & Extraction** panel, then select the **Upload to Source** button:

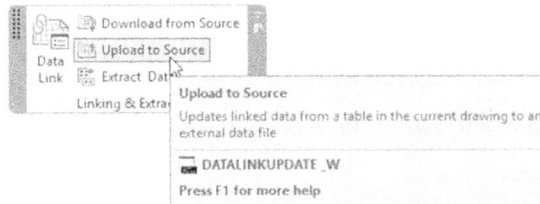

Download from Source
Data Link    Upload to Source
   Extract Dat
Linking & Extra

**Upload to Source**
Updates linked data from a table in the current drawing to an external data file

DATALINKUPDATE _W
Press F1 for more help

- It will ask you to select objects; click the desired table from one of its outside borders
- The following prompt will appear:

```
1 object(s) found.
1 data link(s) written out successfully.
```

- You will see the following bubble appears at the lower right corner of the AutoCAD window:

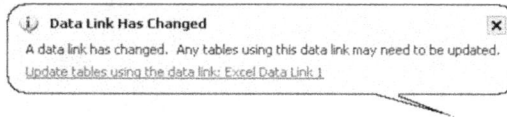

> (i) **Data Link Has Changed** [×]
> A data link has changed. Any tables using this data link may need to be updated.
> Update tables using the data link: Excel Data Link 1

## PRACTICE 12-3    SHARING EXCEL AND WORD CONTENT IN AUTOCAD

**1.** Start AutoCAD 2025

**2.** Open **Practice 12-3.dwg** file

**3.** Using Microsoft Word open the file House General Notes.doc and copy all its contents

**4.** Go to Full Plan layout

**5.** Make layer Text current

**6.** Using Paste option, paste the text in the lower left corner of the layout below the viewport

**7.** Make the text size = 5

**8.** Open the file Door Cost Schedule.xls and copy its contents

**9.** Using Paste Special / Paste Link, select AutoCAD Entities

**10.** Close the Excel sheet

**11.** Change the quantity of Type 02 to be 21, and reflect it back to the Excel sheet

**12.** Update the AutoCAD file with the new value to recalculate the new values

**13.** Make sure that the Excel file has changed

**14.** Save and close the files

## 12.6    HYPERLINKING AUTOCAD OBJECTS

This command will hyperlink any AutoCAD object(s) to a website, an AutoCAD drawing, to an Excel sheet, to a Power Point presentation, etc. To issue this command, go to the **Insert** tab, locate the **Data** panel, then select **Hyperlink** button:

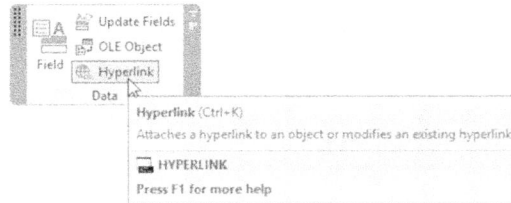

You will see the following prompt:

```
Select objects:
```

Select the desired objects; when done, press [Enter] and you will see the following dialog box:

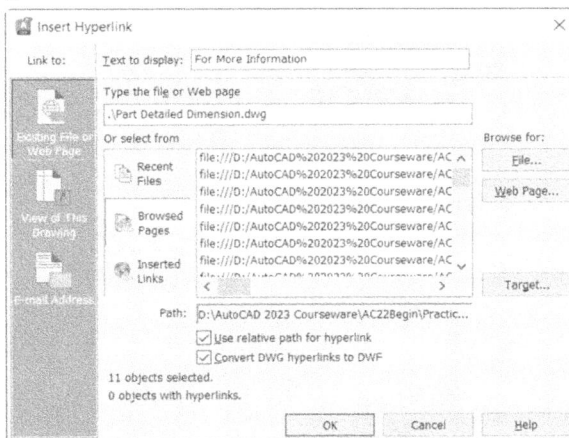

Fill in the **Text to display**, which is like a help note for you, that appears when you get close to the object holding the hyperlink. Then input **Type the file or Web page name**, which is a web site address or file path. Finally click **OK** to end the command.

Using the **Home** tab, locate the **Clipboard** panel and you will find an option called **Paste as Hyperlink**, which will do the job in reverse order, as it will paste the contents as hyperlink to an object.

## PRACTICE 12-4   HYPERLINKING AUTOCAD OBJECTS

**1.** Start AutoCAD 2025

**2.** Open **Practice 12-4.dwg** file

**3.** Hyperlink the 3D shape to a file called Part Detail Dimension.dwg

**4.** Test the hyper link by holding the [Ctrl] key and clicking the 3D shape

**5.** Save and close the files

## 12.7   PURGING UNUSED ITEMS

We normally create a fair amount of content inside the AutoCAD drawing, like layers, blocks, dimension styles, text styles, multileader styles, etc. We use some of them, and others are left without being used. Purge command will help you to eliminate these unused contents, in addition it will remove zero length geometry and empty text objects. To reach this command go to the **Manage** tab, locate **Cleanup** panel, then select the **Purge** command:

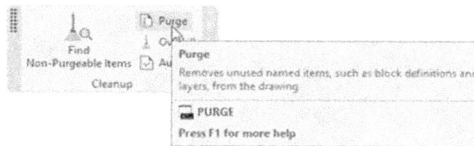

You will see the following dialog box:

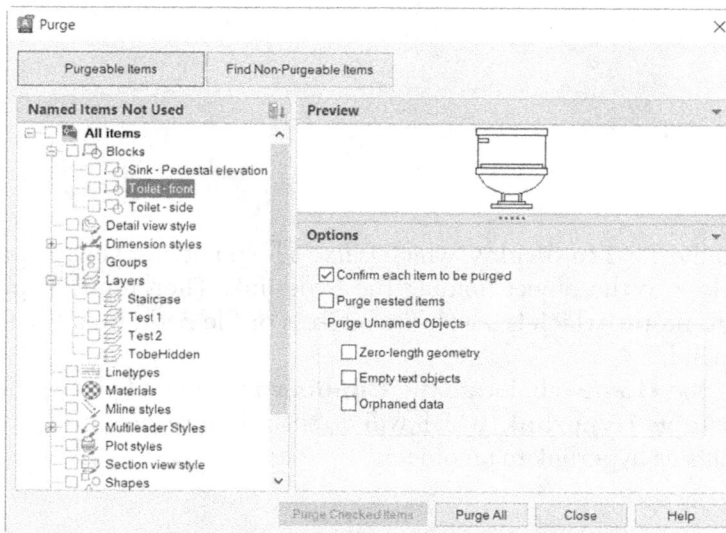

At the top left of the dialog box, if Purgeable Items button is selected AutoCAD will list all the unused items. In the above dialog box, you can see there are unused Blocks, Dimension styles, Layers, Multileader style, and Text styles. If you click

the plus sign beside each item, it will list the names of these item, similar to the following:

You can select to purge all or some items in the same category
Under Options, you can do all or any of the following:

- Confirm each item to be purged; if this is turned on, you have to answer to purge or not each time you want to purge an item, and you will see the following:

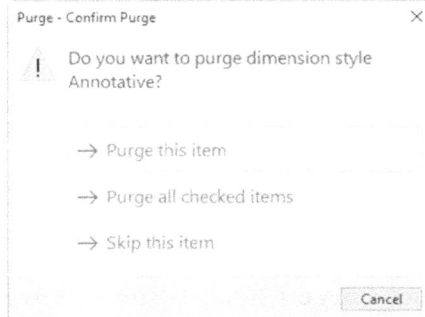

- Purge nested items; this option will help you purge nested blocks. A nested block is a block that contains blocks. If this option is turned off, then you will purge only the big block, and the nested blocks will remain there. But, if you turn it on, it will purge the big block and all nested blocks, barring they are not used in the current file
- Purge zero-length geometry and empty text objects; this option will remove all lines, arcs, and polylines that have a length of zero, and all MTEXT and Text that contains only spaces

If you click Find Non-Purgeable Item, you will see the following:

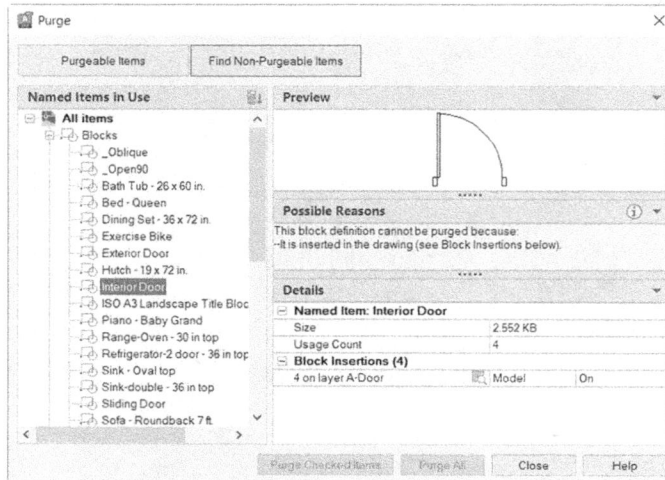

If you select one of the blocks (as the above example suggests) you will see a preview of the block, a possible reason why this block cannot be purged, and finally the size of the block, along with Usage Count, and in which layer this block was inserted.

You can reach the same command, if you go **Manage** tab, locate **Cleanup** panel, and select **Find Non-Purgeable Items** button:

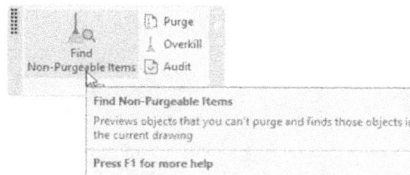

Finally, select either **Purge Checked Items**, **Purge All**, or **Close** button to end the command.

Another command which will help you to tidy up your drawings from any redundancy is **Overkill** command, which will cleanup overlapping geometry by removing duplicated and unneeded objects. Go to **Manage** tab, locate **Cleanup** panel, click **Overkill** command:

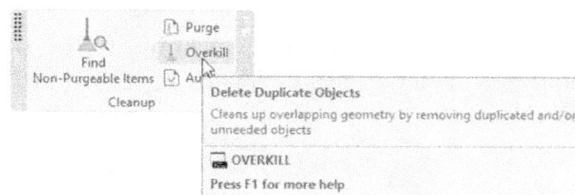

It will ask you to select the objects you think there are redundancy in them. Once you press [Enter], you will see the following dialog box:

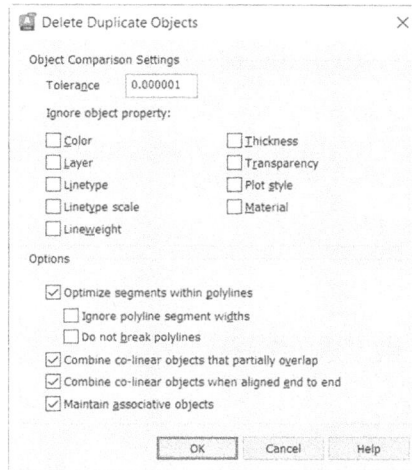

Using Object Comparison Settings, set the Tolerance between the original objects, and the objects overlapping them, and whether to ignore any property of these objects. In Options, select the different options of how to optimize your drawing.

## PRACTICE 12-5    PURGING ITEMS

1. Start AutoCAD 2025

2. Open **Practice 12-5.dwg** file

3. Start Purge command

4. How many blocks to be purged? _____ (3)

5. See the preview of the blocks to be purged.

6. Select only Toilet – front, and Toilet – side only

7. Select all dimension styles

8. Keep only the Staircase layer, and select other three

9. Keep the existing Multileader styles

10. For Text styles select Mine 2

11. Clear the checkbox Confirm each item to be purged, and turn on Nested items checkbox. Make sure that Zero-length geometry, and Empty text both are turned on. Click Purge Checked Items button

12. Click Find Non-Purgeable Items button, click Blocks, then select Interior Door block, see the preview of the block, why it will not be purged, and what is the Usage Count, and in which layer

13. Click Close to end the command

14. If you zoom to the south of the Master Bedroom, the lines representing the wall to the west of the door are overlapping each other. Use Overkill command, to eliminate the overlapping items

15. Save and close the files

## 12.8 USING VIEWS & VIEWPORTS

### 12.8.1 Creating Views

A view in AutoCAD is any rectangular-shaped portion of the drawing that will be saved under a name. There are two ways to define it, either by zooming to the part of the drawing you want to save and then issuing the command. Or alternatively, you can do that inside the View command. To issue this command go to the **View** tab, locate the **Named Views** panel, then click **New View** button:

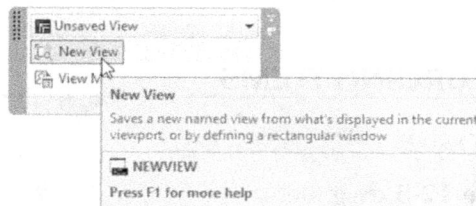

You will see the following dialog box:

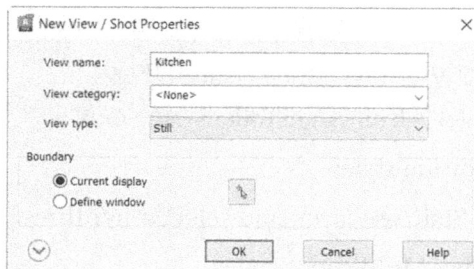

As a first step, input the **View name**. Under **Boundary** select whether your view is the **Current display** or if you want to **Define window** (click the small button at the right to zoom to the desired area). If you click the small button at the lower left part of the dialog box, you will see **Settings**, select whether you want to **Save layer snapshot with view**, or not? Layer snapshot, is the status of layers (on/off, Thaw/Frozen, Unlock/Lock, etc.). When done click **OK** to save the view, take a look of the extended dialog box:

There are multiple ways to retrieve the saved views, which are:

- Using In-canvas View control at the upper left corner of the screen, and using **Custom Model Views** as shown below:

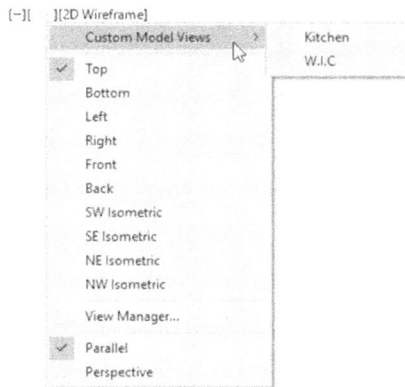

- Using the **View** tab, locate the **Named Views** panel and click the name of the desired view, as shown in the following:

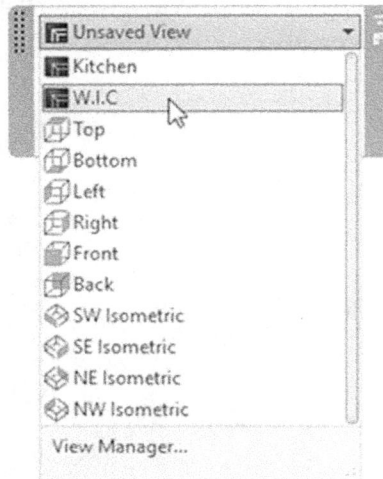

## 12.8.2 Using Views in Viewports

The user can show the saved views in viewports using the **Viewport** dialog box. Do the following steps:

- Be sure you are in layout not at the Model tab
- Click the Layout tab at the rightmost, then locate Layout Viewports panel, click New Viewports:

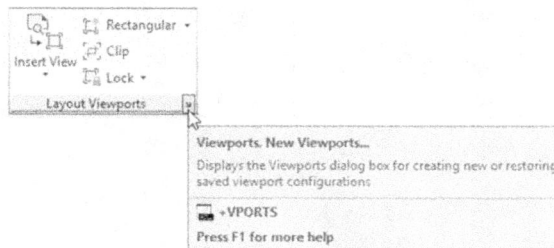

- Select one of the arrangements (the following example is Two: Vertical) using Preview; if you click on one of the viewports it will be current, hence, you can select one of the saved views to show inside it.

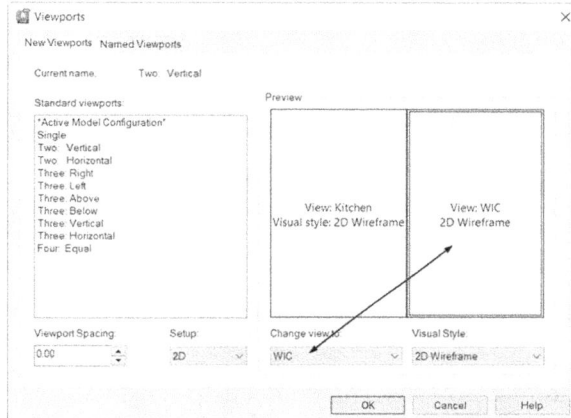

### 12.8.3 Creating a Named Viewport Arrangement – Method (I)

If you do not approve of the viewport arrangement AutoCAD provides and you want to create your own arrangement, take the following steps:

- Make sure you are at Model Space
- Go to the **View** tab and locate the **Model Viewports** panel, then select the **Viewport Configuration** button to see the following:

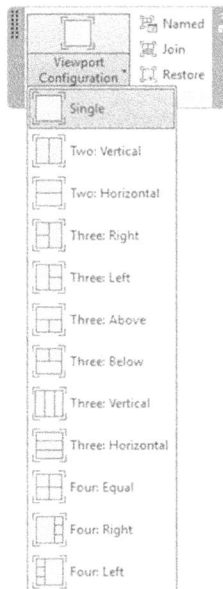

- Or you can reach the same command using In-canvas viewport control

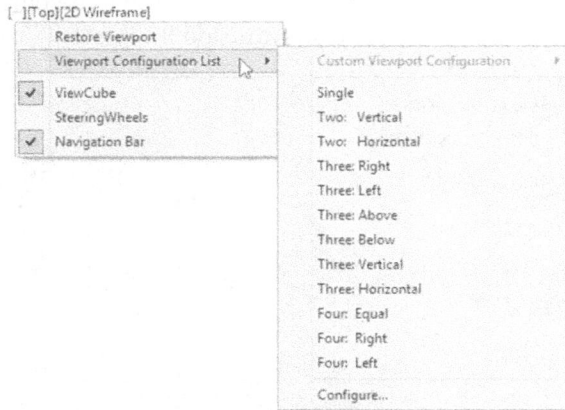

[ ][Top][2D Wireframe]

| Restore Viewport |
| Viewport Configuration List ▶ | Custom Viewport Configuration ▶ |
| ✔ ViewCube | Single |
| SteeringWheels | Two: Vertical |
| ✔ Navigation Bar | Two: Horizontal |
| | Three: Right |
| | Three: Left |
| | Three: Above |
| | Three: Below |
| | Three: Vertical |
| | Three: Horizontal |
| | Four: Equal |
| | Four: Right |
| | Four: Left |
| | Configure... |

- Choose one of the existing arrangements to start with
- To break a viewport to even smaller viewports, select one of the Model viewports, Start Viewport command again, and pick an arrangement
- Make sure that at the lower left corner of the dialog box, under Apply to, you select **Current Viewport**, then click OK

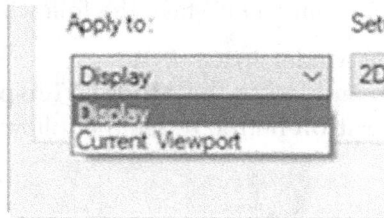

Apply to:                    Setι

| Display ∨ | 2D |
| Display | |
| Current Viewport | |

- This should cut your small viewport to even smaller areas
- Do that to different viewports
- Go to the **View** tab, locate the **Model Viewports** panel, and select the **Join** button. This will enable you to join adjacent viewports (the condition here to form a rectangular shape)

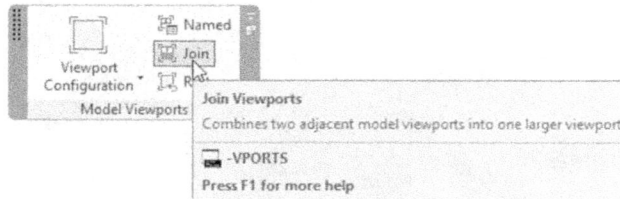

Named
Join
Viewport
Configuration ▾
Model Viewports

**Join Viewports**
Combines two adjacent model viewports into one larger viewport

-VPORTS

Press F1 for more help

- Once you are done joining viewports, you will obtain a new arrangement, and all of you have to do is to save it. Start the **Named** command and type the name of the new arrangement as shown below:

- To retrieve it in a layout, go to the desired layout
- Go to the **View** tab, locate the **Model Viewports** panel, and click the **Named** button:

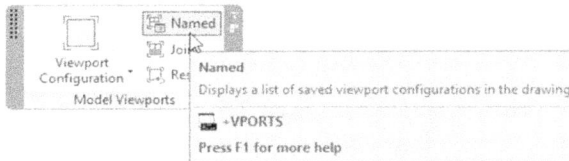

- Select the desired named viewport arrangement and insert it as you do with the existing arrangements

### 12.8.4   Creating a Named Viewport Arrangement – Method (II)

Another way to do the above is:

- Make sure you are at Model Space
- Go to the **View** tab and locate the **Model Viewports** panel, then select the **Viewport Configuration** button to see the following:

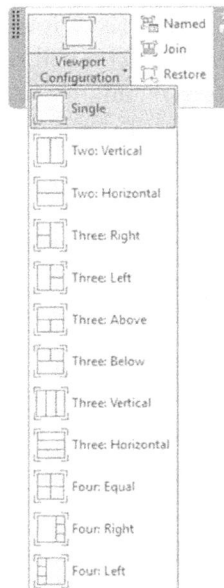

- Or you can reach the same command using In-canvas viewport control

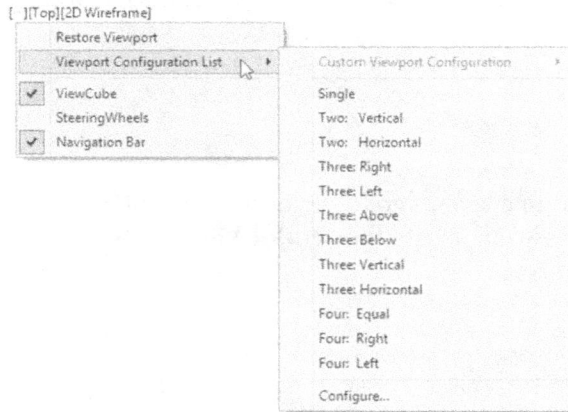

- Once you do this, click inside one of the viewports and you will see a thick blue line as a boundary
- Go to one of the sides; the shape of the mouse will change to double arrows to resize the viewport horizontally or vertically
- To resize it in both directions in one step, go to one of the corners, the shape of the mouse will change to four arrows, click hold and move
- Notice a small (+) in the thick black lines separating any two viewports; if you click it and move it (horizontally will add a horizontal viewport, and if you move it vertically, you will add a vertical viewport). You see a green line which will add a new viewport
- To join viewports together, go to the **View** tab, locate the **Model Viewports** panel, and select the **Join** button. This will enable you to join adjacent viewports (the condition here to form a rectangular shape)

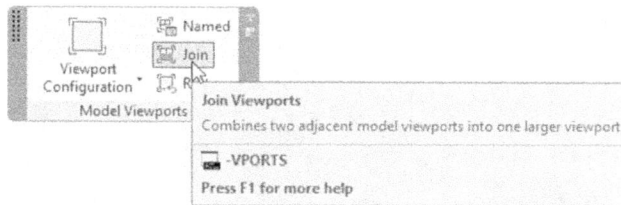

- Once you are done joining viewports, you will acquire a new arrangement, and all you need to do is to save it. Start the **Named** command and type the name of the new arrangement as shown in the following:

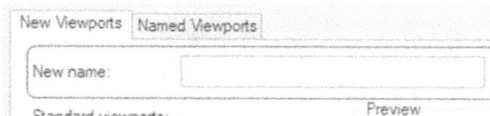

- To retrieve it in layout, go to the desired layout
- Go to the **View** tab, locate the **Viewports** panel, and click the **Named** button:

- Select the desired named viewport arrangement and insert it as you do with the existing arrangements

### 12.8.5    Insert View as Viewport

In Layout, you can insert saved views as viewports. Do the following steps:

- Go to the desired layout
- Click the Layout tab at the rightmost, locate **Layout Viewports** panel
- Click **Insert View** button, and select the desired saved view, as shown in the following:

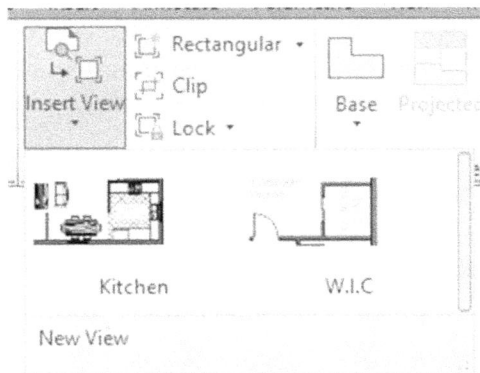

- Drag it to the sheet
- Right-click and select the desired scale
- Insert it in the desired location

## PRACTICE 12-6    USING VIEWS AND VIEWPORTS

1. Start AutoCAD 2025

2. Open **Practice 12-6.dwg** file

3. Create four views making sure that you will save a **layer snapshot** with the view; the four views are:

   **a.** Master Bedroom

   **b.** Master Bedroom Bathroom

   **c.** WIC

   **d.** Kitchen

4. Thaw layer Dimension

5. If you retrieve one of the four views, what happens to the layer Dimension? Why? _____

6. Insert Four: Equal viewports in the Detail 1 layout showing a view in each viewport

7. Using Method (I) or Method (II) using Model Space create Four: Equal

8. At the upper left viewport, split it into Two: Vertical. At the lower right one, split into Two: Vertical

9. At the upper ones join the small to big ones, and at the lower ones join the small to the bigger one, then save this new arrangement under the name **Four Unequal**

10. Go to the Detail 2 layout and insert the new arrangement. Going to each viewport, zoom to a different part of the floor plan

11. Go to Detail 3 layout, and insert the four views as viewports using 1:40 as scale

12. Save and close the files

# NOTES

## CHAPTER REVIEW

**1.** One of the following statements is not correct:

    **a.** Using the Paste option, the content will be always an OLE object in AutoCAD

    **b.** You can insert a table in AutoCAD from an Excel sheet and the editing is two-way street

    **c.** Pasting content from Word will be always MTEXT

    **d.** To make the content linked you have to use Paste Special

**2.** Use _____ in the Offset command to create filleted or chamfered edges while offsetting

**3.** All of the following is true about View and Viewport except one:

    **a.** You can save a layer snapshot with the view

    **b.** You can create a new viewport arrangement in Model and Paper spaces

    **c.** You can insert a new viewport arrangement in Paper space

    **d.** You can show in each viewport a saved view

**4.** You can match properties across files:

    **a.** True

    **b.** False

**5.** While copying objects between files, you can use only drag-and-drop using the left button:

    **a.** True

    **b.** False

**6.** _____ option will trim objects based on extended cutting edges

**7.** Using the Right-click menu and the Clipboard option, you can use the Paste Special option:

    **a.** True

    **b.** False

## CHAPTER REVIEW ANSWERS

**1.** c

**3.** b

**5.** b

**7.** b

# 13

# *ADVANCED PRACTICES — PART II*

## In This Chapter

- How to deal with Fields
- Quick Select, Select Similar, and Add Selected
- Partial Open and Partial Load
- Object Visibility

## 13.1 USING QUICK SELECT

The Quick Select command will select objects in the current drawing based on their properties, which is handy in dense and complicated drawings containing hundreds of thousands of different types of objects. There are multiple ways to issue the Quick Select command:

- Go to the **Home** tab, locate the **Utilities** panel, then select the **Quick Select** button:

- Without issuing any command, right-click, then select the Quick Select option from the menu:

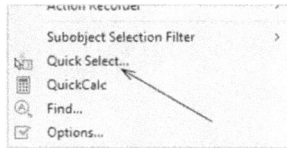

You will see the following dialog box:

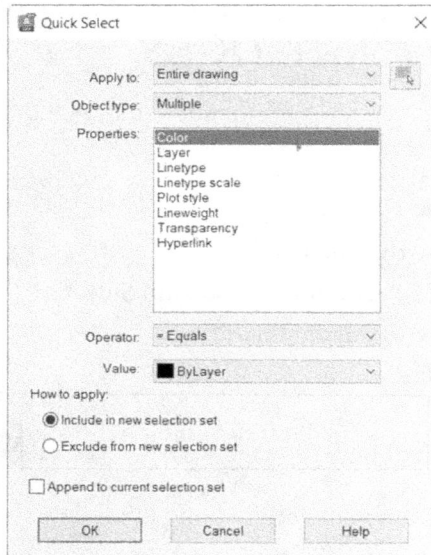

This method is similar to SQL (Structured Query Language), which will apply a filter to find information-based properties.

Select first the **Apply to:** part; it is either for the Entire drawing, or click the buttons beside it to restrict the filter to an area of your choice. Select **Object type** to search for. AutoCAD will list objects found in the current file and the user will see the following image:

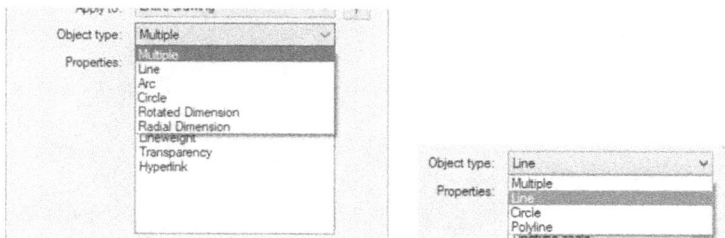

Select the desired object; accordingly, select the Properties of the selected object. If you select **Multiple** in **Object type**, then you will see only general properties; if you select an object type, then you will see both general and specific properties to choose from. The following is a list of properties of a circle:

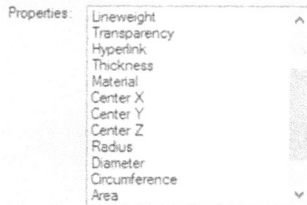

The next step is to select the desired **Operator**. You will see the following:

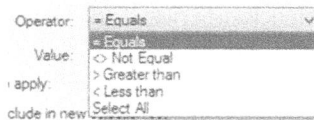

Accordingly specify the proper **Value** for the selected operator.

Finally, select **How to apply** the filter in the drawing. You can create a fresh new selection set from it, or if you selected objects prior to the Quick Select command, you can exclude the objects from it, or you can append it to already selected objects.

As a final note, you can access Quick Select while you are at the Properties palette, as shown in the following:

## PRACTICE 13-1   USING QUICK SELECT

1. Start AutoCAD 2025

2. Open **Practice 13-1.dwg** file

3. Since the drawing says that we have 12 × R 0.06 we want to make sure that this is right. So using Quick Select, select all circles with R = 0.06, how many circles are selected? _____

4. Using Quick Select and Properties, change the radius of all the other eleven circles to have R = 0.06

5. Using Quick Select and Properties move all circles of R = 0.06 to layer Holes

**6.** Using Quick Select and Properties move all dimensions to layer Dimensions (Hint: linear dimensions are called Rotated dimensions, and since we have two types rotated and radial, the user should use append or can do this step twice)

**7.** Save and close the file

## 13.2   USING SELECT SIMILAR AND ADD SELECTED

The Select Similar command will select an object and then select all similar objects that hold the same properties following certain settings. The Add Selected command will allow the user to select an object then initiate the command which will draw the same object that holds the same properties.

### 13.2.1   Select Similar Command

There are two ways to issue this command; the first method involves the following steps:

- Select the desired object
- Right-click, then choose the **Select Similar** option:

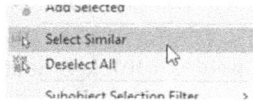

- Based on the current settings, AutoCAD will select the similar objects

The second method involves the following steps:

- Type at the command window SELECTSIMILAR, and the following prompt will appear:

```
Select objects or [SEttings]:
```

- Right-click and select the **Settings** option and you will see the following dialog box:

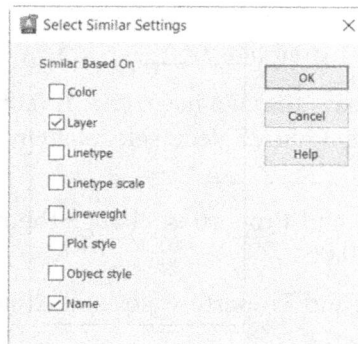

As shown above, AutoCAD will select based on layer and name (anything has a name like block). Other things to choose from are Color, Linetype, Linetype scale, Lineweight, Plot style, Object style (anything has a style like dimension, text, etc.). If you select more than one object (arc and circle), then right-click and chose Select Similar, and AutoCAD will select objects similar to both selected objects.

### 13.2.2　Add Selected Command

Let's assume you have a command that lies in layer Centerline, the color is yellow, and the linetype is dashdot, and you want to create a new line that lies in the same layer and holds the same properties; this command will help you accomplish your mission easily. Select the desired object, then right-click and choose the **Add Selected** option:

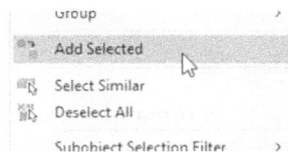

You will see that AutoCAD has started the command you want, and you are ready to start drafting; the new object will reside in the same layer and will hold the same properties as the selected object.

## 13.3　WHAT IS OBJECT VISIBILITY IN AUTOCAD?

The normal practice in AutoCAD to hide an object is to turn the layer off or freeze it. However, this means all the other objects in the same layer will be hidden as well. This command will hide an object or group of objects without hiding all the other objects in the same layer. To do that, select the desired object (or as many objects as you wish) and right-click then select the **Isolate** option, and you will see the following:

As you can see there are three commands: Isolate Objects, Hide Objects, and End Object Isolation. Here is a discussion of each one of these commands:

- **Isolate Objects** will show the selected objects and will hide all other objects
- **Hide Objects** will hide the selected objects and show all other objects
- **End Object** Isolation will cancel the first command, meaning it will show all objects

You can see a button containing a circle, a rectangle, and a triangle at the lower right-hand side of the screen, and it is either on or off; if you click it, you will see the same menu, and hence, you can do everything from down there.

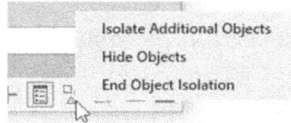

## PRACTICE 13-2    USING SELECT SIMILAR AND ADD SELECTED

1. Start AutoCAD 2025

2. Open **Practice 13-2.dwg** file

3. Type SELECTSIMILAR command, select Settings option, and clear all checkboxes, then select one of the red circles and press [Enter]

4. You will notice that AutoCAD selected all circles in the drawing

5. Right-click and select Isolate/Isolate Objects

6. Make the layer Centerlines current

7. Start Line command, and using OTRACK and the center point of one of the two large circles, take 1.2 from the center to the left to start the line, then draw 2.4 horizontal line passes through the center of the two large circles

8. End Object Isolation

9. Make layer 0 current

10. Select one of the green centerlines at the right or at the left, right-click, and select Add Selected and draw a vertical centerline for the large circles

11. Does it look different than the horizontal line? _____ (yes), why? (the line-type scale of the two lines at the right and the left is 0.5, hence, the new line holds all the properties of the original line)

12. Select one of the two linear dimensions, then right-click and choose Select Similar. Two dimension blocks will be selected, right-click, and select Precision, and select 0.000

13. Select the four centerlines and right-click then select Isolate/Hide Objects

14. Save and close the file

## 13.4   ADVANCED LAYER COMMANDS

In this section, we will discuss some advanced commands related to layers. Most of these commands depend on selecting a tool to perform a certain task related to the layer of the object selected. These commands will make things easier and, hence, using them will lead to a decrease in the time to complete a drawing. You can find these commands if you go to the **Home** tab and locate the **Layers** panel:

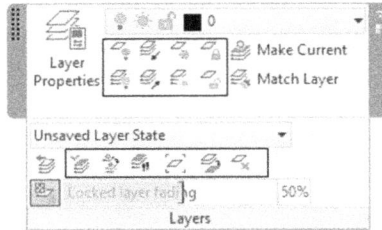

### 13.4.1   Using Isolate and Unisolate Commands

Use the following two buttons:

The Isolate command will ask you to select object(s) and the layer(s) of these objects will be shown, while the other layers will be turned off (or locked). The Unisolate command means to cancel the effects of the Isolate command.

### 13.4.2   Using Freeze and Off Commands

Use the following two buttons:

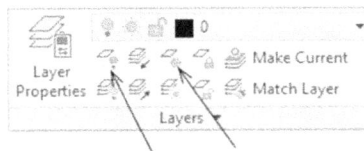

The Freeze and Off commands will ask you to select object(s), and the layer(s) of these objects will be turned off, or frozen.

### 13.4.3 Using the Turn All Layers On and Thaw All Layers Commands

Use these two buttons:

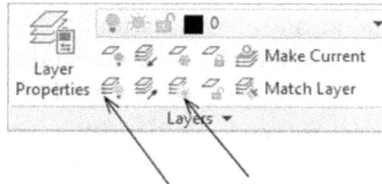

These two commands will turn all layers on and will thaw all layers, a very handy tool for doing this process in one shot.

### 13.4.4 Using the Lock and Unlock Commands

Use these two buttons:

The Lock command will ask you to select the object(s), and the layer(s) of these object(s) will be locked. You can lock one layer at a time. The Unlock command will ask you to select object(s), and the layer(s) of these objects will be unlocked. You can unlock a layer one at a time.

### 13.4.5 Using the Change to Current Layer Command

Use this button:

This command will ask you to select object(s), and then will change the layer of these object(s) to be the current layer.

### 13.4.6 Using Copy Objects to New Layer Command

Use this button:

This command will ask you to select object(s) and then copy them to a new location in the drawing (sounds like a normal copy command) and then change the layer of the new objects to a new layer either by selecting an object residing in the desired layer or by typing its name.

### 13.4.7 Using the Layer Walk Command

Use this button:

This command will show a dialog box listing all layers in the current file. To show the contents of a layer, click its name in the list (by default all layers are selected). A checkbox at the bottom says "Restore on Exit," which suggests whenever you will close this dialog box, all layers will be restored to their previous status. If you this layer is turned off, this command will maintain its effects:

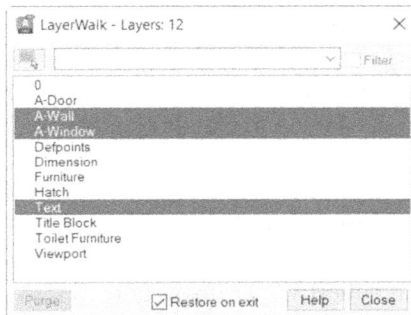

### 13.4.8 Using the Isolate to Current Viewport Command

Use this button:

This command will ask you to select an object and then freeze the layer of this object in all viewports except the current viewport.

### 13.4.9 Using the Merge Command

Use this button:

This command will merge a layer or more into a target layer. First AutoCAD will ask you to select an object (AutoCAD will list the name of the layer of the selected object), and AutoCAD will keep reminding you to select objects until you are done and press [Enter]. The last step will be to select an object in the target layer and AutoCAD will report the deleted layers.

### 13.4.10 Using the Delete Command

Use this button:

This is a very practical command, as we know that AutoCAD will not delete a layer unless it is empty. This command will delete (purge) a layer by selecting any objects residing in it, except for the current layer.

## 13.5 LAYER'S TRANSPARENCY

You can set the visibility of a layer. The default value for all layers is 0 (zero) and can be as much as 90. If you are making a test plot for a drawing with lots of solid hatching, set the visibility to the minimum so you do not lose plotter ink. You can use the same color for numerous layers, then control the visibility of the layer to give each layer a different tone of the color. When you start the Layer Properties Manager, you can see a column called Transparency, and if you are at a layout, you will see another column called VP Transparency:

| S.. | Name | ▲ | VP Linew... | Transparency | VP Transparency | Plot St... | VP Plo... | N. | Desc |
|---|---|---|---|---|---|---|---|---|---|
| ✓ | 0 | | —— Defa... | 0 | 0 | Normal | Normal | | |
| | A-Door | | —— Defa... | 0 | 0 | Normal | Normal | | |
| | A-Wall | | —— Defa... | 0 | 0 | Normal | Normal | | |
| | A-Window | | —— Defa... | 0 | 0 | Normal | Normal | | |
| | Defpoints | | —— Defa... | 0 | 0 | Normal | Normal | | |
| | Dimension | | —— Defa... | 0 | 0 | Normal | Normal | | |
| | Furniture | | —— Defa... | 0 | 0 | Normal | Normal | | |
| | Hatch | | —— Defa... | 0 | 0 | Normal | Normal | | |
| | Text | | —— Defa... | 0 | 0 | Normal | Normal | | |
| | Title Block | | —— Defa... | 0 | 0 | Normal | Normal | | |
| | Toilet Furniture | | —— Defa... | 0 | 0 | Normal | Normal | | |
| | Viewport | | —— Defa... | 0 | 0 | Normal | Normal | | |

You can control the visibility of new objects; to do that go to the **Home** tab and locate the **Properties** panel:

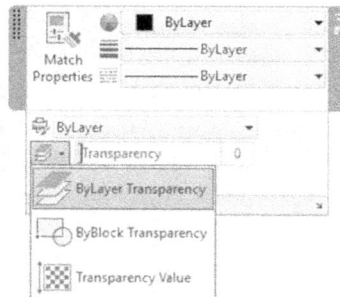

Select either to set the **Transparency** to be ByLayer, ByBlock, or Transparency Value (which will set it to 0 (zero)), or move the slider to any desired value.

Also, set the value for object(s) using the Properties palette, as in the following:

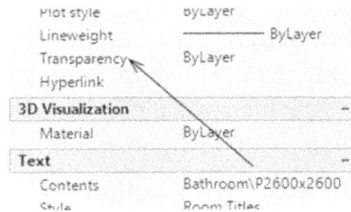

## PRACTICE 13-3    USING ADVANCED LAYER COMMANDS

1. Start AutoCAD 2025

2. Open **Practice 13-3.dwg** file

3. Using the Freeze command select one of the dimension blocks; what happened to the Dimension layer? _____

4. Start the Isolate command and select a line of the walls; what happened to other objects in other layers? _____

5. Unisolate

6. Start the Layer Walk command, and you will notice that all layers are selected

7. Click any of the layer names, then select A-Door layer, hold [Ctrl] and select A-Wall, and A-Window. Uncheck Remove on exit checkbox, and click Close. On the warning message click Continue

8. Start the Layer Properties Manager palette. What is the state of the layers you did not select in the previous step? _____

9. Start the Layer Walk again, and while holding the [Ctrl] key, select Furniture and Toilet Furniture layers. Close Layer Walk command

10. Merge layer Toilet Furniture to layer Furniture

11. What happened to layer Toilet Furniture after the merging process?

12. Save and close the file

## 13.6 USING FIELDS IN AUTOCAD

AutoCAD stores many data in the drawing database. Some of them are constant, and some of them are variable. The user can utilize these types of data by inserting them in the drawing, to benefit from the updating feature that AutoCAD will perform on the variable data. This approach is better than writing text using Text and Mtext commands, which will need to be updated manually. There are four methods to insert a field in your drawing:

- Using Field command. To issue this command, go to the **Insert** tab, locate the **Data** panel, then select the **Field** button:

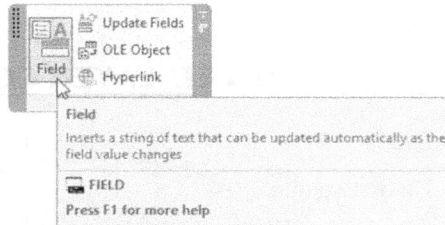

- Using the Text command, and after you specify the starting point, height, and rotation, and before you start writing, right-click and select the **Insert Field** option:

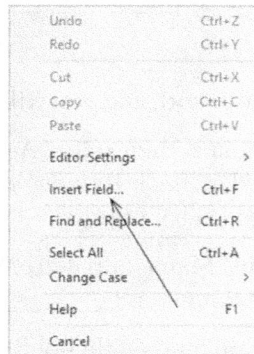

- Using the Mtext command, and after you specify the text area, you will find the **Text Editor** context tab and the **Insert** panel, and then select the **Field** button:

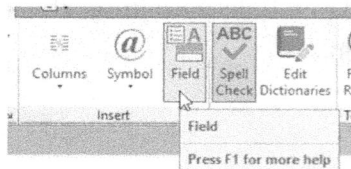

- Also, you can insert fields in Table and Attribute commands

Regardless of the method used to reach the **Field** command, the following dialog box will appear:

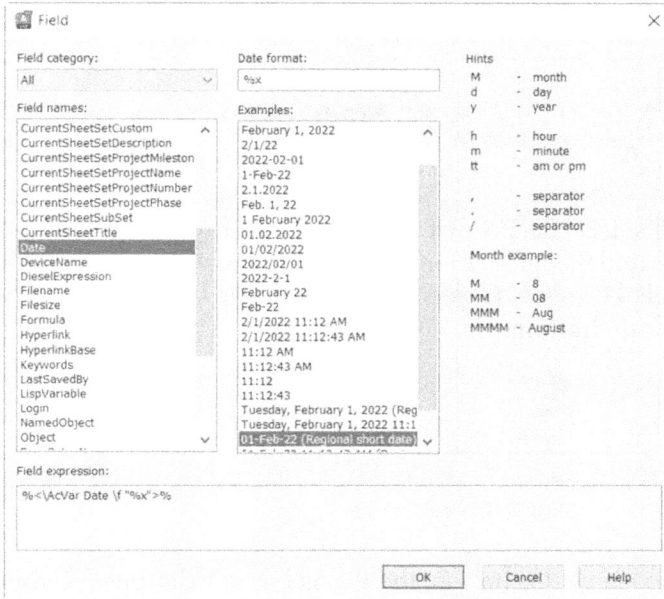

The first thing to control is the Field category, which will help you find the desired data quickly. Clicking at the pop-up list will show the following list:

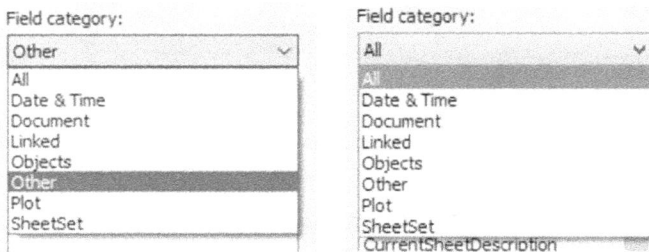

There are seven field categories and each one will show related field names. For instance, the Objects category will show four field names, which are:

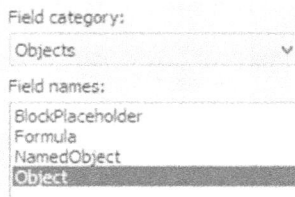

Based on the field category and field name, you will find at the right-hand side of the dialog box things to help you control the appearance of the field. For instance, if you select Field category Document, and Field name is File name, then you will see the following at the right:

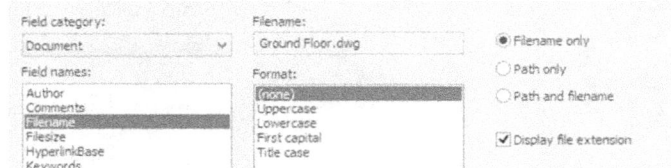

You will select the Format first, then select whether to show Filename, Path only, or Path and filename. Finally, control to show or hide the file extension. While if you use Filed category Objects and Field name Object, you will need to select an object from the drawing:

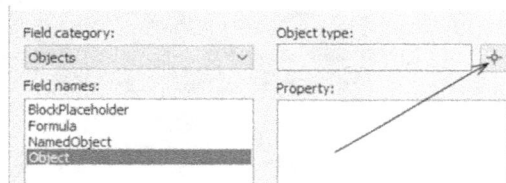

When you click this button, the dialog box will disappear temporarily to let you select the desired object; when done, you will see the following:

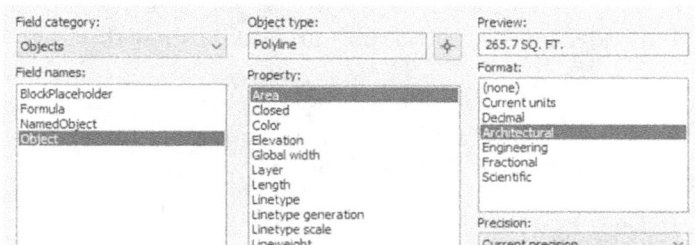

As you can see, we selected a circle, and we chose to use the Area with decimal format.

When you insert a field in the drawing, the default settings will display it with a background, similar to the following:

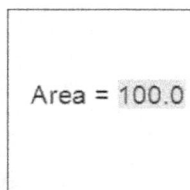

Area = 100.0

To remove the background, go to the Application menu and select the **Options** button. Select the **User Preferences** tab, then locate **Fields**, and uncheck the **Display background of fields** checkbox, similar to the following:

Fields
☑ Display background of fields

Field Update Settings

Click off the **Display background of fields** checkbox if you do not want to display the background. Click **Field Update Settings** to see the following dialog box:

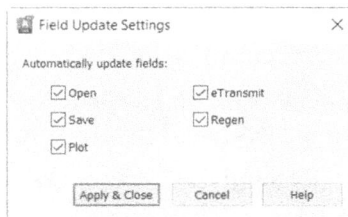

Field Update Settings ✕

Automatically update fields:

☑ Open        ☑ eTransmit
☑ Save        ☑ Regen
☑ Plot

Apply & Close    Cancel    Help

This updates the field automatically each time you save, open, plot, eTransmit, or Regen. But if you need to update manually, then go to the **Insert** tab, locate the **Data** panel, then select the **Update Fields** button:

Update Fields
Manually updates the fields in selected objects to the current value

UPDATEFIELD
Press F1 for more help

You need to select the desired fields to be updated.

## PRACTICE 13-4   USING FIELDS IN AUTOCAD

**1.** Start AutoCAD 2025

**2.** Open **Practice 13-4.dwg** file

**3.** Make sure that the current layer is Polyline

**4.** Using the Boundary command, click inside the master bedroom to add a polyline

**5.** Make layer Text current

**6.** Go to the Annotate tab and make sure that Room Titles is the current text style

7. Start the Single Line text command and specify a point almost at the middle of the room, with rotation angle = 0. Type "Area = " then right-click and select the Insert Field option, and insert the area of the polyline showing 266 SQ. FT.

8. Go to Full Plan layout

9. Zoom to the lower right portion of the title block

10. Using the Application Menu, go to Drawing Utilities, then select Drawing Properties, go to Summary tab, and at Title input: Munir Hamad Villa, at Author input your name, then click OK to end the command

11. Make Standard the current text style

12. Using the Field command, and under the title Project Name, insert Title field using upper case. Under Designed By insert Author, using First capital

13. Under Date insert the date of today using MMMM d, yyyy format

14. Under Filename insert the filename only without the extension using lower case

15. Remove the background for fields

16. Save and close the file

## 13.7  USING PARTIALLY OPENED FILES

Users may deal with a huge drawing containing plenty of views and layers. Large files tend to take a significant amount of time to open. To eliminate this obstacle, the user can use the Partial Open command, and afterwards use Partial Load to add more contents to the partially opened file.

### 13.7.1  How to Open a File Partially

Select the normal Open command, select the desired file, then DO NOT click Open as you always do; instead click the small arrow at its right to see the following list of options and select Partial Open:

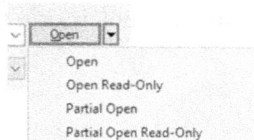

You will see the following dialog box:

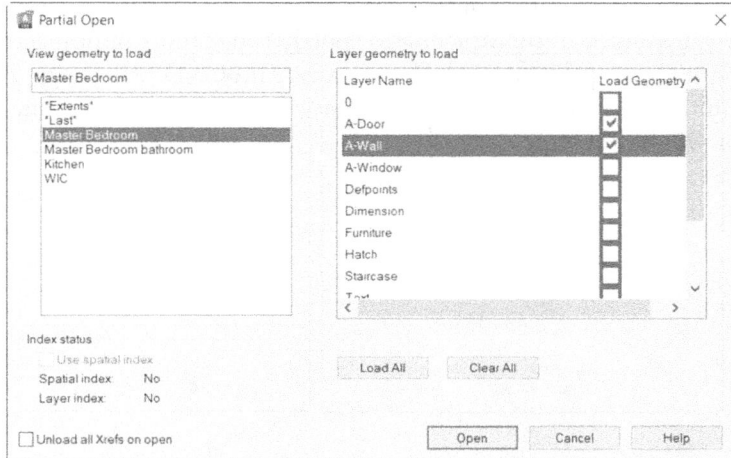

The user can do all or any of the following:

- Select the desired layers to open
- Select the desired views to open, or you can use Extents or Last views
- Then click Open to open the file with the settings you selected

### 13.7.2   Using Partial Load

This command is not applicable for normal files; instead, you can use it only on the files that are partially opened. To use this command, make sure the menu bar is shown, then select **File / Partial Load**, or you can type **PARTIALLOAD** on the command window to see the following dialog box:

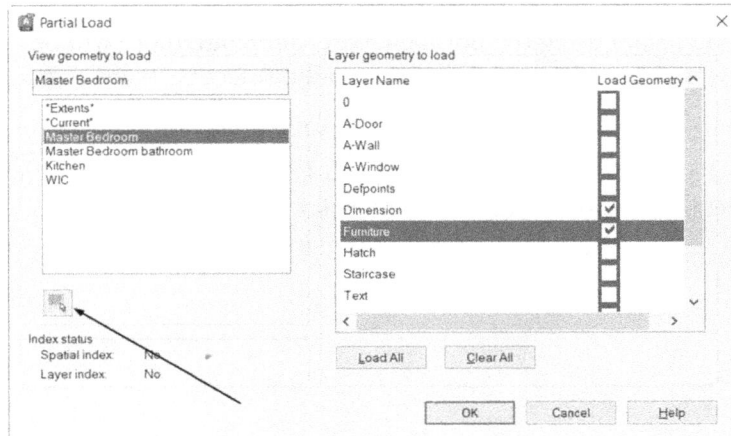

As you can see this dialog box matches the Partial Open dialog box, with the exception of the small button at the lower left portion of the dialog box. This button will specify a window in the drawing to specify the extents of objects to be loaded.

Saving the partially loaded file means AutoCAD will keep these settings working until you change them. So, when you try to open it again, the following dialog box will be shown:

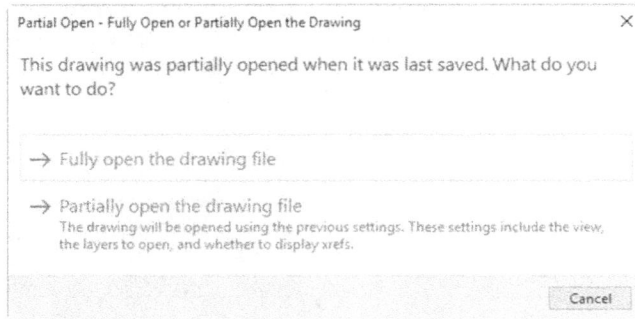

Partial Open - Fully Open or Partially Open the Drawing                                    ×

This drawing was partially opened when it was last saved. What do you want to do?

→ Fully open the drawing file

→ Partially open the drawing file
The drawing will be opened using the previous settings. These settings include the view, the layers to open, and whether to display xrefs.

Cancel

## PRACTICE 13-5    USING PARTIALLY OPENED FILES

1. Start AutoCAD 2025

2. Using the Partial Open command, open Practice 13-5.dwg partially, using Master Bedroom view, and A-Door, A-Wall, and A-Window layers

3. Using Partial Load, the select layer Furniture

4. Save and close the file

5. Open it again using the normal Open command, there will be a message asking you either to open it fully or partially, select partially

6. Zoom out to see that all of the furniture is shown

7. Use Partial Load again, select Hatch layer, click Pick a window button, and select the area of the kitchen, then click OK

8. Save and close the file

# NOTES

## CHAPTER REVIEW

1. Using Fields in AutoCAD:

   a. If you used Date, you could set the format of the Date

   b. You add the Length of the Polyline

   c. If you want to add Area field, you can set the format of the Area

   d. All of the above

2. _____ command will work only on partially opened files

3. In the Select Similar command

   a. You can select all circles

   b. You can select all circles in the same layer

   c. You can select all circles with the same color

   d. All of the above

4. One of the following statements is not true:

   a. You can isolate objects and you can isolate layers

   b. You can control field update settings

   c. You need to empty the layer in order to delete using the Delete in Layers panel

   d. You can hide objects rather than hiding layers

5. You can insert the area of a polyline as a field without a background:

   a. True

   b. False

6. _____ will show a dialog box listing all layers in the current file; to show the contents of a layer, click its name in the list

## CHAPTER REVIEW ANSWERS

1. d

3. d

5. a

# 14

# *USING BLOCK TOOLS AND BLOCK EDITING*

## In This Chapter

- How to use Automatic Scaling
- How to use the Design Center
- How to use and customize the Tool Palette
- How to edit a block

## 14.1 AUTOMATIC SCALING FEATURE

When you create a block, you input the Block Unit, as if you are telling AutoCAD that each unit you used in this block will represent a certain unit. In order for this circle to be completed, you have to control the drawing file unit. Go to the **Application menu**, select the **Drawing Utilities**, then select **Units**, and you will see the following dialog box:

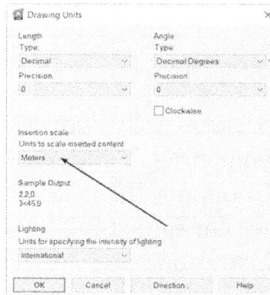

Under **Units to scale inserted content,** select the desired unit, which represents your drawing's file unit. AutoCAD will convert the block unit to the drawing unit.

## 14.2 DESIGN CENTER

In Design Center, you can share blocks, layers, dimension styles, text styles, table styles, etc. created in other files and put them in yours, whether the file is in your computer, network, or even on the Internet.

To issue the command go to the **View** tab, locate the **Palettes** panel, then click the **Design Center** button:

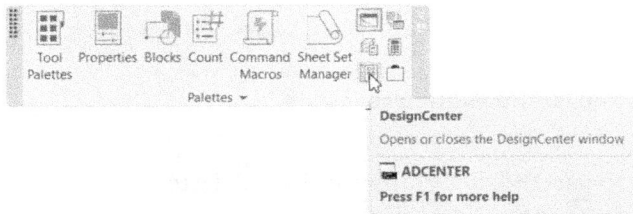

You will see the following palette:

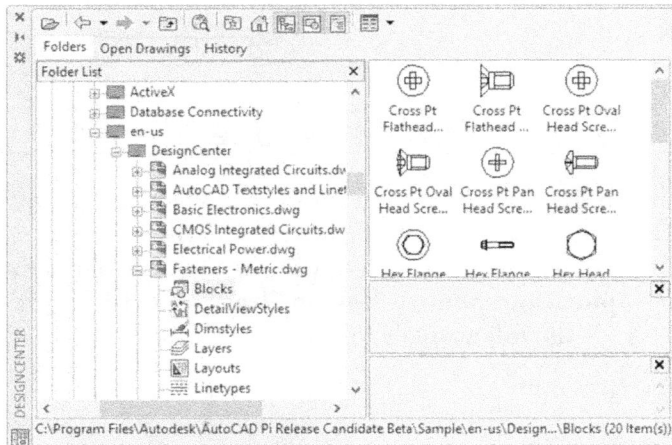

The Design Center left pane is like My Computer in Windows OS, which contains a listing of all of your drives, folders, and files. This will enable you to locate the desired file, which contains the needed blocks, layers, etc. Whenever you locate the file and click at the plus sign at the left of the file name, a list will appear, as shown in the following illustration:

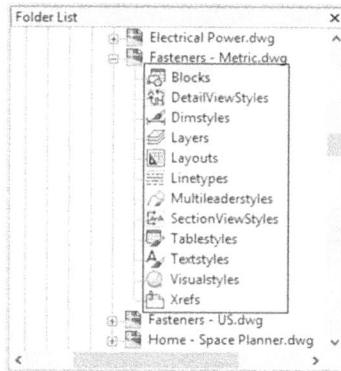

Select the desired content; if you select Blocks, you will see the shapes and the names at the right pane. There are three methods to copy blocks to your current drawing using Design Center, which are:

- Drag-and-Drop using the left button to insert the block in the current file
- Drag-and-Drop using the right button; when you release it, you will see the following menu (self-explanatory)

- Double-click and the Insert dialog box will appear immediately

## PRACTICE 14-1    USING DESIGN CENTER

1. Start AutoCAD 2025

2. Open **Practice 14-1.dwg** file

3. Make layer Furniture the current layer

4. Start Design Center

5. Locate your AutoCAD 2025 folder, then go to the following path: \Sample\ en-us\Design Center (en-us folder is assumed for the English language Auto-CAD, this folder may change depending on the language chosen)

6. Locate file Home-Space Planner.dwg, and choose Blocks to put in the furniture shown below

7. Make layer Toilet the current layer

**8.** Locate file House Designer.dwg, and choose Blocks to put in the toilet as shown below

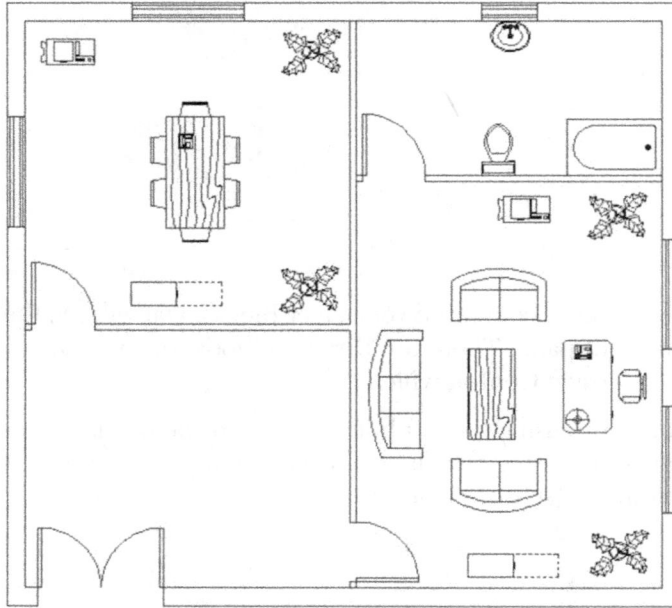

**9.** Save and close the file

## 14.3    TOOL PALETTES

Using Design Center to share data between files was a gift. There are, however, still some difficulties to deal with. You have to be careful of the current layer, the scale, the rotation angle, and the agonizing idea of searching for your desired file each time you need to copy something from it. We waited until AutoCAD 2004, when AutoCAD introduced us to Tool Palettes, which will take blocks to the next level.

Tool Palettes will help you to store any type of object in them, and then retrieve them in any opened file; moreover, you can control the object's property, so the next time you drag it in your file, you will not worry about layers, scale, or rotation angles. You can keep several copies of the same block, each holding different properties.

To start the command, go to the **View** tab, locate the **Palettes** panel, then click the **Tool Palettes** button:

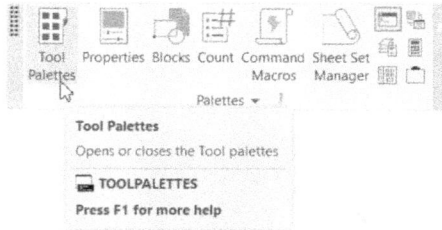

You will see the following:

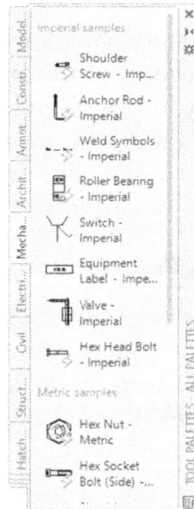

By default, there are some tool palettes that accompany AutoCAD, but the possibility to create your own palette is there.

### 14.3.1 How to Create a Tool Palette from Scratch

This method will create an empty palette, so you will be able to fill it using different methods. To do this, right-click over the name of any existing tool palette, and you will see the following menu:

Pick the **New Palette** option, and a new empty palette will be created, allowing you to name it immediately. Type in the name of the new palette as shown below:

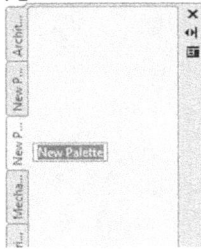

### 14.3.2 How to Fill the New Palette with Content

The user can fill the new empty palette with content using the drag-and-drop method from the different drawings to the palette. Say you opened one of your colleague's drawing files, and you discovered that he or she created several new blocks; you could simply click the block, avoiding the grips, hold, drag it to the palette and there you go. You can do the same with normal objects (lines, arcs, circle, polylines, hatches, tables, dimensions, etc.).

### 14.3.3 How to Create a Palette from Design Center Blocks

AutoCAD allows users to create a palette from all blocks in a file using Design Center. In order for this to work, make sure that both Design Center and Tool Palettes are displayed in front of you. Proceed to your desired file in the left pane of the Design Center and right-click and you will see the following:

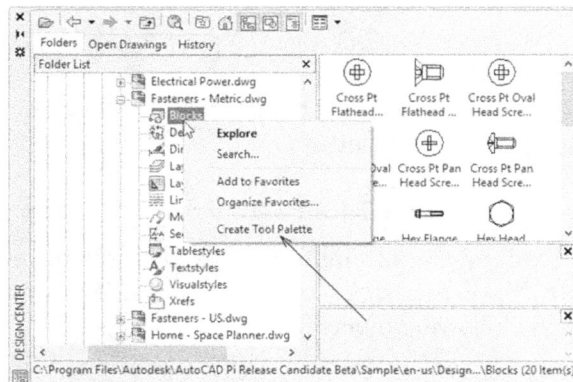

Select the **Create Tool Palette** option, and a new tool palette holding the same name of the file will be created containing all the blocks.

**NOTE** *The user has the ability to make a drag-and-drop for any block in any file from the Design Center to any tool palette.*

### 14.3.4 How to Customize Tools Properties

You can create several copies of blocks and hatches in tool palettes using the normal Copy/Paste procedure. Once you have several copies of the same block/hatch, you can change the properties of these copies according to your needs. Follow these steps:

- Right-click on the copied block/hatch and you will see the following menu:

- Pick the **Properties** option and you will see the following dialog box:

Type a new **Name** and a new **Description**. Properties of block/hatch are cut into two categories:

- **Insert** type of properties
- **General** type of properties

By default, the **General** properties are all **use current**. For a block or a hatch, you can specify that whenever you drag it from the tool palette it will reside in a certain layer (regardless of the current layer) and it will hold a certain color, linetype, lineweight, etc.

With Tool Palettes, CAD Managers can create for each user his or her own tool palettes, which hold all the needed blocks and hatches, customized according to the company standard, so it will become a simple drag-and-drop process. If we know that 30-40% of any drawing is blocks, and 10-20% is hatches, this means your time savings in these two areas will be enormous, if we use tool palettes effectively.

## 14.4 HATCH AND TOOL PALETTE

Just like blocks, you can store hatch patterns in tool palettes. Storing a hatch in tool palettes is not as important as storing hatches with altered properties, like layer, scale, color, etc. to ensure simple and fast hatch insertion in the drawing. Issue the Tool Palettes command and select the Hatches tab as shown in the following:

Use the Drag-and-drop technique from the drawing and to the drawing. Inside the tool palette right-click and select the **Properties** option to change the properties of the saved hatch.

## PRACTICE 14-2 USING TOOL PALETTES

1. Start AutoCAD 2025

2. Open **Practice 14-2.dwg** file

3. Open Design Center, and locate /Sample/en-us/Design Center/Fasteners-US.dwg

**4.** Locate the Block underneath it, right-click, then select the Create Tool Palette option

**5.** If Tool Palettes are not displayed, they will be displayed with a new palette called Fasteners-US. Close Design Center

**6.** Make sure that the current layer is 0

**7.** Using the newly created tool palette, locate Hex Bolt ½ in. -side, right-click, select Properties, and change the layer to be Bolts

**8.** Create another copy from it, and make the Rotation angle = 270

**9.** Using drag-and-drop, drag the two blocks to the proper places as shown below

**10.** Using the newly created tool palette, locate Square nut ½ in. -top, and set its layer to be Nut

**11.** Create a copy of Square nut ½ in -top and set the scale to 1.5

**12.** Using drag-and-drop, drag the new block to the proper places as shown below

**13.** Erase the lines to get the following shape:

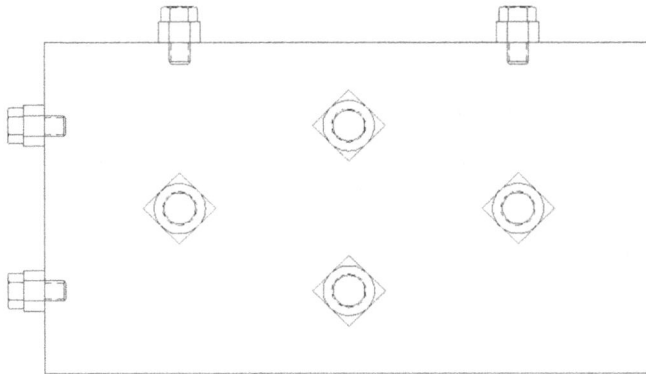

**14.** Close tool palettes, Save and close the file

## 14.5   CUSTOMIZING TOOL PALETTES

Users can customize the appearance of tool palettes to satisfy their needs, set the transparency, set view options, and so on. All of these functions will be available at the right-click menu.

### 14.5.1 Allow Docking

Users can be allowed to dock the tool palette to the right or the left of the screen to make the tool palette permanent. (The default mode is that the tool palette is floating.) To do this, make sure that the tool palette is shown, right-click any empty space within the tool palette (avoid icons), and you will see the following menu; select Allow Docking (if you see (✓) at its left, this means it is on):

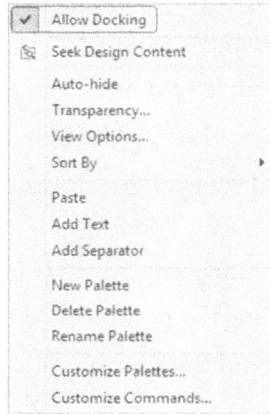

| |
|---|
| ✓ Allow Docking |
| 🔊 Seek Design Content |
| Auto-hide |
| Transparency... |
| View Options... |
| Sort By ▶ |
| Paste |
| Add Text |
| Add Separator |
| New Palette |
| Delete Palette |
| Rename Palette |
| Customize Palettes... |
| Customize Commands... |

Now you can drag the tool palette to the right or left of the screen.

### 14.5.2 Transparency

The user can set the transparency value of the tool palette. The default value is 100% opacity. To do this, make sure that the tool palette is shown, right-click any empty space within the tool palette (avoid icons), and you will see the following menu; select the Transparency option:

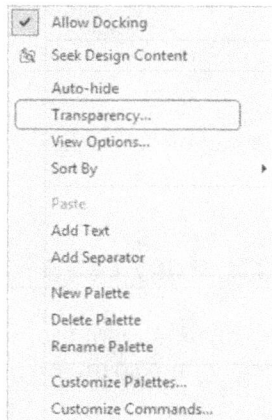

| |
|---|
| ✓ Allow Docking |
| 🔊 Seek Design Content |
| Auto-hide |
| Transparency... |
| View Options... |
| Sort By ▶ |
| Paste |
| Add Text |
| Add Separator |
| New Palette |
| Delete Palette |
| Rename Palette |
| Customize Palettes... |
| Customize Commands... |

You will see the following dialog box:

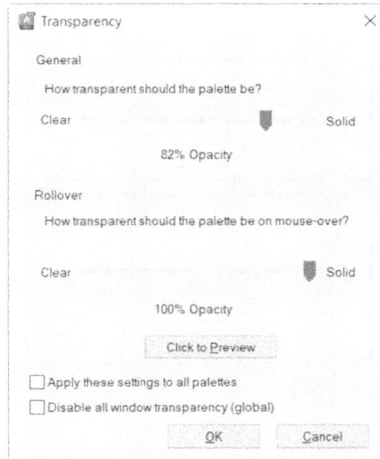

The user is asked to change the value of the transparency for the palette (use the slider under General), and for Rollover (when your mouse is rolled over the palette). The value of transparency of Rollover should be always equal to or more than the general value.

Also, select whether these settings affect the current palette or all palettes, and whether to disable all window transparency or not.

### 14.5.3   View Options

The user can set how the icons will appear inside the palette. To do this, make sure that the tool palette is shown, right-click any empty space within the tool palette (avoid icons), and you will see the following menu; select the View Options option:

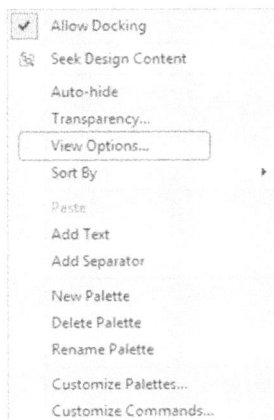

You will see the following dialog box:

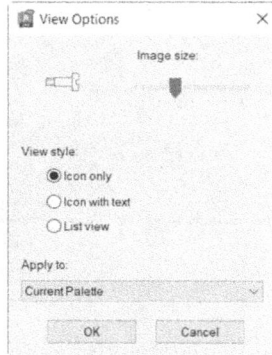

Using this dialog box, you can:

- Set the size of the image
- Set the view style: Icon only, Icon with text, or List view
- Decide whether to apply these changes for the current palette or for all palettes

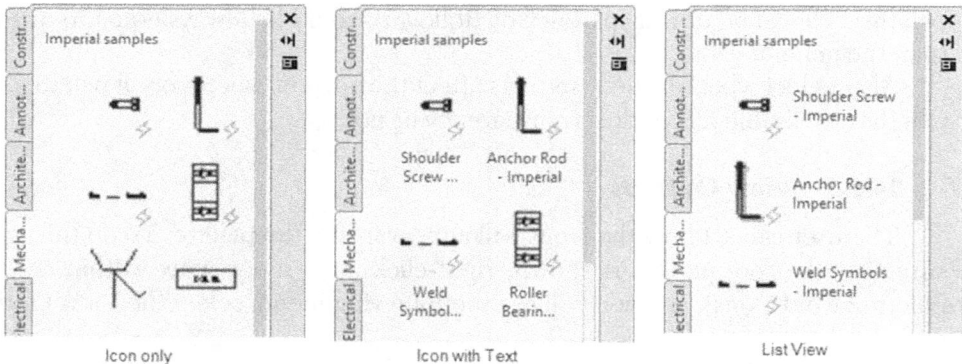

Icon only          Icon with Text          List View

### 14.5.4 Add Text and Add Separators

The user can add text to create an internal grouping for each palette, and also add some lines which will work as a separator. To do this, make sure that the tool palette is shown, right-click any empty space within the tool palette (avoid icons), and you will see the following menu; select Add Text and Add Separator options:

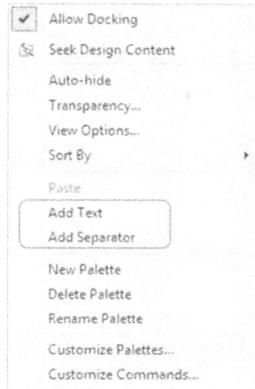

This is what you will receive:

### 14.5.5   New / Delete / Rename Palette

These three commands will create a new palette, delete an existing palette, or rename an existing palette. To do this, make sure that the tool palette is shown, right-click any empty space within the tool palette (avoid icons), and you will see the following menu: select New Palette, Delete Palette, and Rename Palette options:

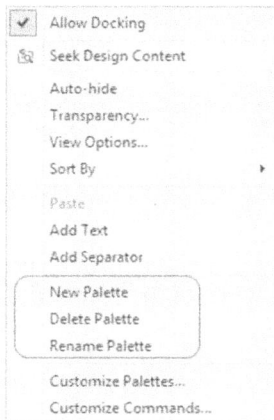

### 14.5.6 Customize Palettes

The user can create palette groups to organize palettes. The user can then set one of the groups to be the current palette group. To do this, make sure that the tool palette is shown, right-click any empty space within the tool palette (avoid icons), and you will see the following menu; select the Customize Palettes option:

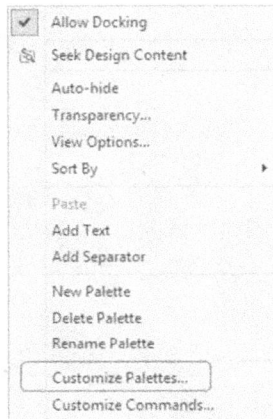

You will see the following dialog box:

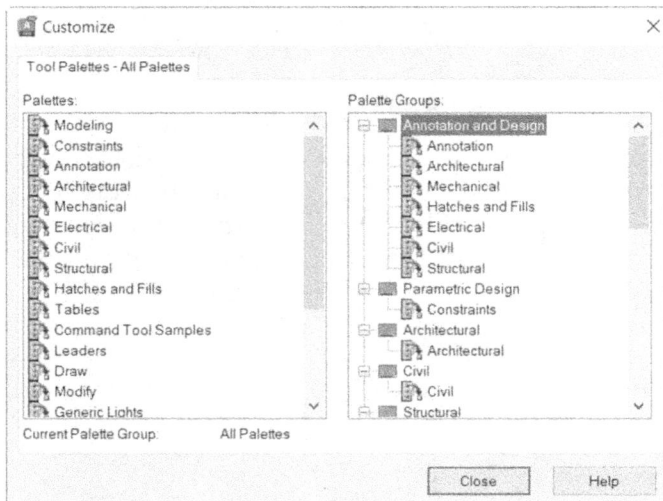

At the right part, you will see the predefined palette groups along with the palettes belonging to them. At the left, you will see the defined palettes. In order to create a new group, at the right-part right click anywhere, and you will see the following menu:

```
New Group
Rename
Delete
Set Current
Export...
Export All...
Import...
```

- Select the New option and type the name of the new group. You may move the group from any place to any place using drag-and-drop technique
- To fill this group with palettes, use the drag-and-drop technique again from the left part to the right part
- To make this group the current group, right-click it, then select Set Current option:

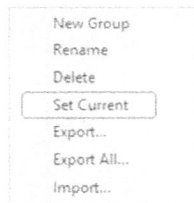

```
New Group
Rename
Delete
Set Current
Export...
Export All...
Import...
```

Using this menu you can Rename a group, Delete a group, Export one group, Export all groups, and finally Import groups. The Palette group file extension is *.XPG.

## PRACTICE 14-3   CUSTOMIZING TOOL PALETTES

1. Start AutoCAD 2025

2. Open **Practice 14-3.dwg** file

3. Start Tool Palette command

4. Using the tool palette created from Practice 14-2, rearrange the tools in Fasteners to see all nuts together, all screws together, and all bolts together

5. Add separators between the three types, and add text as title for each type

6. Change the transparency of the tool palette to be 50% for general

7. Using View Options, change the size of the tool to be less than maximum with one degree, showing Icons with text

8. Using Customize Palettes create a new group and call it My Group

9. Drag-and-drop Fasteners – US to it

10. Make My Group the current group (you can see that this group contains only one tool palette)

11. Export My Group to be My Group.xpg (you can go to another computer and try to import the same)

12. Save and close the file

## 14.6 EDITING BLOCKS

We can edit the original block in Block Editor, which is used to create dynamic features of a block (this is an advanced feature of AutoCAD). Block Editor will redefine the block by adding/removing/modifying the existing objects.

To issue the command, go to the **Insert** tab, locate the **Block Definition** panel, then click the **Block Editor** button:

You will see the following dialog box:

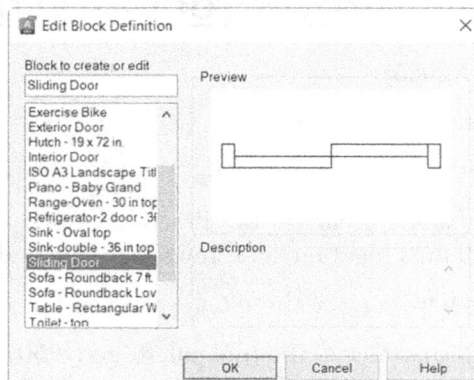

Select the name of the desired block to edit and then click **OK** to start editing. Another easy way is to double-click one of the block incidences.

When the Block Editor opens, you will see many things take place, such as: the background color will be different, the new context tab "Block Editor" will appear, and lots of new panels will appear as well. Ignore all of this, and start adding, removing, or modifying objects of your block. Once you are done, click the **Save Block** button at the **Open/Save** panel (you can find it at the left), then click the **Close Block Editor** button in the **Close** panel to end the command.

## PRACTICE 14-4   EDITING BLOCKS

**1.** Start AutoCAD 2025

**2.** Open **Practice 14-4.dwg** file

**3.** Double-click the incidences of block Single Door

**4.** Change the arc linetype to be Dashed2

**5.** Save the block with the new changes

**6.** What happened to the other incidences of block Single Door?

**7.** You should have the following shape:

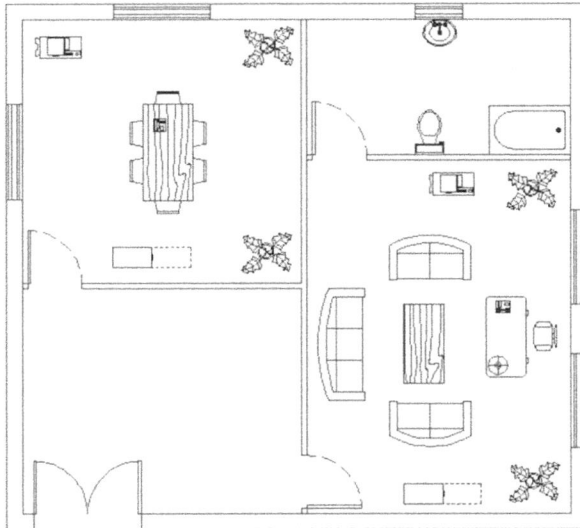

**8.** Save and close the file

## 14.7   BLOCKS SEARCH AND CONVERT

AutoCAD offers smarter block solutions to modernize your design creation methods.

If you have in your drawings shapes that consist of the same objects and are repeated several times without being created as a block, you can select them, and ask AutoCAD to search for similar instances to convert them to blocks

Do the following steps:

- Using Command Window, type the BCONVERT, you will see the following:

```
Select objects to convert to blocks:
```

- Select the desired objects, and press [Enter]. You will see that AutoCAD highlighted all similar objects and you will see the following:

```
Select instances or [Source objects only]:
```

- Select Instances means; AutoCAD will convert all selected instances including the source objects into blocks
- Source objects only means: AutoCAD will convert the source objects only into blocks.
- Once you specify your choice, you will see the following dialog box:

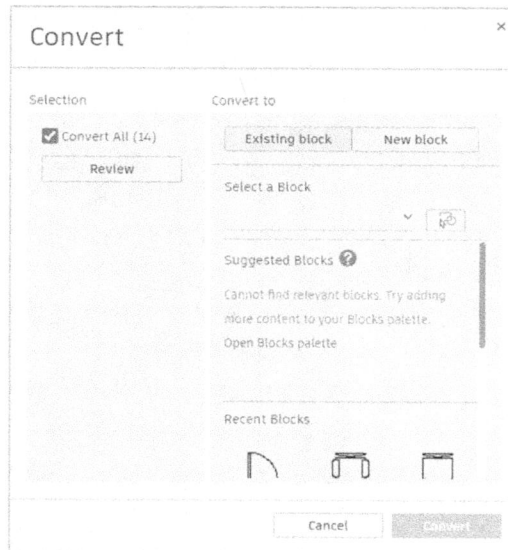

- Choose whether you want to convert it into New block or an Existing Block. Either way, you need to specify a name, once done, click Convert button. Then Accept Placement, or Move, Rotate, and Scale

## PRACTICE 14-5   BLOCKS SEARCH AND CONVERT

1. Start AutoCAD 2025

2. Open **Practice 14-5.dwg** file

3. Check the single door shape, you will discover they are objects

4. Start BCONVERT command, and select one of the single door objects, press [Enter]. AutoCAD will identify all the other similar objects

5. How many instances of the single door shape? _____ (14)

6. Press [Enter] to answer for all instances

7. Click Existing, and from the drop-down list select the block name: Single Door

8. Click Convert

9. Move the Single Door to layer Door

10. Do the same process for the 3-seat Sofa, creating a new block, and name it Sofa

11. Save and close the file

# NOTES

## CHAPTER REVIEW

    **1.** The Tool Palette can store commands like line, polyline, etc.

        **a.** True

        **b.** False

    **2.** _____ is the tool to allow users to share blocks, layers, etc.

        **a.** Insert Center

        **b.** Design Office

        **c.** Design Center

        **d.** The internet

    **3.** Using Tool Palettes, you can customize all tools by changing their properties

        **a.** True

        **b.** False

    **4.** Double-clicking a block will show _____ dialog box

    **5.** If both the Design Center and Tool Palettes are open, you can drag any block from the Design Center to a Tool Palette:

        **a.** True

        **b.** False

    **6.** Using _____ of the right-click menu will give you Icon only, Icon with text, and List view:

        **a.** Transparency

        **b.** View Options

        **c.** Auto Hide

        **d.** Allow Docking

## CHAPTER REVIEW ANSWERS

    **1.** a

    **3.** a

    **5.** a

# CREATING TEXT AND TABLE STYLES AND FORMULAS IN TABLES

**In This Chapter**

- How to create a Text Style
- How to create a Table Style Then using the Table command to insert Table
- In Tables, how to use Formulas
- In Tables, how to utilize the cell functions

## 15.1 STEPS TO CREATE TEXT AND TABLES

Writing text in AutoCAD involves two simple steps:

- Create a text style (normally will be created once) which will hold the size and the shape of the text
- Write text using either Single Line Text or Multiline Text

Normally creation of any style is tedious and lengthy. It is not the mission of novice users to create these styles; it falls on the shoulders of the CAD managers, who should always think about standardization—and of course text styles are part of this process.

The same applies to tables because users will perform two steps:

- Create a table style
- Insert and fill a table

Text Style, Table Style, and all the other styles should be part of Template, which will hold the company standards. If you do not have a template, you still can share these styles using Design Center, mentioned in Chapter 14.

## 15.2   HOW TO CREATE A TEXT STYLE

The first step in writing text in AutoCAD is to create a text style. Text style is where you define the characteristics of your text. To start the Text Style command, go to the **Annotate** tab, locate the **Text** panel, and click the small arrow at the lower right part:

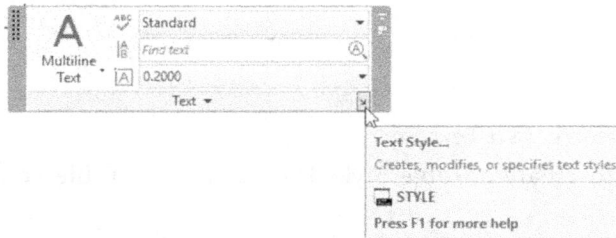

The following dialog box will appear:

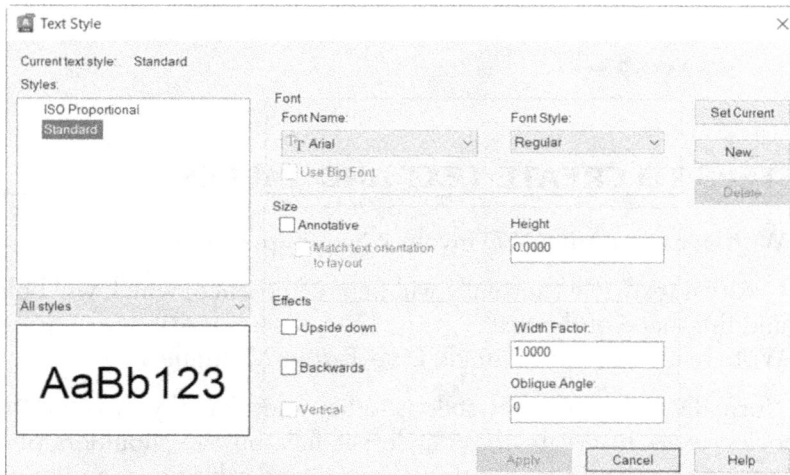

There will be two predefined text styles: one called **Standard**, and the other called **Annotative**. Virtually, they are the same except the latter uses the Annotative feature (which will be discussed in Chapter 17). Both styles use Arial as a font. Professional users will be advised not to use these two, but to create their own. To create a new text style, click the **New** button, and you will see the following dialog box:

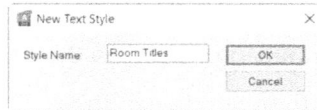

| New Text Style | × |
|---|---|
| Style Name: Room Titles | OK |
| | Cancel |

Input the name of the new text style, and when you are done click **OK**. The first step you will take is to select the **Font**. There are two types of fonts that you can use in AutoCAD:

- Shape files (*.shx), the very old method of fonts (out-of-date)
- True Type fonts (*.ttf)

See the illustration below to identify that TTF files are more accurate and look better:

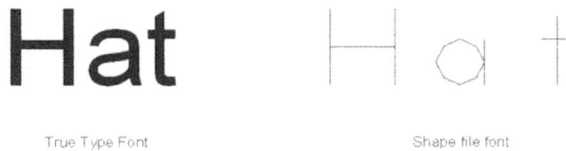

## Hat

True Type Font

Shape file font

Next, the user will select the **Font Style** (applicable if true type font); you will choose one of the following:

- Regular
- Bold
- Bold Italic
- Italic

Keep **Annotative** off for now (we will discuss later). Next, you must specify the **Height** of the text, which is the height of the capital letters; small letters will be 2/3 of the height. See the following illustration:

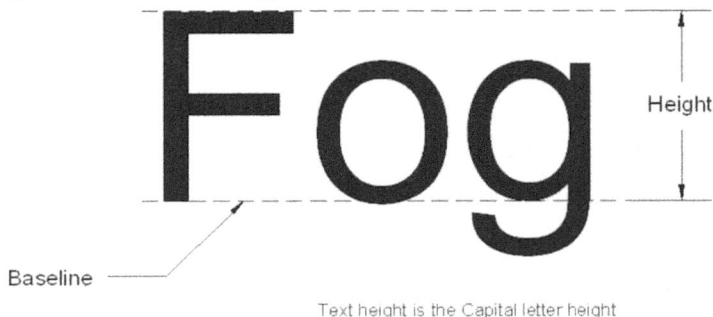

## Fog

Height

Baseline

Text height is the Capital letter height

So, how to set the height of the text?

- Setting height to be 0 (zero) means it will be variable (you must input the value each time you use this style)
- Setting the height to a value greater than 0 (zero) means the height is fixed

Finally set the effects, there are five of them:

- **Upside down**, to write text upside down
- **Backward**, to write text from right to left
- **Vertical**, to write from top to bottom. Good for Chinese words, but only for Shape files
- **Width Factor**, to set the relationship between width of the letter and its length. If the value>1.0, the text is wide. If the value is <1.0, the text is long
- **Oblique Angle**, to set the angle to italicize either to the right (positive), or to the left (negative)

Whenever you are done, click the **Apply** button, then the **Close** button.

You can show **All styles** or show **Styles in use** by checking the pop-up list at the left as shown below:

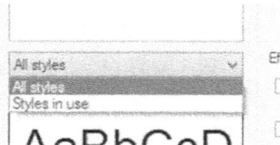

# PRACTICE 15-1   CREATING TEXT STYLE AND SINGLE LINE TEXT

1. Start AutoCAD 2025

2. Open **Practice 15-1.dwg** file

3. Create a text style with the following specs:

   **a.** Name = Room Names

   **b.** Font = Tahoma

   **c.** Annotative = off

   **d.** Height = 0.3

   **e.** Leave the rest to default values

4. Make Room Names text style current

5. Make layer Text current

6. Type the room names as shown below

**7.** Make Standard text style current (in this text style the height = 0, hence you should set it every time you want to use this style)

**8.** Make layer Centerlines current

**9.** Zoom to the upper left centerline, check the letter A is missing

**10.** Start Single Line text, right-click, and select the Justify option and from the list pick MC (Middle Center); select the center of the circle as the Start point, for the height set it to be 0.25, rotation angle = 0, then type A and press [Enter] twice.

**11.** You should receive the following results:

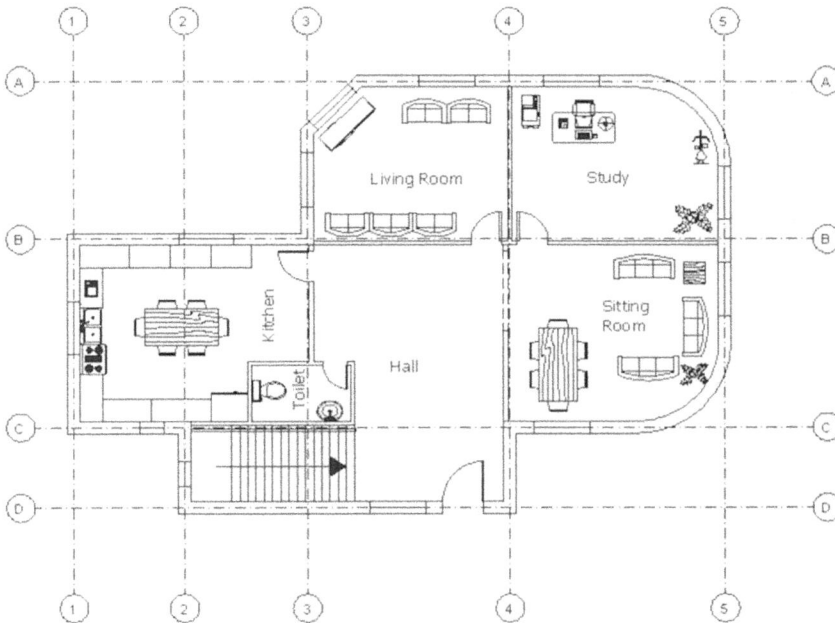

**12.** Save and close the file

## 15.3   CREATING TABLE STYLE

In order to create a professional table, the user should create a Table Style, which holds the features of the table specifying the Title, Header, and the Data rows. Using this style, you can insert as many tables as you prefer. Table Style can be shared using the Design Center.

To issue the command, go to the **Annotate** tab, locate the **Tables** panel, and then select the **Table Style** button:

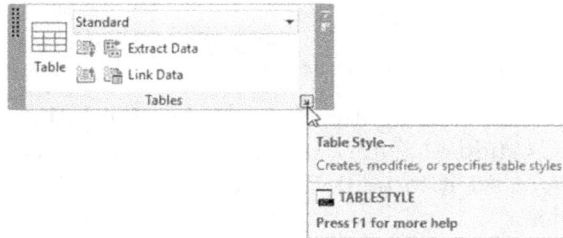

The following dialog box will be displayed:

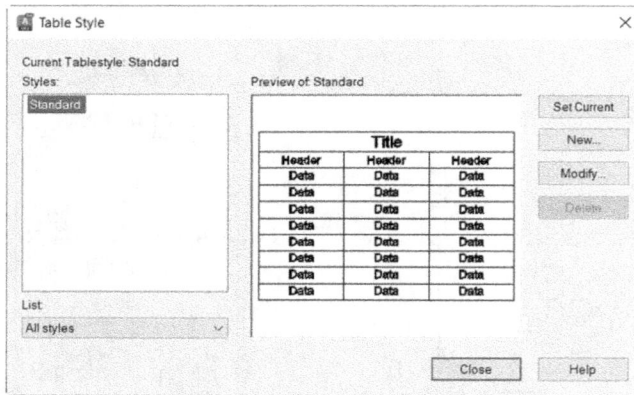

Just like text style, there is a predefined table style called Standard. Click the **New** button to create a new table style and you will see the following dialog box:

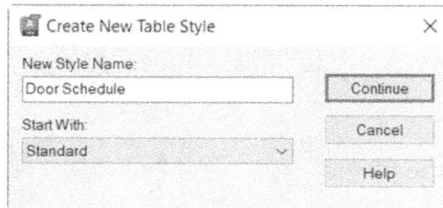

Input the name of the new table style and click **Continue**.

You will see the following dialog box:

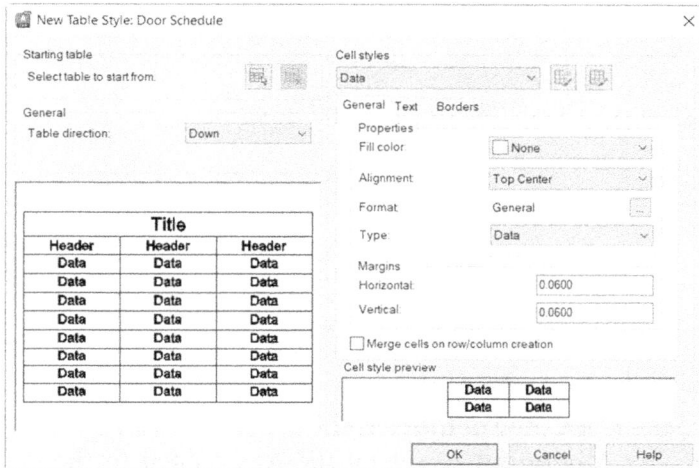

AutoCAD will allow you to start a new table style based on an existing table in your current drawing; click the button shown.

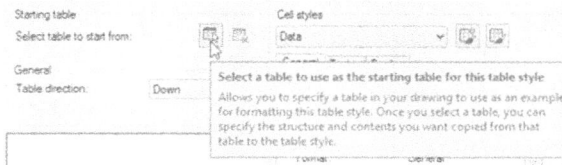

If not, you need to specify the characteristics of your table by selecting the **Table direction**, whether **Down** or **Up**, as in the following illustration:

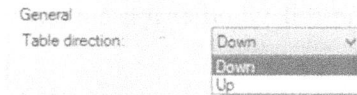

AutoCAD has three portions of any table: Title, Header, and Data. Using the Table Style command, you can set these three parts by selecting the desired portion, then by setting the features using the **General** tab, the **Text** tab, and the **Border** tab.

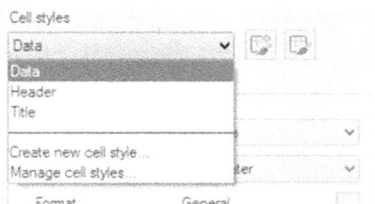

### 15.3.1 General Tab

The General tab is shown below:

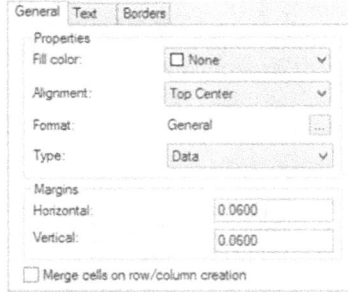

Edit the following features:

- Change the **Fill Color** of the cells (by default it is None)
- Change the **Alignment** to set up the text related to the cell borders. For instance, if you say Top Left, this means the text will reside in the top left part of the cell. See the following illustration:

- Change the **Format**. You will see the following dialog box, which will allow you to set the format of the cell, whether it is currency, percentage, date, etc.:

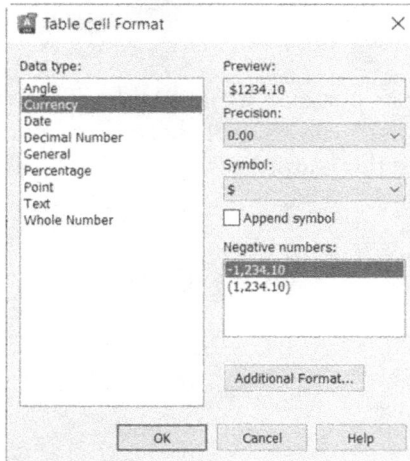

- Change the **Type** of the contents of the cell, whether **Data** or **Label**. This portion is very important, as some of the cells may hold numbers, but these numbers should not be included in a mathematical formula, hence, the type of the data will be Label and not Data
- Change the **Margins** left for the text away from the border of the cell, the **Horizontal** margin and **Vertical** margin

### 15.3.2 Text Tab

The Text tab is shown below:

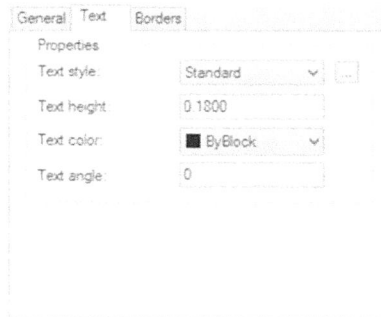

Edit the following features:

- Change the **Text style** of the text filling the cells
- Change the **Text height** of the text filling the cells (noting that if the used Text style has a height > 0.0, the number here is meaningless)
- Change the **Text color**
- Change the **Text angle**; see the following illustration:

### 15.3.3 Borders Tab

The Borders tab is shown below:

Edit the following features:

- Change the **Lineweight**, **Linetype**, and **Color** of the border lines
- Change the border to Double line instead of Single line (default value) and if yes, what is the Spacing between the lines?
- Change whether you want lines representing column separators, row separators, or not

## 15.4  INSERTING A TABLE IN THE DRAWING

This command will insert a table in the current drawing. To issue this command, go to the **Annotate** tab, locate the **Tables** panel, then select the **Table** button:

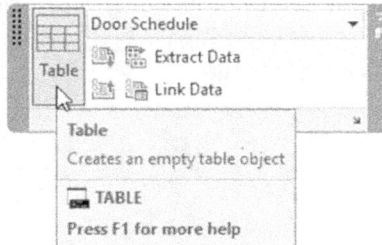

You will see the following dialog box:

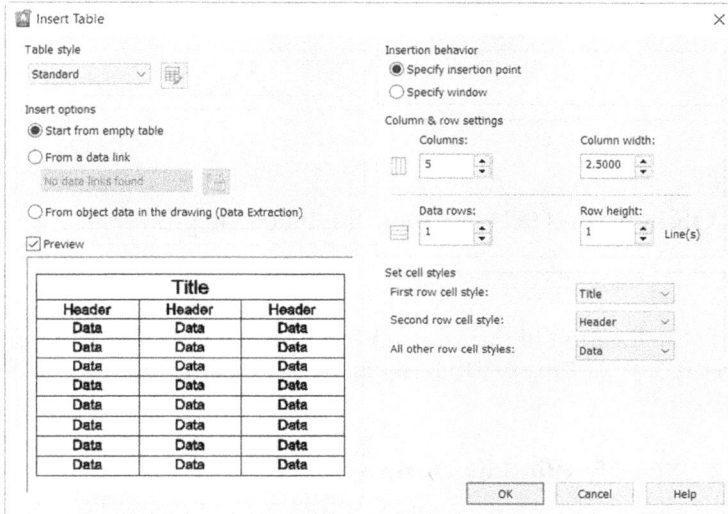

The first step is to select the desired predefined **Table style** name from the available list. If you did not create your table style yet, click the small button beside the list to go and define it right now.

Select the proper **Insert options**. To insert the table in your drawing, select from the following three choices:

- Start from empty table
- From a data link – this is an advanced feature of AutoCAD
- From object data in the drawing (Data Extraction) – this is also an advanced feature of AutoCAD

Most likely, we will click the first choice which is Start from empty table. Now select the **Insertion behavior** and select one of the two available choices:

- Specify insertion point
- Specify window

### 15.4.1 Specify Insertion Point Option

The insertion point meant here is the upper left corner of the table. You will hold the table from this point. To fulfill the rest of information, fill in the following data:

- Columns (which means the number of columns)
- Column width
- Data rows (which means the number of rows, but without Title and Heads)

■ Row height (in lines)

Click **OK** and AutoCAD will show the following prompt:

```
Specify insertion point:
```

Specify the location of the table and start filling the cells; you can use arrows at the keyboard or the [Tab] key to jump from one cell to another. Use [Shift] + [Tab] to go backward.

### 15.4.2 Specify Window Option

Using this option you will specify a window, which means you will give AutoCAD the total length and the total width of the table. In order to fulfill the rest of the information, fill in the following:

■ Input the number of **Columns**, and AutoCAD will calculate the column width. Or input the **Column width**, and AutoCAD will calculate the number of columns

■ Input the number of **Data rows**, and AutoCAD will calculate the row height. Or input the **Row height**, and AutoCAD will calculate the number of rows

Click **OK** and AutoCAD will show the following prompt:

```
Specify first corner:
Specify second corner:
```

Specify the two opposite corners of the table, then start inputting the cell contents using arrow, [Tab], and [Shift] + [Tab].

## PRACTICE 15-2   CREATING TABLE STYLE AND INSERTING TABLES IN THE CURRENT DRAWING

**1.** Start AutoCAD 2025

**2.** Open **Practice 15-2.dwg** file

**3.** Create a new Table Style with the following features:

**a.** Name = Door Schedule

**b.** Title Text style = Notes

**c.** Header Fill color = Yellow

**d.** Header Text style = Notes

**e.** Data Alignment = Middle Left

**f.** Data horizontal margin = 10

**g.** Data Text style = Notes

**h.** Data Text color = Blue

**4.** Create a table using the above Table style and the frame drawn, using the Window option to insert the table:

| Door Schedule | |
|---|---|
| Door No. | Size |
| Door 01 | 2'-6" x 6'-8" |
| Door 02 | 3'-0" x 6'-8" |

**5.** After finishing the table input, erase the frame

**6.** Save and close the file

## 15.5   USING FORMULAS IN TABLE CELLS

In this section, we will discuss how to create formulas in AutoCAD table cells, along with more functions like merging cells, using premade functions, and others. If you know how to use Excel, you can skip this part; if not read it because we will be needing it in the coming parts:

■  We call the intersection of column and row in a table a **Cell**

- The cell address comes from the column letter and row number. For example, B15 means this cell is in column B and row 15
- The cell address comes from the column letter and row number. For example, B15 means this cell is in column B and row 15
- To make AutoCAD know that you will be writing a formula, start with the equals sign. Along with the cell addresses, the user can use arithmetic functions like add, subtract, multiply, and divide. For example: =(B3/2)+A9
- AutoCAD provides some premade functions like SUM, COUNT, and AVERAGE
- To allow users to copy formulas from one cell to another, AutoCAD opted to make the cell address relative. For example, suppose we have a formula in cell C3 that says =A3+B3; if we copy it to C4, what is the result? AutoCAD will change the formula to =A4+B4. AutoCAD provides a tool called Autofill grip in order to copy formulas (or any cell content); click and drag it upward or downward

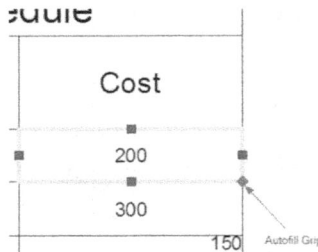

- Adding a dollar sign $ will attach the formula either to a column or to a row or to both. As an example: $F12 means column F is the column to use while copying formulas, with variable rows. While if we type F$12, this means row 12 is the row to use while copying, with variable columns. $F$12 will mean a fixed address

## 15.6   USING TABLE CELL FUNCTIONS

There are three different modes for table editing, which are:

- If you click the outer frame, the user will edit the whole table by adjusting the total width/height for the whole table or columns and rows, similar to the following:

- If you click a cell or group of cells (to select a group of cells, click and drag) a new context tab called Table Cell will be added to control the cells
- If you double-click inside a cell, you will be able to input data or edit existing data, hence the Text Editor context tab will appear to help you finish your job

In the following discussion, we will cover the second case which is the Table Cell context tab and its contents.

### 15.6.1 Using the Rows Panel

The user will use the below panel:

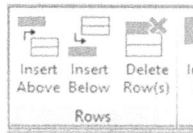

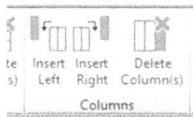

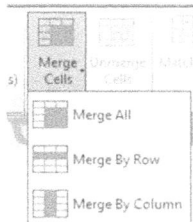

You can do all or any of the following:

- **Insert** a new row **Above** the selected cell(s)
- **Insert** a new row **Below** the selected cell(s)
- Delete the selected Row(s)

### 15.6.2 Using the Columns Panel

The user will use the below panel:

You can do all or any of the following:

- **Insert** a new column to the **Left** of the selected cell(s)
- **Insert** a new column to the **Right** of the selected cell(s)
- Delete the selected Column(s)

### 15.6.3 Using the Merge Panel

The user will use the below panel:

You can do all or any of the following:

- Merge all selected cells in a single cell
- Merge selected cells using rows
- Merge selected cells using columns
- If you have cells containing different types of data (numbers, text, dates, etc.) which one will be used? The answer is the content of the first cell. You will receive the following message

- Unmerge cells to bring them back to the previous condition before merging

### 15.6.4   Using the Cell Styles Panel

The user will use the below panel:

You can do all or any of the following:

- **Match cell**, to match the properties of a selected cell with other cells
- Select the **Alignment** using one of the nine available alignments for the content relative to the cell borders
- Select one of the existing cell styles for the selected cells:

- Select one color to be the background for the selected cells:

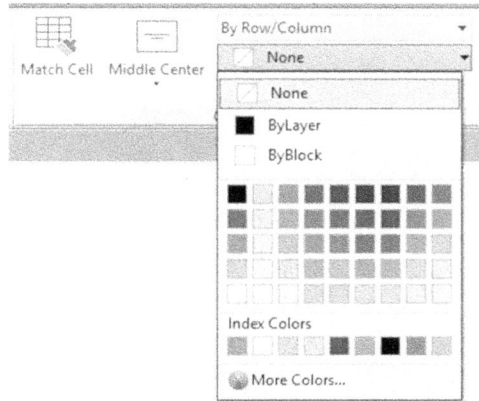

- Click the **Edit Borders** button and the following dialog box will appear:

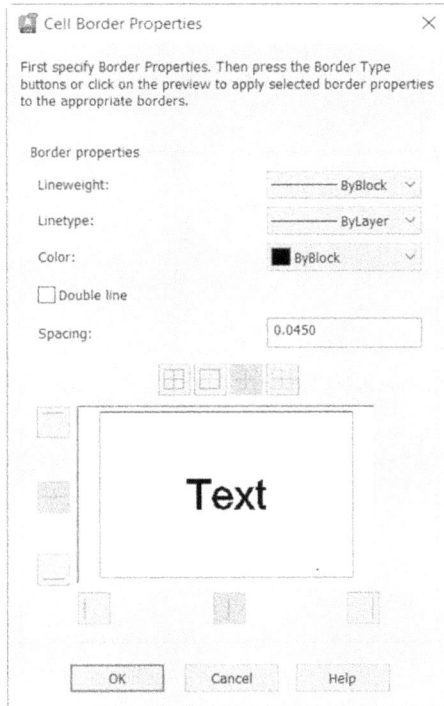

- The user will control all properties of the border like: lineweight, linetype, color, double line and spacing, and showing border lines in a different location

### 15.6.5 Using the Cell Format Panel

The user will use the below panel:

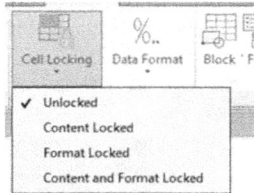

You can do all or any of the following:

■ Select whether to lock the contents or format, or both, of the selected cells. The user can unlock the locked cells as well

■ Select Data Format of the selected cells, choosing one of the following:

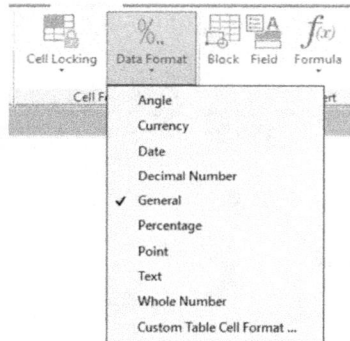

■ You can also customize the data format to something else. Select the Custom Table Cell Format option and you will see the following dialog box:

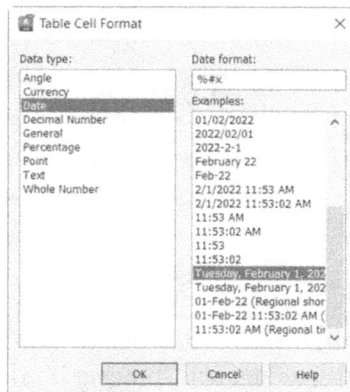

■ Select the Data type you want to change and then make the necessary changes. For some Data Types like Currency, you can click the Additional Format button

to see the following dialog box, and to specify what to use for a Decimal and Thousands separator, and whether or not to suppress leading and trailing zeros:

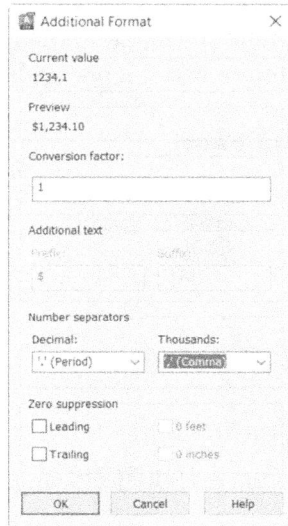

### 15.6.6 Using the Insert Panel

The user will use the below panel:

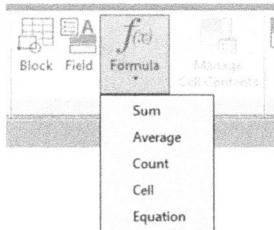

You can do all or any of the following:

- Click the **Block** button, and the following dialog box will appear:

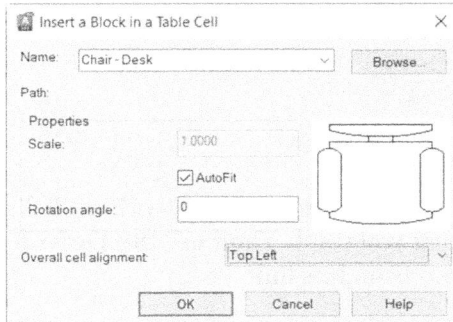

- Select the block name or click the Browse button to bring in any block or file, then specify the block's **Scale** or **AutoFit**, and the **Rotation angle**. At last, specify the Overall cell alignment of the block reference to the cell border
- Click the **Field** button to insert a field in the cell
- Click the **Formula** button to insert a premade formula

### 15.6.7 Using the Data Panel

The user will use the below panel:

You can do all or any of the following:

- Click the **Link Cell** button, and the following dialog box will appear:

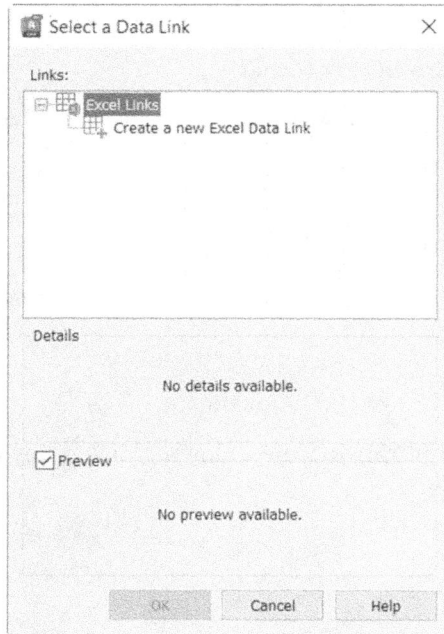

- This is the same dialog box that we dealt with while linking the date with Excel. Create the link that you want. Then use the **Download from Source** button to bring in the data from this link

# PRACTICE 15-3    FORMULAS AND TABLE CELL FUNCTIONS

1. Start AutoCAD 2025

2. Open **Practice 15-3.dwg** file

3. Insert the block in the first column using the second column as your guide (for the first two blocks use AutoFit, and for the rest use Sale = 0.01)

4. Select all Data cells and make them Middle Center

5. Select one of the Cost column cells

6. Insert a new column to its left and name it Quantity

7. Input the following values in it from top to bottom: 4, 1, 1, 6, 4

8. Select all the Data cells of the Cost column

9. Select the Custom Table Cell Format (from Cell Format panel) and change the currency format to show a number like the following: $ 999.99

10. Select one of the Cost column cells

11. Insert a new column to its right and call it Total Cost

12. Input a formula: Quantity * Cost

13. Using AutoFill grip copy it to the other cells

14. Select all Data cells of Total Cost and make sure to add a comma as the thousands separator

15. Select any cell in the lower row and add a new row below it

16. Merge by row all cells except the rightmost cell, and type in it Total Cost, and make it Middle Right

17. Add at the new cell a Sum function using the range of cells containing the Total Cost

18. Select the Title cell and make sure that border lines at the top, right, and left are removed

19. Set the background color of the Header to be Cyan

20. Save and close the file

# NOTES

## CHAPTER REVIEW

1. In _____ the user will define the background color of the table cell

2. There are two methods to insert a table in drawing:

   **a.** True

   **b.** False

3. By default the first row cell style is _____

4. Height in Text style is for:

   **a.** Small letters

   **b.** Capital letters and small letters

   **c.** Capital letters only

   **d.** Everything above the baseline

5. While inserting a table using a window, specifying the number of columns is enough, you do not need to specify also the column width:

   **a.** True

   **b.** False

6. In Table style, you can specify different text styles for header and title:

   **a.** True

   **b.** False

## CHAPTER REVIEW ANSWERS

**1.** Table Style

**3.** Title

**5.** a

## CHAPTER 16

# DIMENSION AND MULTILEADER STYLES

## In This Chapter

- How to create a Dimension Style
- More dimension advanced commands
- How to create Multileader Style
- How to insert Multileader in the drawing

## 16.1 WHAT IS DIMENSIONING IN AUTOCAD?

Dimensioning in AutoCAD is just like Text and Tables; the user should prepare dimension style as the first step, and then use it to insert dimensions. Dimension Style will control the overall outcome of the dimension block generated by the different types of dimension commands.

To insert a dimension, depending on the type of the dimension, the user should specify points or select objects, and then a dimension block will be added to the drawing. For example, in order to add a linear dimension, the user will select two points representing the distance to be measured, and a third point will be the location of the dimension block. Refer to the illustration below:

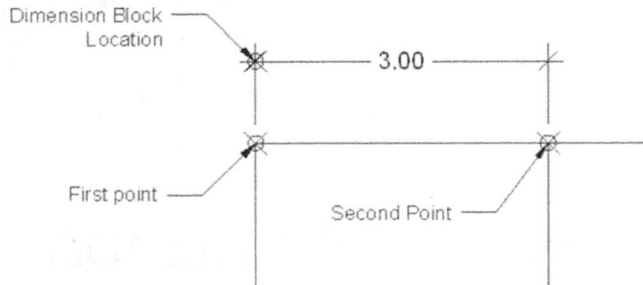

The generated block consist of three portions, which are:

- Dimension line
- Extension lines
- Dimension text

See the following illustration:

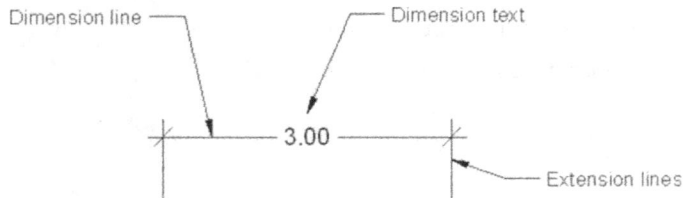

## 16.2   HOW TO CREATE A NEW DIMENSION STYLE

This command will create a new dimension style or modify an existing one. To issue this command, go to the **Annotate** tab, locate the **Dimensions** panel, then select **Dimension Style** button:

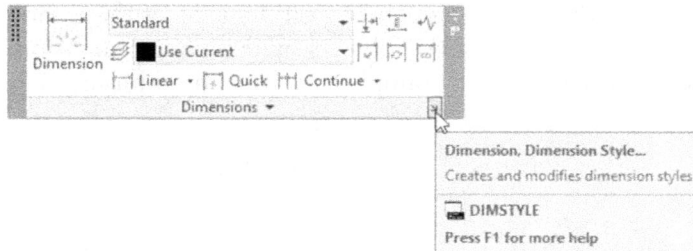

You will see the following dialog box:

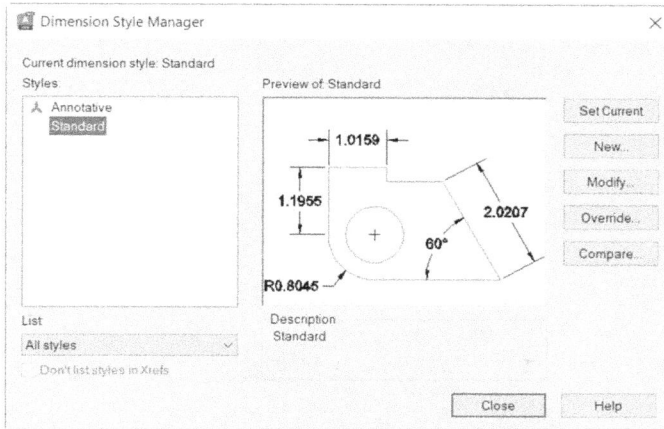

As you can see you will find two predefined dimension styles, Standard and Annotative. Click the **New** button to create a new style. You will see the following dialog box:

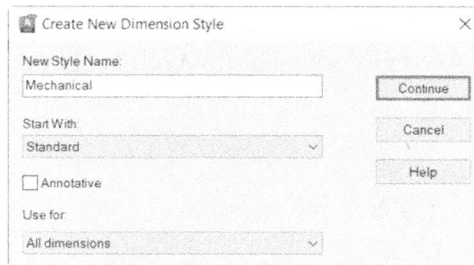

Input the name of the new style to be made, and under **Start With**, select the existing style, which will be your starting point. Leave the Annotative checkbox off, as we will discuss it in Chapter 19. Make sure that under **Use** for is **All dimensions** (we will cover it at the end of our discussion); to start creating the new style click the **Continue** button.

## 16.3   DIMENSION STYLE: LINES TAB

As a rule of thumb and while we are discussing the different tabs of the dimension style command, we will leave Color, Linetype, and Lineweight to their default settings, because we want these things to be controlled by layers rather than the individual dimension block.

The first tab in the dimension style dialog box is Lines; it will allow you control of dimension lines and extension lines:

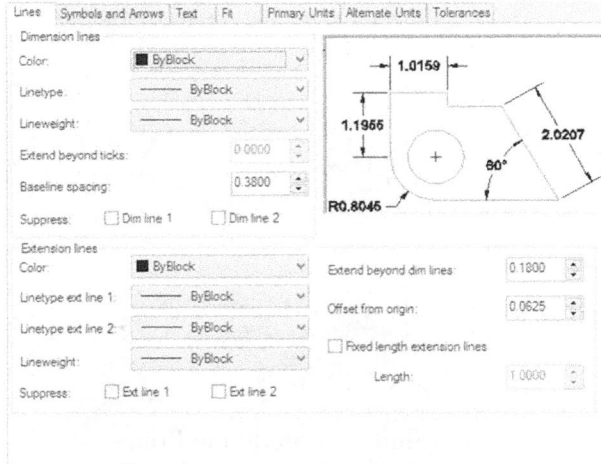

Under **Dimension lines,** change all or any of the following settings:

- Control **Extended beyond ticks,** which is as illustrated below (this option works only with **Arrowhead** equals to **Architectural tick** or **Oblique**):

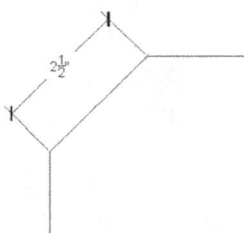

Extended beyond ticks = 0                Extended beyond ticks = 3/8"

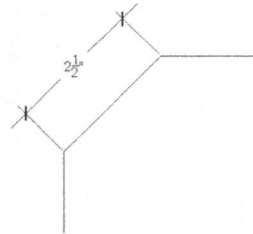

- As we will see in this chapter, when you add Baseline dimension you will not control the spacing between one dimension and another, or control **Baseline spacing**, as illustrated below:

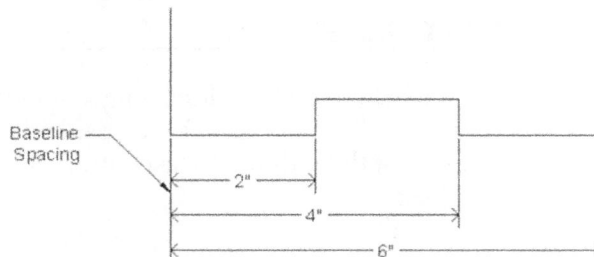

**NOTE** *From now on, if we say First, this means nearest to the first point picked, hence Second means nearest to the second point picked.*

- Choose whether to **Suppress Dim line 1, Dim line 2**, or leave them as is. Check the picture below:

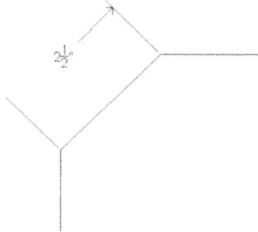

Suppress Dim Line 1                    Suppress Dim Line 2

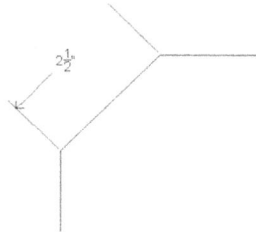

Under **Extension lines,** change all or any of the following settings:

- Choose whether to **Suppress Ext line 1, Ext line 2,** or leave them as is. Refer to the picture below:

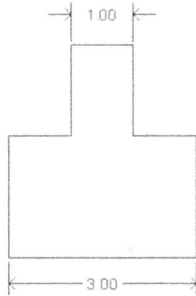

Both Extension lines are displayed          Suppress Both Extension Lines

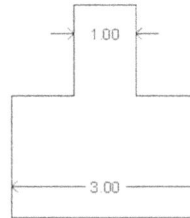

- Input Extend beyond dim lines and Offset from origin. See the picture below:

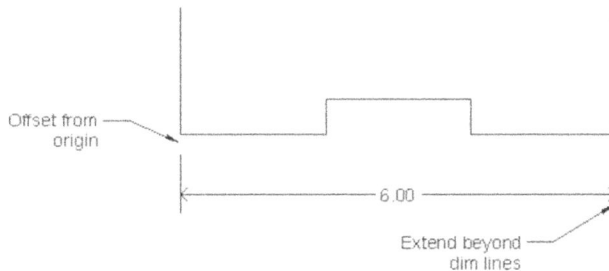

Offset from origin

6.00

Extend beyond dim lines

- Choose whether to fix the length of the extension lines or not; if yes, what will be the length? See the following example:

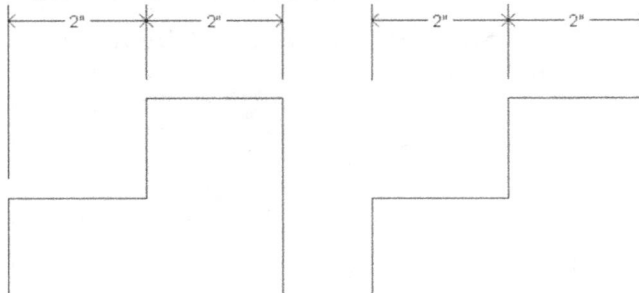

Fixed length = Off           Fixed Length = On

## 16.4 DIMENSION STYLE: SYMBOLS AND ARROWS TAB

This tab will control the arrowheads and related features. Below is the picture of the **Symbols and Arrows** tab:

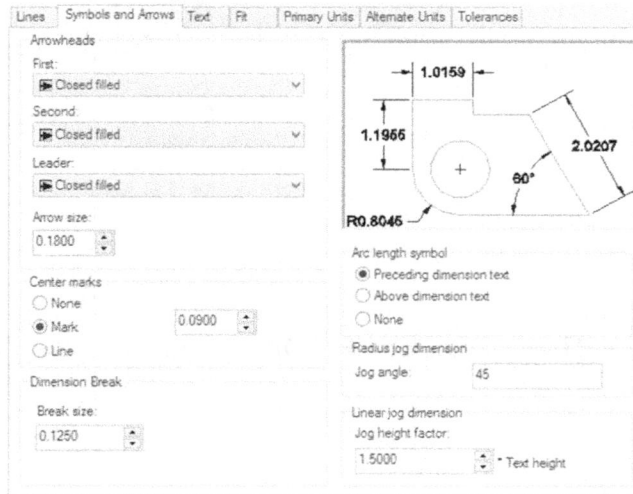

Under **Arrowheads**, change all or any of the following:

- The shape of the **First** arrowhead. When you set the shape of the first, automatically the **Second** arrowheads will change. If you want them to be different change the second.

- The shape of the arrowhead to be used in the **Leader** (Radius and Diameter are not leaders)
- The **Size** of the arrowhead

Under **Center marks**, choose whether to show or hide the Center mark of arcs and circles as shown below, then set the **Size** of the center mark:

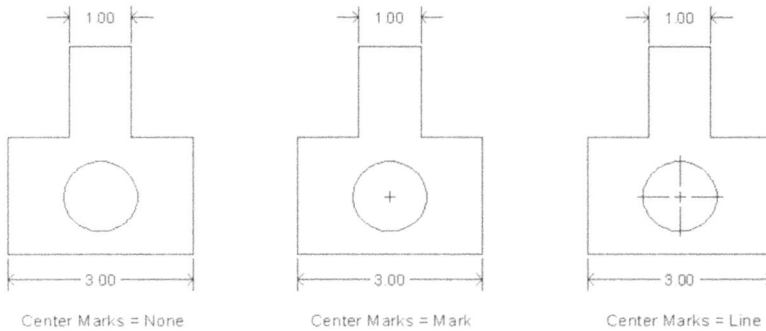

| Center Marks = None | Center Marks = Mark | Center Marks = Line |

Under **Dimension Break**, input the **Break size**; break size is defined as the distance of void left between two broken lines of dimension. Refer to the picture below:

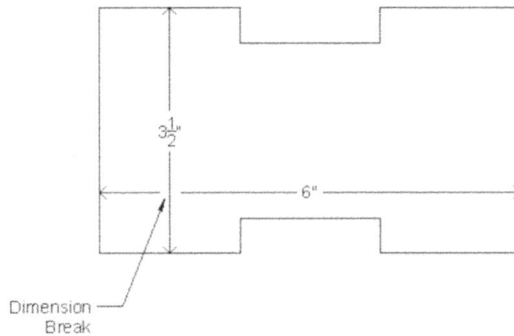

Under **Arc length symbol**, choose whether to show (as shown in the picture below) or hide the arc length symbol:

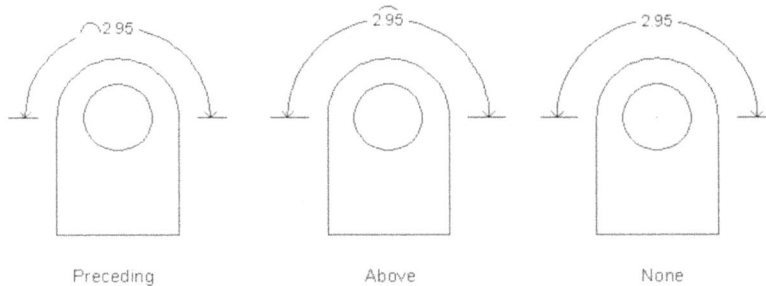

| Preceding | Above | None |

Under **Radius dimension jog**, input the **Jog angle** as shown in the picture as shown below:

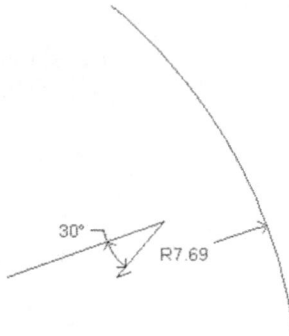

Jog Angle

Under **Linear Jog dimension**, input the **Jog height factor** as shown in the below picture. Jog height factor is defined as a factor to multiply the height of the text used in dimension:

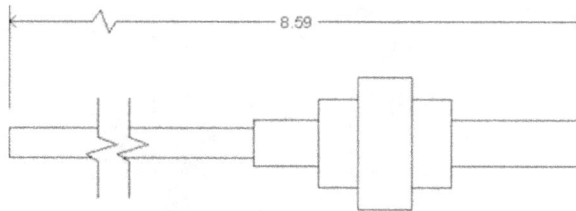

Jog height factor = 2.0

## 16.5   DIMENSION STYLE: TEXT TAB

This tab will control the text appearing in the dimension block. Below is the picture of the **Text** tab:

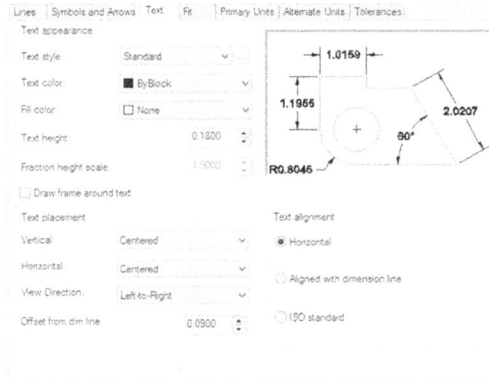

Under **Text appearance** change all or any of the following:

- Select the desired Text Style or create a new one.
- Specify the **Text color** and the **Fill color** (text background color).

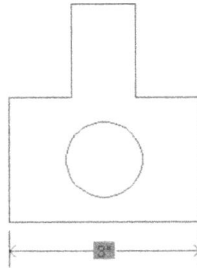

Fill Color

- If your text style has a text height = 0 (zero) then input the **Text height**.
- Depending on the primary units (discussed in a moment), set **Fraction height scale**.

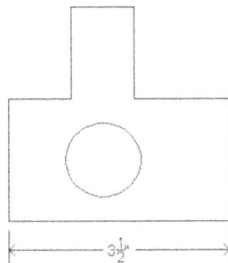

Fraction height scale = 1.0    Fraction height scale = 0.7

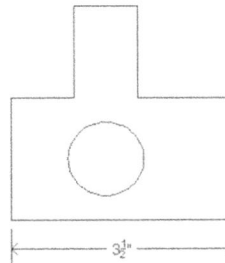

- Select whether your text will be with frame or without frame, see the picture below:

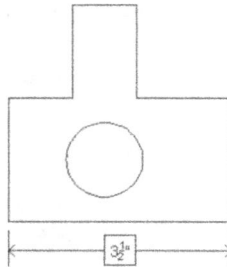

Draw frame around text

Under **Text placement** change all or any of the following:

- Choose the **Vertical** placement of your text related to the dimension line; there are five available choices: Centered, Above, Outside, JIS (Japan Industrial Standard), and Below. See the following images:

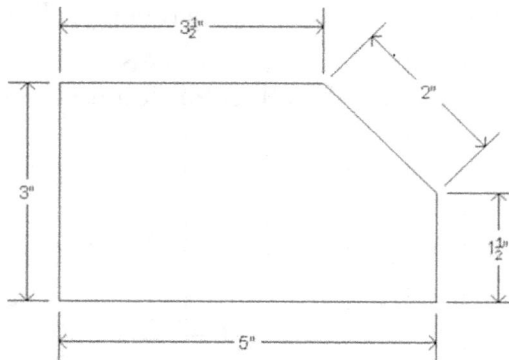

Centered

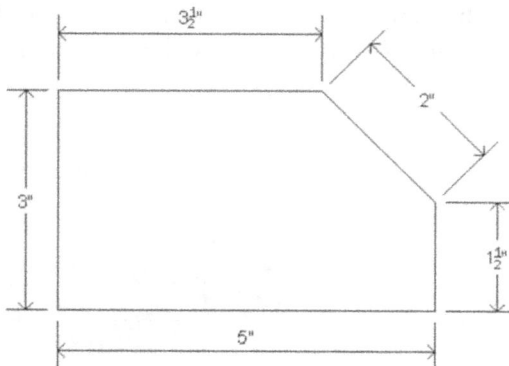

Above

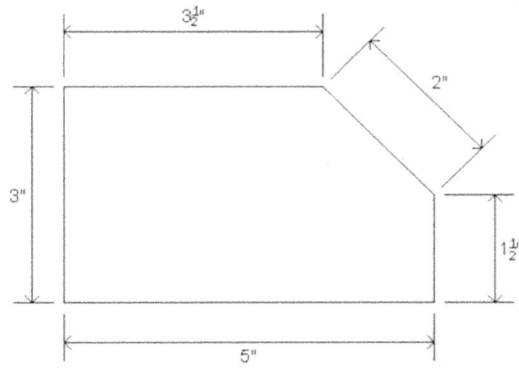

Outside

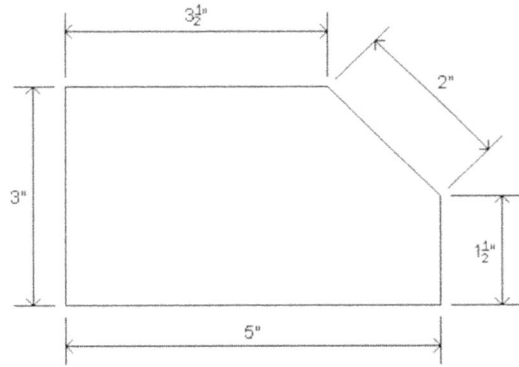

JIS

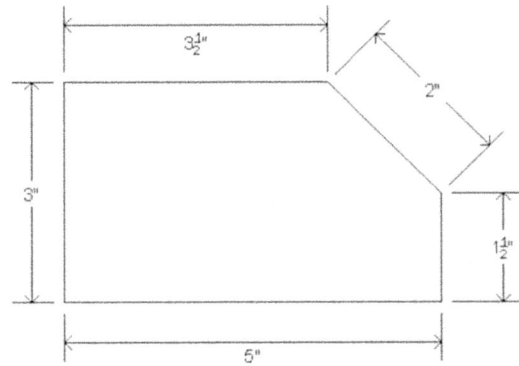

Below

- Choose the **Horizontal** placement; the user has four choices to select from. See the following images:

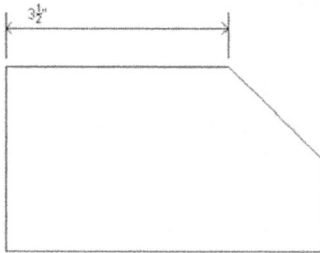

At Ext Line 2

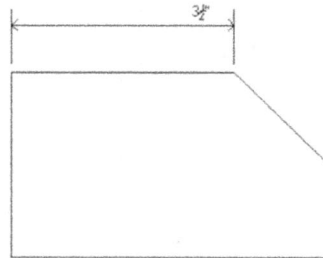

At Ext Line 1

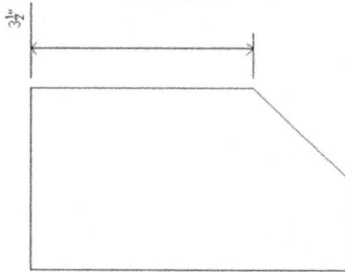

Over Ext Line 2

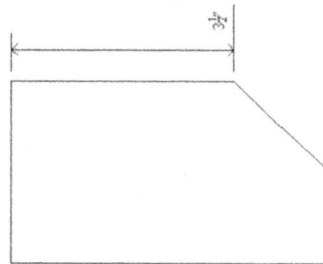

Over Ext Line 1

- Choose the View Direction of the dimension text: Left-to-Right or Right-to-Left. Refer to the following illustration:

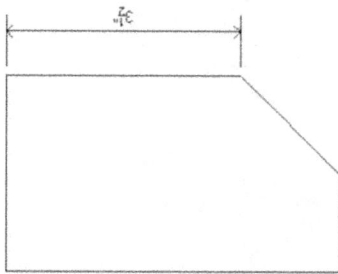

View Direction Right-to-Left

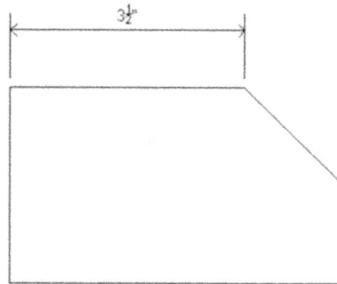

View Direction Left-to-Right

- Input the **Offset from dim line**, as shown below:

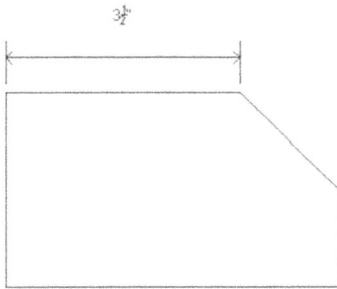

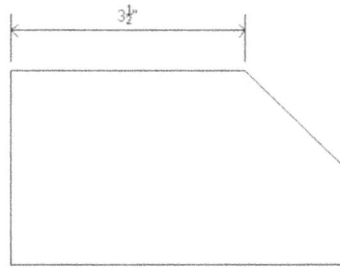

Offset from dim line = 0.50                    Offset from dim line = 0.09

Under **Text alignment**, control the alignment of the text related to the dimension line, whether horizontal always regardless of the alignment of the dimension line, or aligned with the dimension line. ISO will influence only Radius and Diameter dimension; all of the dimension types will be aligned except for those two.

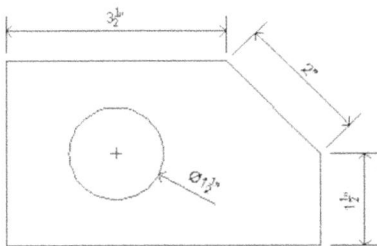

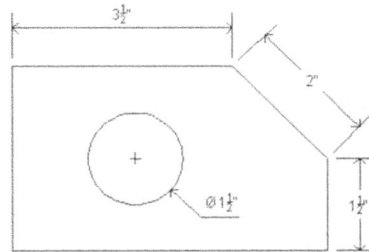

Aligned with dimension line                    Horizontal

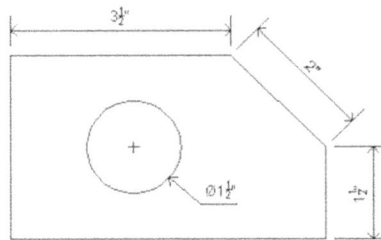

ISO standard

## 16.6 DIMENSION STYLE: FIT TAB

This tab will control the relationship between the dimension block components. Below is the picture of the **Fit** tab:

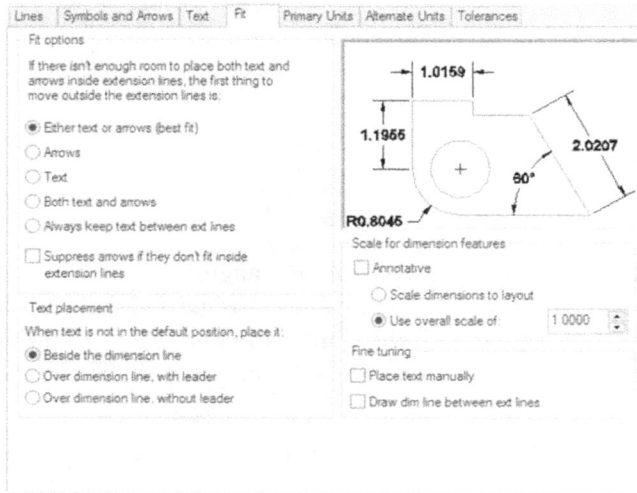

Under **Fit options**, change all or any of the following:

- If there is no room for the text and/or the arrowheads inside the extension lines, what do you want AutoCAD to do? Select proper option

- If there is no room for arrows to be inside the extension lines, do you want AutoCAD to suppress them? Or not?

Under **Text placement**, if the text to be sent outside the extension lines, select one of the following illustrated cases:

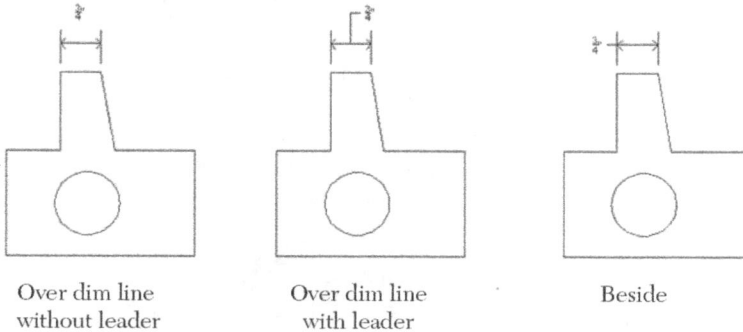

Over dim line without leader

Over dim line with leader

Beside

Under **Scale for dimension features**, , the user will control the size of the text features (length, sizes, etc.). Will it be scaled automatically if it was annotative?

Or it will follow the viewport scale if was input in the layout, or if you want to input it in the Model space, you can set the scaling factor.

Under **Fine tuning**, select whether you want to place your text manually rather than leaving it to AutoCAD. Also, select to force text always inside extension lines.

## 16.7 DIMENSION STYLE: PRIMARY UNITS TAB

In this tab, the user will be allowed to control everything related to numbers, which will appear at the dimension block. Below is the image of the **Primary Units** tab:

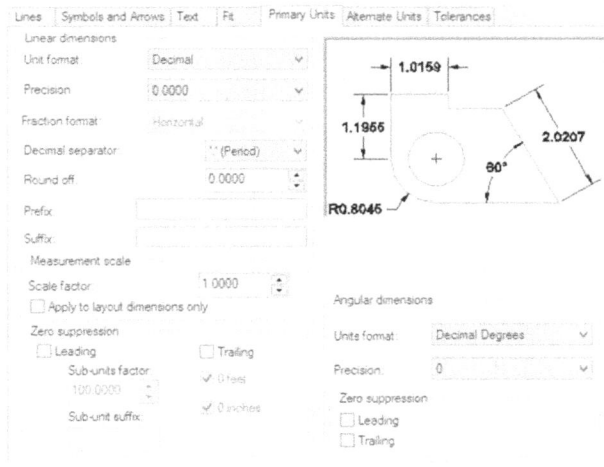

Under **Linear dimensions**, change all or any of the following settings:

- Select the desired **Unit format**, then select its **Precision**
- If your selection was **Architectural** or **Fractional**, then choose the desired **Fraction format**. You will have three choices to pick from: **Horizontal**, **Diagonal**, and **Not Stacked**

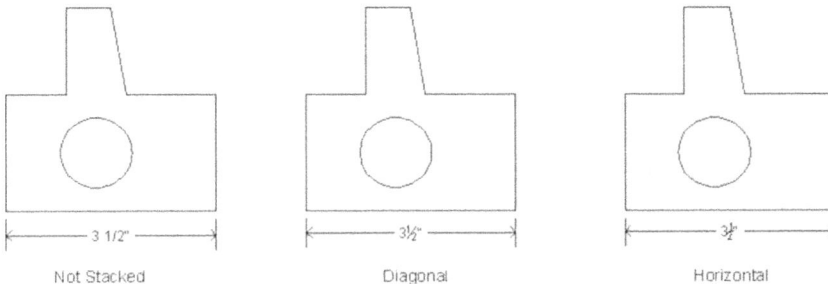

Not Stacked          Diagonal          Horizontal

- If your selection was **Decimal**, then choose the **Decimal Separator**; you will have three choices: **Period**, **Comma**, and **Space**
- Input the **Round off** number
- Input the **Prefix** and/or the **Suffix**, like below:

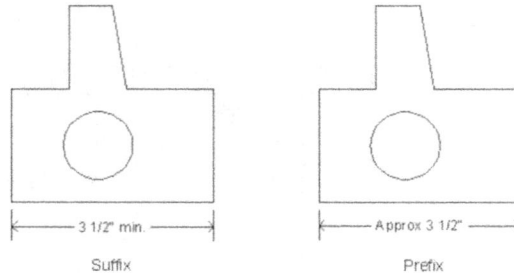

Suffix            Prefix

Under **Measurement scale**, change all or any of the following:

- By default, AutoCAD will measure the distance between the two points specified by you (that is if it was linear) and input the text in the format set by you. But what if you want to show a different value than the measured value? At this moment, input the **Scale factor**.
- Choose whether this scale will affect only the dimension input in the layout

Under **Zero suppression**, choose whether to suppress the **Leading**, and/or the **Trailing** zeros as shown below:

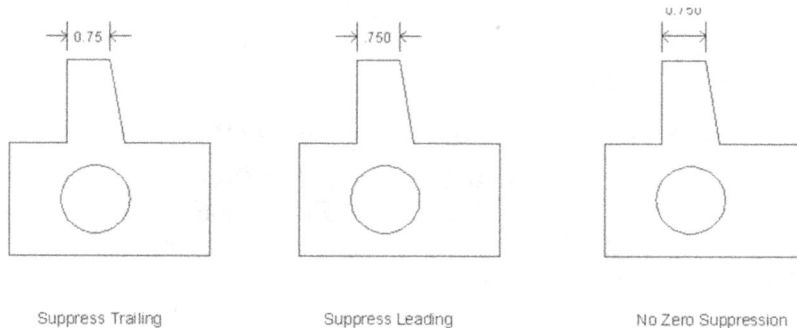

Suppress Trailing       Suppress Leading       No Zero Suppression

If you have meters as your unit, and the measured value was less than one (which is cm) we call this sub-unit. Select the sub-unit factor and the suffix for it (in this example it is cm).

Under **Angular dimensions**, choose the Unit format and the Precision. Control **Zero suppression** for angles as well.

## 16.8   DIMENSION STYLE: ALTERNATE UNITS TAB

This tab will show two numbers in the same dimension block, one showing primary units, and the other showing alternate units. Below is the **Alternate Units** tab:

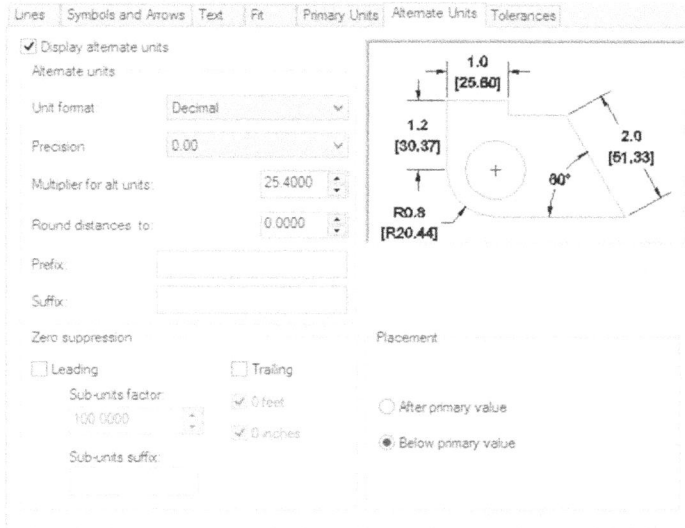

Click on the **Display alternate units** option, then change the following:

- Choose the Alternate **Unit format** and its **Precision**
- Input the Multiplier for all units value
- Input the Round distance
- Input the **Prefix** and the **Suffix**
- Input the **Zero suppression** method/m,
- Choose the method of displaying alternate units, whether **After primary value**, or **Below primary value**. See below:

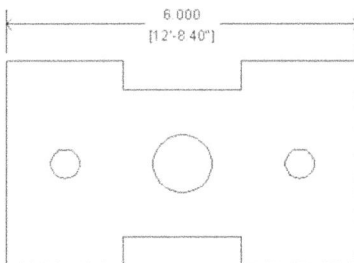

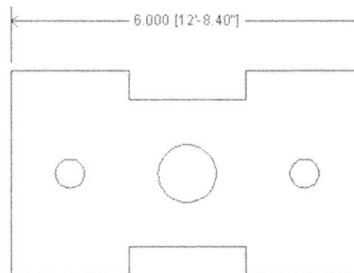

Below primary value                              After primary value

## 16.9 DIMENSION STYLE: TOLERANCES TAB

This tab will control whether or not to show tolerances and what is the used method. Below is an image of the **Tolerances** tab:

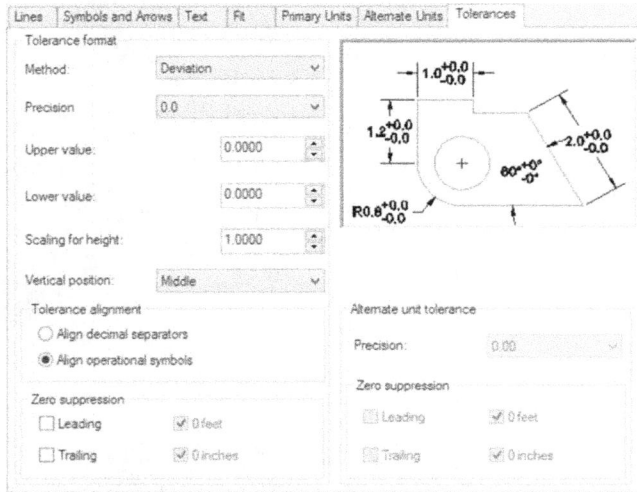

The four methods to show the tolerance are listed below:

- Symmetrical
- Deviation
- Limits
- Basic

The following is an illustration for each of the four choices:

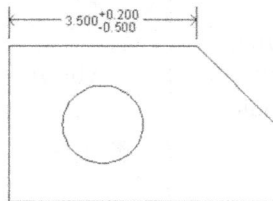

Deviation

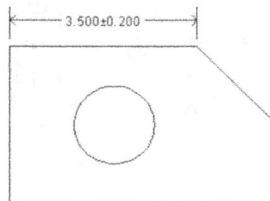

Symmetrical

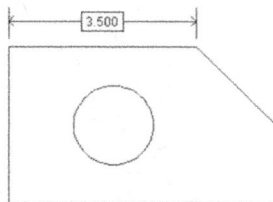

Basic

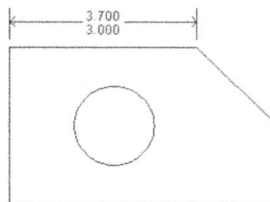

Limits

Under **Tolerance format**, change all or any of the following:

- Select the proper **Method**, and then select its **Precision**
- Depending on the method, specify the **Upper value** and **Lower value**
- Input **Scaling for height** for the tolerance values if desired
- Choose **Bottom**, **Middle**, or **Top** vertical position for the dimension text with reference to the tolerance values. Check the following illustration:

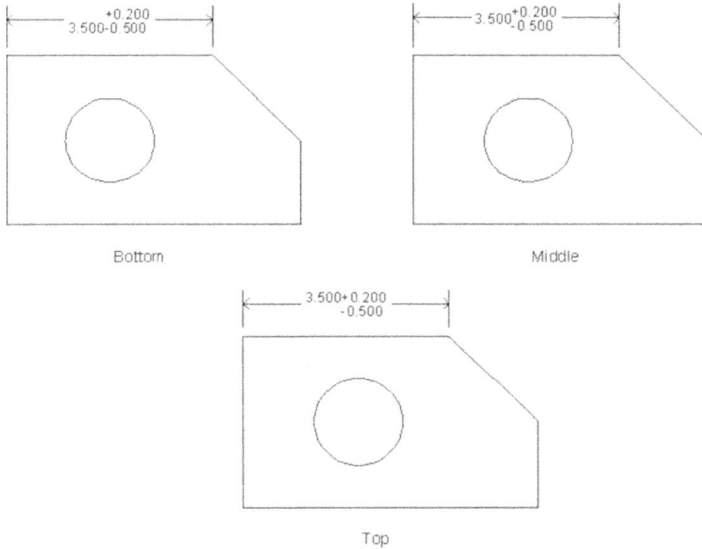

Bottom

Middle

Top

If either **Deviation** method or **Limits** method is selected, then you have to choose whether to **Align decimal separators**, or **Align operational symbols**. As shown below:

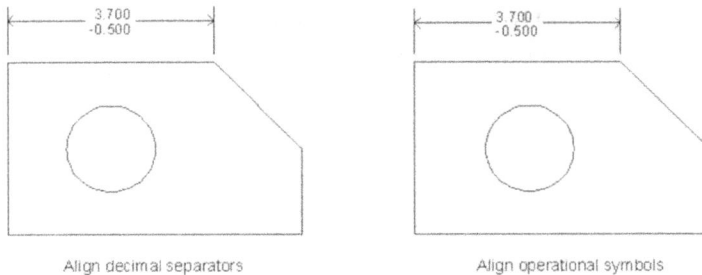

Align decimal separators

Align operational symbols

Under Alternate units tolerances, and if the **Alternate units** option is turned on, then specify the **Precision** of the numbers. Consequently, choose the **Zero suppression** for both the **Primary units** tolerance and the **Alternate units** tolerance.

## 16.10   CREATING A DIMENSION SUB STYLE

By default, the dimension style created will affect all types of dimensions. If you want the dimension style to affect only a certain type of dimension and not the others, you have to create a sub style. Complete the following steps:

- Select an existing dimension style
- Use the **New** button to create a new style and you will see the following dialog box:

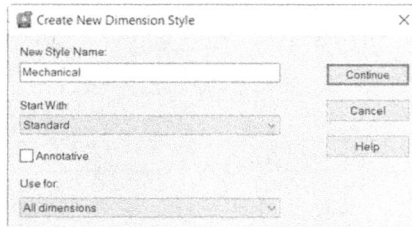

- Do not input anything; go to **Use for** and select the type of dimension (in the following figure we selected Angular diameter) and the dialog box will change to:

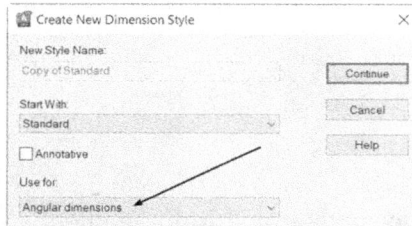

- Click the **Continue** button and make the changes you want; these changes will affect diameter dimensions only
- The picture of the **Dimension Style** dialog box will allow you to differentiate between the style and the sub style and in what features

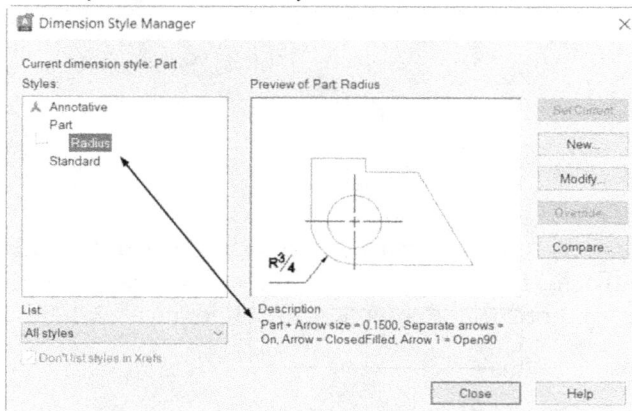

## PRACTICE 16-1    CREATING DIMENSION STYLE

**1.** Start AutoCAD 2025

**2.** Open **Practice 16-1.dwg** file

**3.** Create a new dimension style based on Standard, using the following information:

    **a.** Name: Part

    **b.** Extend beyond dim line = 0.3

    **c.** Offset from origin = 0.15

    **d.** Arrowhead = Right angle

    **e.** Arrow size = 0.25

    **f.** Center mark = Line

    **g.** Arc length symbol = Above dimension text

    **h.** Jog angle = 30

    **i.** Text placement vertical = Above

    **j.** Offset from dim line = 0.2

    **k.** Text Alignment = ISO Standard

    **l.** Text placement = Over dimension line with leader

    **m.** Primary unit format = Fractional

    **n.** Primary unit precision = 0 ¼

    **o.** Fraction format = Diagonal

**4.** Click OK to end the creation process

**5.** Select Part, and create a sub style for Radius

    **a.** Arrowhead = Closed filled

    **b.** Arrow size = 0.15

**6.** Make Dimension the current layer

**7.** Make Part the current dimension style and add the dimensions to create a shape to look similar to the following:

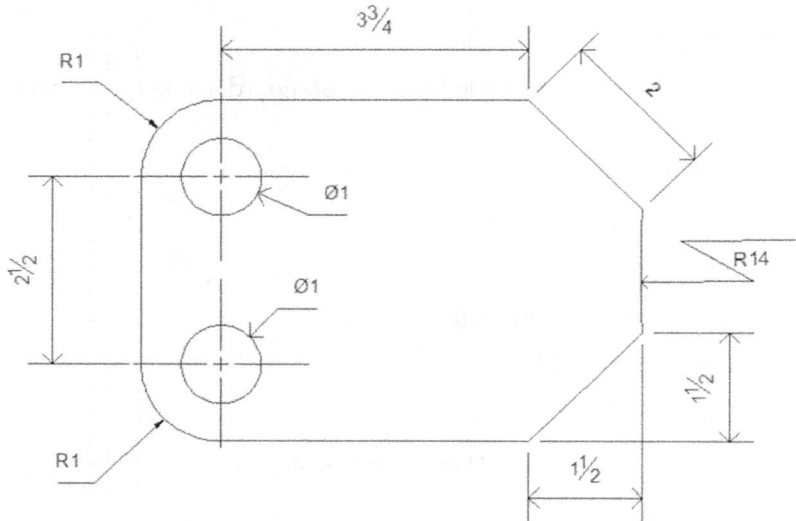

**8.** Save and close the file

## 16.11  MORE DIMENSION FUNCTIONS

In this part, we will discuss more dimension functions, which will help you produce a better look for the final drawing. These functions are as follows:

- Dimension Break
- Dimension Adjust Space
- Dimension Jog Line
- Dimension Center Mark
- Dimension Oblique
- Dimension Text Angle
- Dimension Justify
- Dimension Override

### 16.11.1    Dimension Break

If two or more dimension blocks intersect in one or more points, this command will break one of the blocks at the intersection point. To issue this command, go to the **Annotate** tab, locate the **Dimensions** panel, then select the **Break** button:

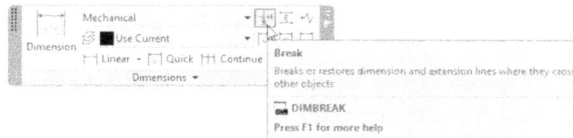

You will see the following prompts:

```
Select dimension to add/remove break or [Multiple]:
Select object to break dimension or [Auto/Manual/
Remove] <Auto>:
```

The first prompt will ask you to select the dimension block, which will be broken; the second prompt will ask you to select the dimension block, which stays as is. These prompts will remove a break if it exists; simply right-click at the second prompt and select the Remove option. These two prompts will be repeated until you press [Enter] to end the command.

The final result will look similar to the following:

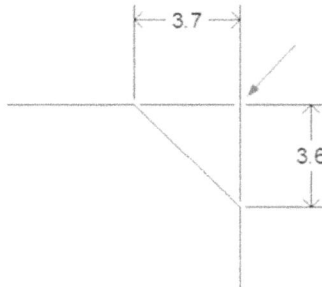

### 16.11.2    Dimension Adjust Space

This command will adjust either the spaces between dimension blocks to be aligned or having equal spaces between them. To issue this command, go to the **Annotate** tab, locate the **Dimensions** panel, then select the **Adjust Space** button:

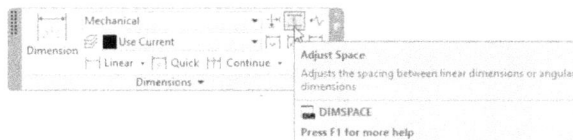

You will see the following prompts:

```
Select base dimension:
Select dimensions to space:
Select dimensions to space:
Enter value or [Auto] <Auto>:
```

The first step is to select the base dimension block, which the other blocks will follow; then select all the other dimension blocks, and when done, press [Enter]. AutoCAD will ask for the value and there are three options:

- Value = 0 (zero), this means all the other blocks will be aligned with the base dimension block
- Value > 0, this will be the distance which will separate the base dimension block from the nearest block, and the others as well
- Value = Auto, which means AutoCAD will try to figure out the best arrangement for the selected blocks

Let's view the following example:

- We have the following situation:

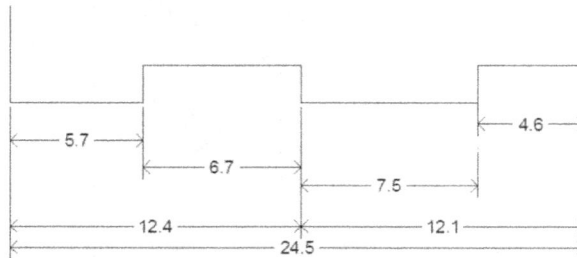

- As the first step, we started the Adjust Space command and selected the dimension block at the left reading 5.7 as our base dimension block, then selected the adjacent 6.7, 7.5, and 4.6, and we set the value to 0:

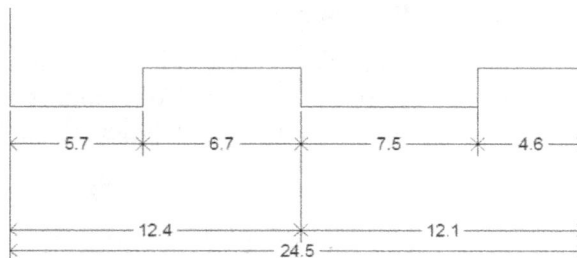

■ We started the command again, and we selected the dimension block reading 5.7 as our base dimension block, then we selected the other three dimension blocks reading 12.4, 12.1, and 24.5 setting the value to 1.0:

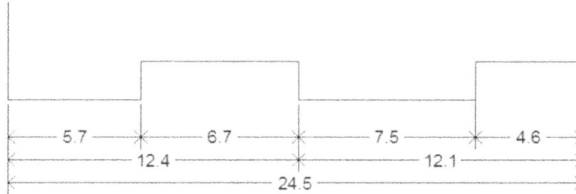

### 16.11.3 Dimension Jog Line

This command will add/remove a jog line to an existing linear or aligned dimension. To issue this command, go to the **Annotate** tab, locate the **Dimensions** panel, then select the **Jog line** button:

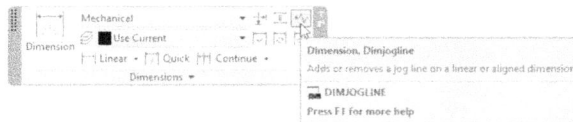

You will see the following prompts:

```
Select dimension to add jog or [Remove]:
Specify jog location (or press ENTER):
```

The first prompt is to ask you to select the desired linear or aligned dimension block. The user can select the position of the jog, but you can press [Enter] to let AutoCAD locate it automatically; you will see the following:

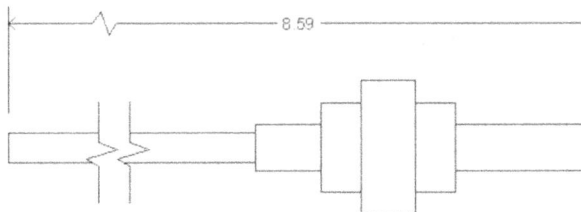

Jog height factor = 2.0

### 16.11.4   Dimension Oblique

This command will change the angle of the extension lines to any angle you want. To issue this command, go to the **Annotate** tab, locate the **Dimensions** panel, then select the **Oblique** button:

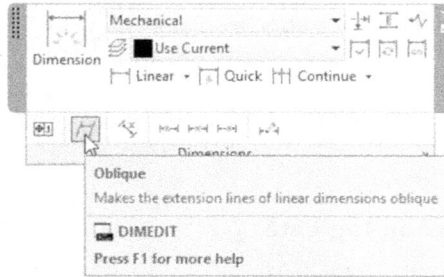

You will see the following prompts:

```
Select objects:
Select objects:
Enter obliquing angle (press ENTER for none):
```

Select the desired dimension blocks, then press [Enter]; finally, input the oblique angle, which can be positive or negative. Refer to the following example:

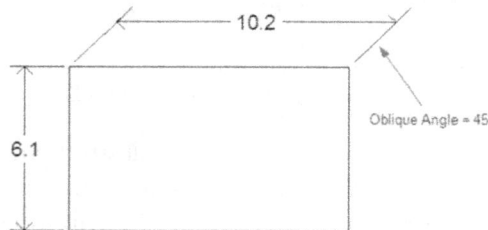

### 16.11.5   Dimension Text Angle

This command will change the angle of the dimension text to any angle you want. To issue this command, go to the **Annotate** tab, locate the **Dimensions** panel, then select the **Text Angle** button:

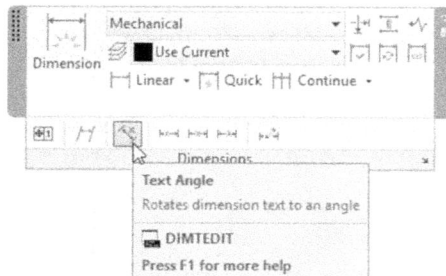

You will see the following prompts:

```
Select dimension:
Specify angle for dimension text:
```

Select the desired dimension blocks and input the text angle, which can be positive or negative. Refer to the following example:

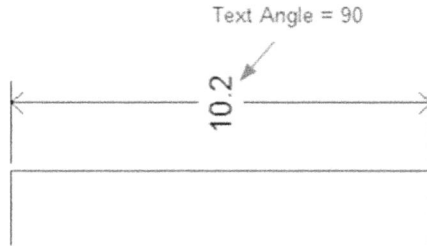

Text Angle = 90

### 16.11.6 Dimension Justify

This command will change the horizontal position of the dimension text. To issue this command, go to the **Annotate** tab, locate the **Dimensions** panel, then click one of the following buttons:

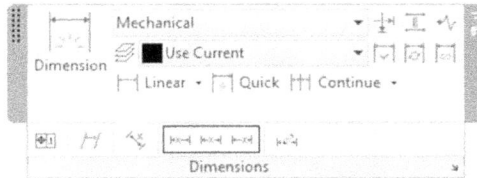

Each one of the functions will move the dimension text either to the left, center, or to the right. Refer to the following example (it moved it to the left):

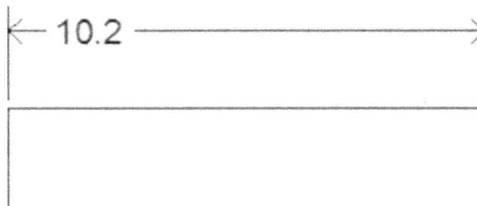

### 16.11.7 Dimension Override

This command will override a dimension system variable (the user should memorize the system variable) or simply remove the override. To issue this command, go to the **Annotate** tab, locate the **Dimensions** panel, then select **Override** button:

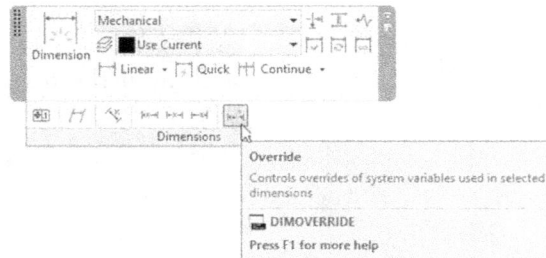

You will see the following prompts:

```
Enter dimension variable name to override or
[Clear overrides]:
```

TThis command is very useful if you want to remove (clear) all the overriding steps you make on a dimension block using the right-click menu. To answer to the above prompt, either type the name of the dimension system variable, or type C to clear the override. Refer to the following illustration:

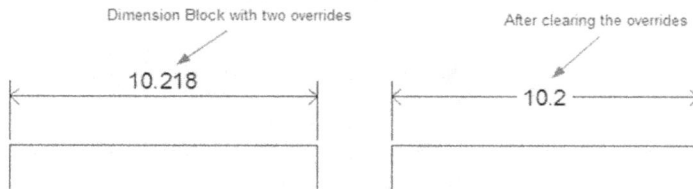

Dimension Block with two overrides

10.218

After clearing the overrides

10.2

## PRACTICE 16-2    MORE DIMENSION FUNCTIONS

1. Start AutoCAD 2025

2. Open **Practice 16-2.dwg** file

3. Using Adjust Space, adjust the space between the six continuous dimensions to be all at the same alignment with 2 ¾ dimension

4. Using Adjust Space, adjust the space between the six continuous dimensions and the total single dimension to 1.5

5. Make sure that Dimension layer is the current layer; if not, make it current

6. Using Dimension Break, break the horizontal 1 ½ using the vertical 1 ½

7. Rotate the text of the vertical 1 ½ to be 45

8. What is the upper horizontal dimension read? _____

**9.** Using Dimension Override, clear the override, what does it read now? _____

**10.** Make the same dimension left justified

**11.** You should receive the following:

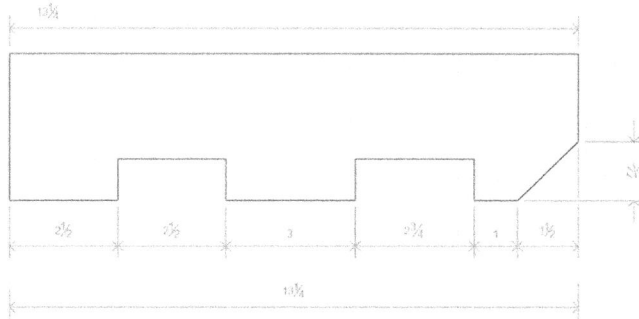

**12.** Save and close the file

## 16.12   ASSOCIATIVE CENTERLINES AND CENTER MARKS

Associative Center Marks for circles, arcs, and polygonal arcs, and Centerlines for lines and polylines are powerful tools offered by AutoCAD for creating and editing center marks and centerlines.

Associative means, if the geometry change, both Centerline and Center Mark will react accordingly.

### 16.12.1   Center Mark

To add a new Center Mark, complete the following steps:

■  Go to **Annotate** tab, locate **Centerlines** panel, click **Center Mark** button:

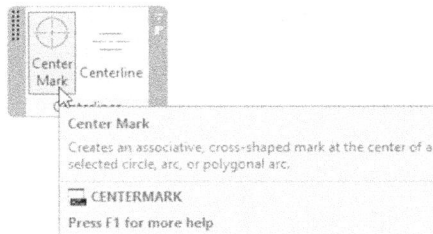

■  The following prompt will appear:

```
Select circle or arc to add centermark:
```

■  Click on the desired arc, circle, or polygonal arc, a center mark will be added.

- You will receive the following:

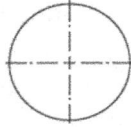

- You can continue adding for other circles, arcs

### 16.12.2   Centerline

To add a new Centerline, complete the following steps:

- Go to **Annotate** tab, locate **Centerlines** panel, click **Centerline** button:

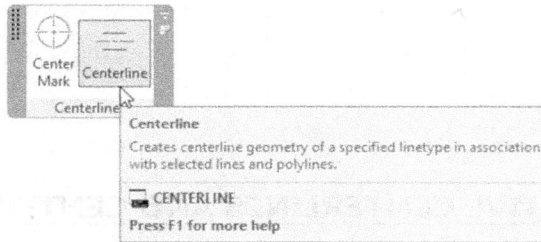

- The following prompts will appear:

```
Select first line:
Select second line:
```

- Select the first line/polyline, then select the second line/polyline, the command will end. You will receive the following

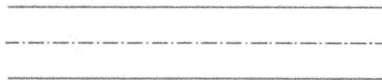

### 16.12.3   Centerline and Center Mark Properties

You can control everything related to Centerline and Center Mark using Quick Properties. All properties are self-explanatory:

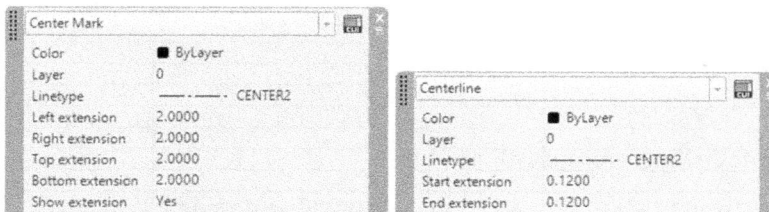

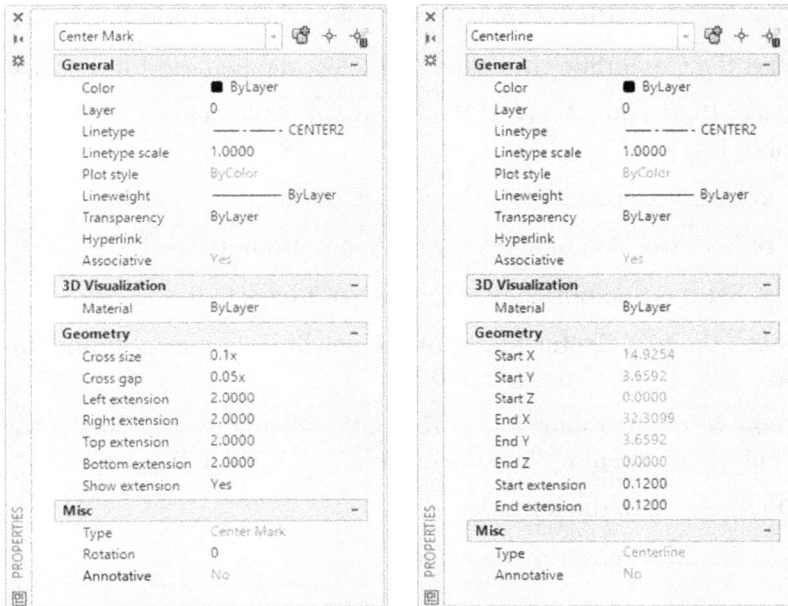

Also, you can use the grips to modify both Centerline and Center Mark, as shown below:

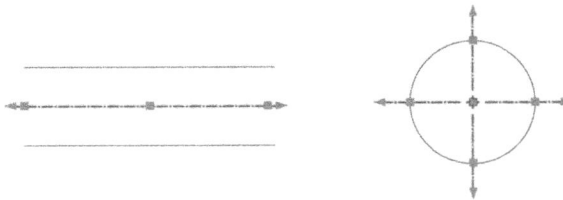

# PRACTICE 16-3    CENTERLINES AND CENTER MARKS

1. Start AutoCAD 2025

2. Open **Practice 16-3.dwg** file

3. Set the layer Center to be the current layer

4. Start Center Mark command, and click the big green circle

5. Click the two arcs of polyline

6. End the command, by pressing [Enter]

**7.** Start the Centerline command and select the two short lines at the right

**8.** Start the Centerline command and select the two short lines at the left

**9.** Select the Center Mark of the big green circle, and show Properties, do the following:

    **a.** Under Geometry, set Cross Size = 0.2x

    **b.** Set extension fields (left/right/top/bottom) to be 0.5

    **c.** Set the Rotation to be 45, then get it back to 0

**10.** Select the two Center Marks of the polyline, and using Properties, set Cross Size = 0.2x, and Cross Gap = 0.1x

**11.** Zoom to the two short line at the right. Select the short line at the left of the Centerline, and move it to the outside by 0.5. What happen to the Centerline?

**12.** Do the same to the left side

**13.** You should receive the following:

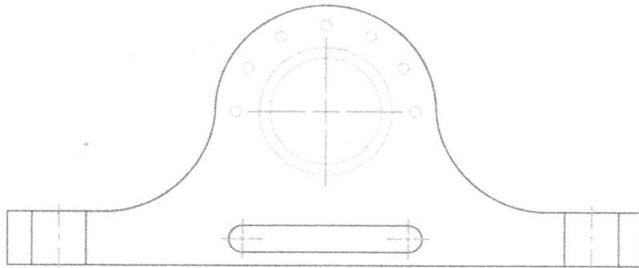

**14.** Save and close the file

## 16.13 HOW TO CREATE A MULTILEADER STYLE

Multileader is a replacement for the normal leaders that once existed in AutoCAD. Leaders used to follow the current dimension style and they were always single. Multileader has its own style and the single leader can point to a different location in the drawing.

Multileader style will set the characteristics of the Multileader block. To start this command, go to the **Annotate** tab, locate the **Leaders** panel, then select the **Multileader Style** button:

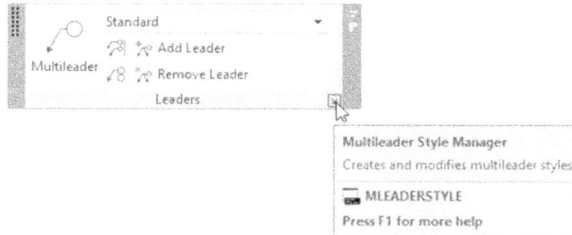

You will receive the following dialog box:

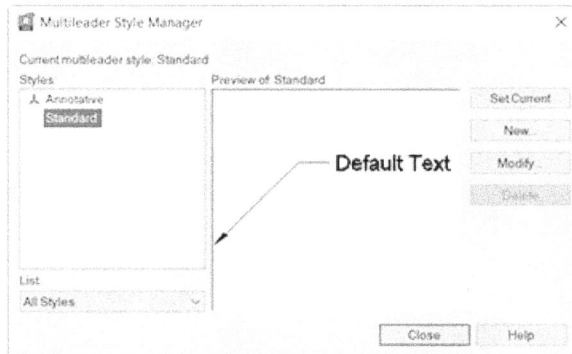

As you can see there are two predefined styles, one called Standard (default) and the other called Annotative. Click the **New** button to create a new multileader style, as you will see in the following dialog box:

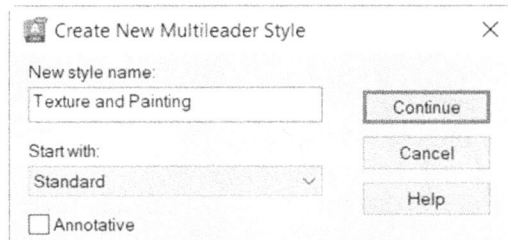

Input the name of the new style, then click the **Continue** button. There are three tabs and each one will control part of the multileader block. These are:

- Leader Format
- Leader Structure
- Content

### 16.13.1   Leader Format Tab

- Below is the image of the Leader Format tab:

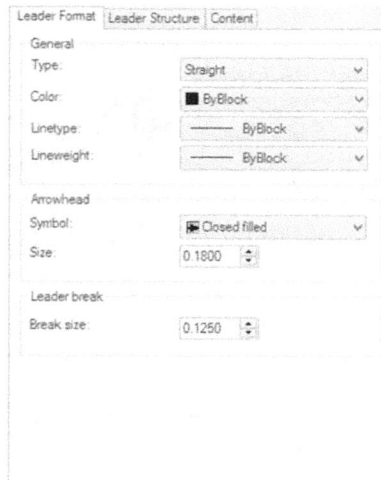

Change all or any of the following:
- Edit the type of the leader; choose one of the three choices: Straight, Spline, or None. The following is an example of both straight and spline options:

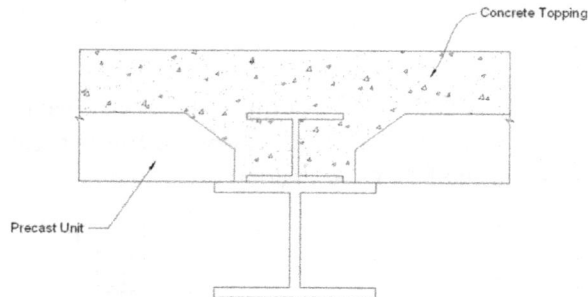

- Edit the Color, Linetype, and Lineweight
- Select the arrowhead shape and its size
- Set the distance of the dimension break from any two blocks that will intersect

## 16.13.2    Leader Structure Tab

Below is the image of the Leader Structure tab:

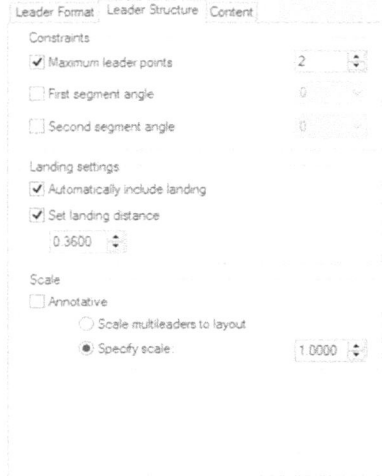

You can change all or any of the following:

- Specify if you want to change the **Maximum leader points** or not. Then input the desired value. By default this value is 2, meaning the first point pointing to the geometry and the second point in the end of the multileader.

Max leader points = 2          Max leader points = 4

- Specify whether you want to change the **First segment angle** and the **Second segment angle** or not. If yes, what are the angle values?
- Specify whether you want AutoCAD to **Automatically include landing** or not. If yes, what is the **landing length**
- Specify whether the multileader will be Annotative or not (we will discuss annotative in the coming chapters)

### 16.13.3 Content Tab

Below is the image of the Content tab:

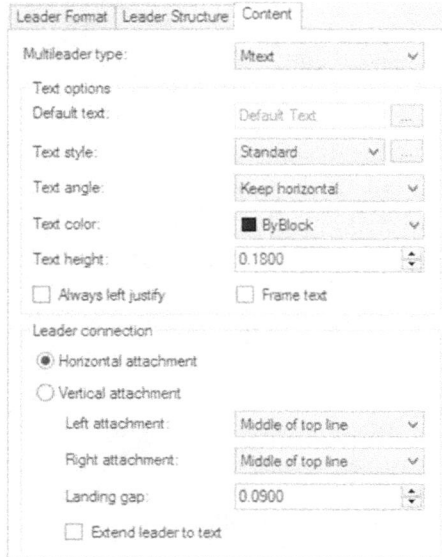

In AutoCAD, there are two Multileader types:

- Mtext
- Block (Either predefined, or user-defined)

The following image displays the two types:

If you select the **Mtext** option, you will be able to change all or any of the following:

- If there is **Default text**
- Select **Text style**, **Text angle**, **Text Color**, and **Text height** (if Text Style's height = 0)
- Select whether the text is **Always left Justify** or not, and with **Frame** or not

- Choose whether the leader connection is horizontal or vertical. If vertical, then edit the position of the text relative to the landing for both left and right leader lines, then control the gap distance between the end of landing and the text

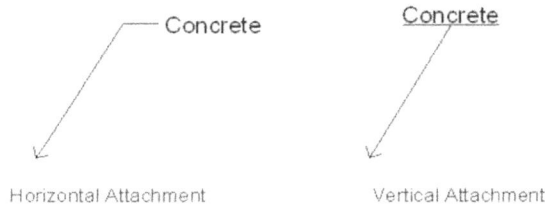

Concrete

Concrete

Horizontal Attachment

Vertical Attachment

If you select the **Block** option, you can change all or any of the following:

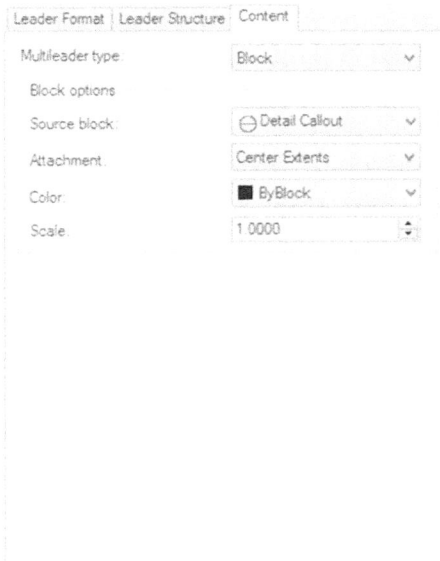

- Specify the **Source block**:

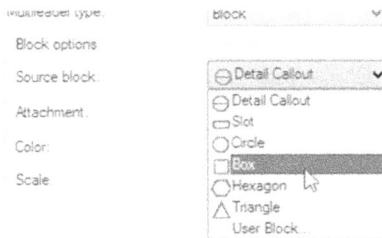

- Specify the **Attachment** position, **Color** of attachment, and finally, the **Scale** of the attachment

## 16.14  INSERTING A MULTILEADER DIMENSION

This group of commands will add a single multileader, then add a leader to an existing multileader, remove a leader from an existing multileader, and align and group an existing multileader. You will always start with the **Multileader** command, which will insert a single leader. To start this command, go to the **Annotate** tab, locate the **Leaders** panel, then select the **Multileader** button:

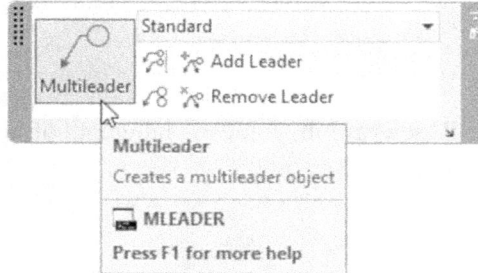

You will see the following prompts:

```
Specify leader arrowhead location or [leader Landing
first/Content first/Options] <Options>:
Specify leader landing location:
```

First, specify the leader arrowhead location, then specify the leader landing location, and then type the text you want to appear beside the leader. To add a leader to an existing multileader, go to the **Annotate** tab, locate the **Leaders** panel, then select the **Add Leader** button to add more leaders.

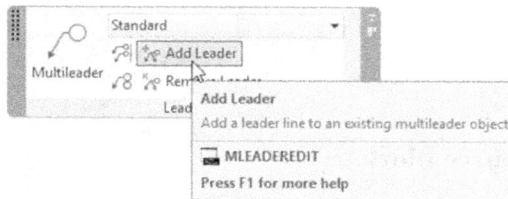

You will see the following prompts:

```
Select a multileader:
1 found
Specify leader arrowhead location:
Specify leader arrowhead location:
```

To remove a leader from an existing multileader, go to the **Annotate** tab, locate the **Leaders** panel, then select the **Remove Leader** button to remove leaders:

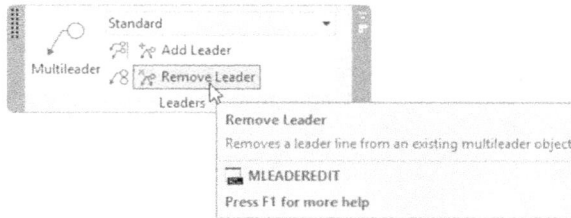

You will see the following prompts:

```
Select a multileader:
1 found
Specify leaders to remove:
Specify leaders to remove:
```

To align a group of multileaders, go to the **Annotate** tab, locate the **Leaders** panel, then select the **Align** button:

- You will see the following prompts:

```
Select multileaders: 1 found
Select multileaders: 1 found, 2 total
Select multileaders:
Current mode: Use current spacing
Select multileader to align to or [Options]:
Specify direction:
```

To collect a group of similar multileaders to be a single leader, go to the **Annotate** tab, locate the **Leaders** panel, then select the **Collect** button. This command works only with leaders containing blocks:

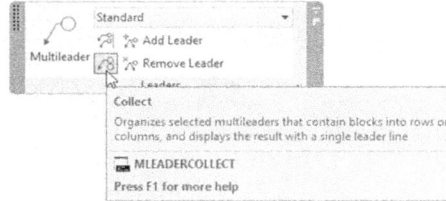

You will see the following prompts:

```
Select multileaders:
Select multileaders:
Specify collected multileader location or
[Vertical/Horizontal/Wrap] <Horizontal>:
```

**NOTE**

- *You can change the wrap of the multileader text while editing using the two arrows pointing to the right and left.*
- *Multileader command has an option to select an existing MTEXT object and add a leader for it using the current Multileader style.*

# PRACTICE 16-4  CREATING MULTILEADER STYLE AND INSERTING MULTILEADERS

**1.** Start AutoCAD 2025

**2.** Open **Practice 16-4.dwg** file

**3.** Create a new multileader style based on Standard:

    **a.** Name = Texture and Painting

    **b.** Arrowhead symbol = Dot small

    **c.** Arrowhead size = 0.35

    **d.** First segment angle = 0

    **e.** Automatically include landing = Off

    **f.** Multileader type = Block

    **g.** Source block = Circle

**4.** Create a new multileader style based on Standard:

    **a.** Name = Material

    **b.** Leader format = Spline

    **c.** Arrowhead symbol = Right angle

    **d.** Arrowhead size = 0.25

**5.** Make layer Dimension current

**6.** Using both styles, insert the following multileaders:

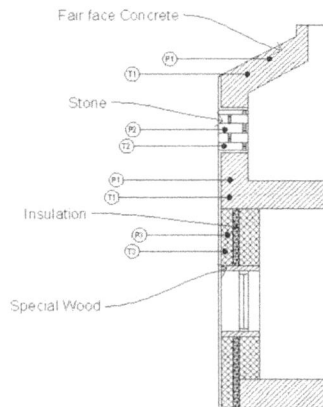

**7.** Using Add leader, Align, and Collect, try to get the following image:

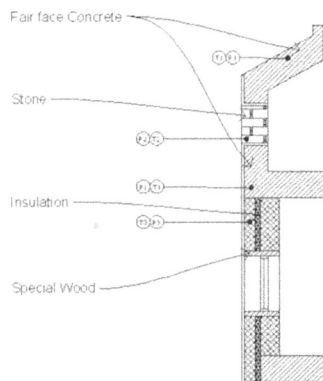

**8.** Start Multileader command, right-click, choose the option select Mtext, select the Mtext "Special Window" and point the arrow to the window

**9.** Save and close the file

# NOTES

## CHAPTER REVIEW

**1.** Symmetrical and Deviation are two types of _____ in dimension style

**2.** There are two types of multileader blocks, multiline text or blocks

    **a.** True

    **b.** False

**3.** When creating a new dimension style, which of the following is NOT correct:

    **a.** You can create a dimension style affecting all types of dimensions

    **b.** You have to select the existing dimension style to start with

    **c.** You cannot create a dimension style affecting only one type of dimension

    **d.** You can create a sub style

**4.** You can show _____ units, and _____ units in a dimension block

**5.** Using the Multileader style there should be always a landing in my block:

    **a.** True

    **b.** False

**6.** Collect and Align are _____ commands

## CHAPTER REVIEW ANSWERS

**1.** Tolerance

**3.** c

**5.** b

## CHAPTER 17

# PLOT STYLE, ANNOTATIVE, AND EXPORTING

## In This Chapter

- How to create and use the two types of Plot Styles
- What is Annotative in AutoCAD?
- Exporting to a DWF and PDF file

## 17.1   PLOT STYLE TABLES – FIRST LOOK

Plot styles will be used to convert colors used in the drawing to printed colors. The default setting is to keep matching color at the printer. Before AutoCAD 2000, there was only one type of conversion method, dependent on the colors used in the drawing. After AutoCAD 2000, a new concept called Plot Style was introduced. There are two types of plot styles; these are:

- Color-dependent Plot Style Table
- Named Plot Style Table

## 17.2   COLOR-DEPENDENT PLOT STYLE TABLE

This plot style table is a simulation for the only method that existed before AutoCAD 2000. The essence of this method is simple: for each used color in your drawing, specify the color to be used in the printer. AutoCAD will provide the user with the ability to set the lineweight, linetype, etc. for each color. The problem of this method is its limitation, because there are only 255 colors to be used.

Each time you create a Color-dependent Plot Style Table, AutoCAD will create a file with extension **\*.ctb**. You can create Plot Style tables from outside AutoCAD using the Control Panel of Windows, or from inside AutoCAD using the menu bar.

From outside AutoCAD, start the Windows Control Panel, double-click the **Autodesk Plot Style Manager** icon or show menu bar, then select the **Tools/ Wizards/Add Plot Style Table**, and you will see the following dialog box:

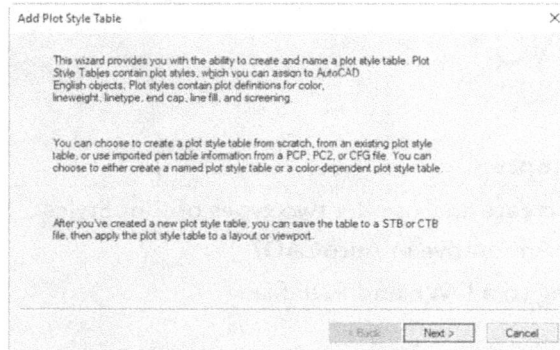

The first screen is an introduction to the whole concept of plot styles; read it to understand the next steps. When done, click the **Next** button, and you will see the following dialog box:

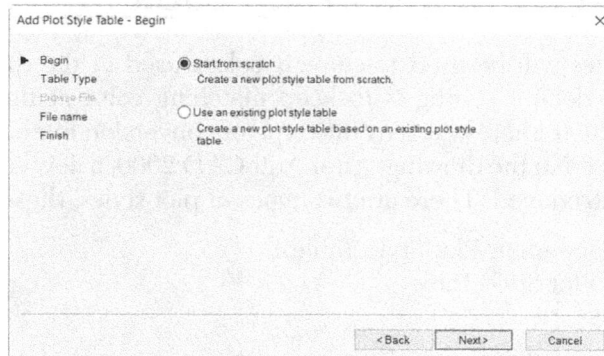

AutoCAD will list the four possible choices, which are:

- Start from scratch
- Use an existing plot style

Select **Start from scratch**, then click the **Next** button, and you will see the following dialog box:

Select the **Color-Dependent Plot Style Table** and click **Next** button, and you will see the following dialog box:

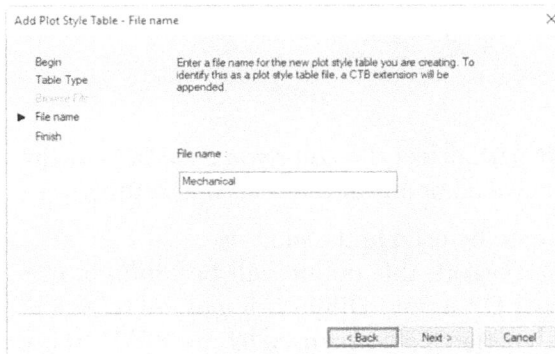

Input the name of the new plot style and click **Next** button and you will see the following dialog box:

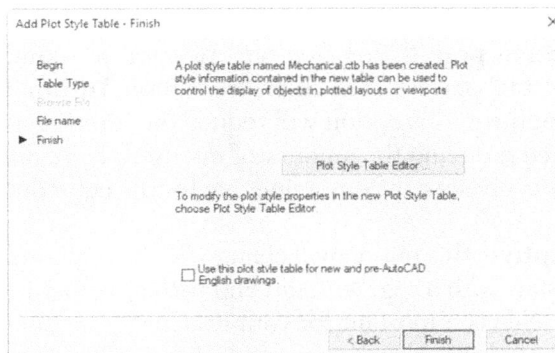

Select the **Plot Style Table Editor** button and the following dialog box will be displayed:

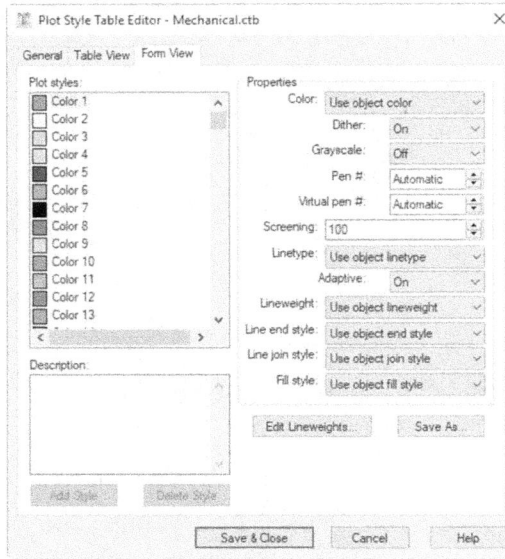

From the left part, select the color you used in your drawing file, and then at the right part, change all or any of the following settings:

- Change **Color** to be used in the plotter
- Switch **Dither** on/off; this option will be dimmed if your printer or plotter doesn't support Dithering. Dither is a method to give the impression of using more colors than the 255 colors used by AutoCAD. It is preferable to leave this option off. It should be on if you want **Screening** to work
- Change **Grayscale**; this method is good for laser printers
- Change **Pen #**, this option is valid for the old types of plotter – pen plotters which are not used these days
- Change **Virtual pen #**, for non-pen plotters to simulate pen plotters by assigning a virtual pen for each color; leave it on **Automatic**
- Change **Screening**; this option will reduce the intensity of the shading and fill hatches, hence reducing the amount of ink used. Depends on **Dithering**
- Change **Linetype**; set a different linetype for the color or leave it to the object's linetype
- Change **Adaptive**; this option will change the linetype scale of all objects using the color to start with a segment and end with a segment Turn this option off if the linetype scale is important for your drawing
- Change **Lineweight**; this option will change the lineweight for the color selected

- Change **Line end style**; this option will allow the user to select the end style for lines, pick one of the following: Butt, Square, Round, and Diamond
- Change **Line join style**; to select the line join shape, the available options are: Miter, Bevel, Round, and Diamond
- Change **Fill style**; this option will set the fill style for filled areas in the drawing (good for trial printing)

Click **Save & Close**. Then click **Finish.**

Your last step should be linking your plot style with a layout. Complete the following steps:

- Select the desired layout, then start the **Page Setup Manager**
- Select the name of the current Page Setup and click **Modify**
- At the upper right part of the dialog box, and under **Plot style table (pen assignment)**, select the desired plot style table:

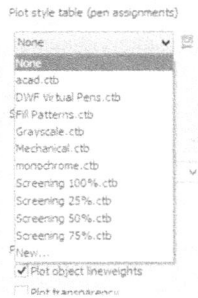

- Click Display plot styles checkbox on

You can assign one **ctb** file for each layout. In order to see the linetype and lineweight of the objects, you have to switch the **Show/Hide Lineweight** button at status bar on, as shown below:

## 17.3    NAMED PLOT STYLE TABLE

This method does not depend on colors. The created plot style tables will be linked later on with layers, hence, you may have two layers holding the same color, yet they will print with different colors, linetypes, and lineweights.

The Named Plot Style Table has a file extension of *.stb. The creation procedure of Named Plot Style is identical to Color-dependent Plot Style, except the last step, which is configuring the **Plot Style Table Editor**. You can create it from outside AutoCAD using the Control Panel, then double-clicking the **Autodesk Plot Style Manager** icon; or you can show the menu bar, then select **Tools/Wizards/Add Plot Style Table**. You will see the same screen you saw while creating a Color-dependent plot style table.

Do the same steps until you reach the **Plot Style Table Editor** button; select it and you will see the following dialog box, click the **Add Style** button:

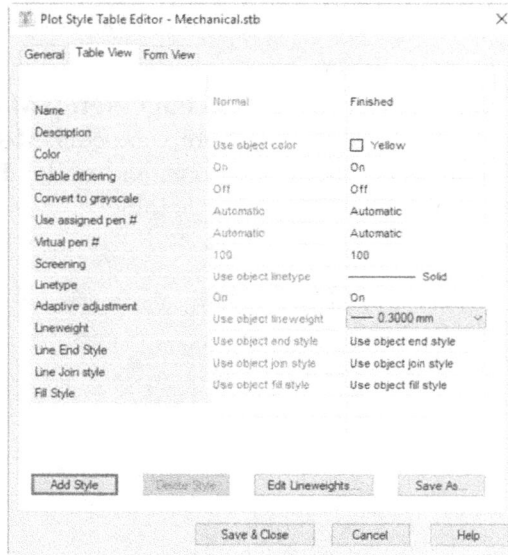

As you can see, you have to change all or any of the following:

- Input the Name of the style and a brief Description
- Change the Color to be used in the plotter
- We explained the rest of features in the Color-dependent Plot Style Table

You can add as many styles as you wish in the same Named Plot Style. Click **Save & Close**. Then click **Finish**.

Linking a Named Plot Style Table with any drawing is a bit more complicated than linking color-dependent plot style tables; complete the following steps:

- The first step is a precautionary step. You may want to print a drawing, and then discover it takes only ctb files. To solve this problem, you have to convert one of the **ctb** files to **stb** file. At the command window type **convertctb**, and a dialog box listing all ctb files will appear. Select one of them, keeping the same name, or giving a new name, then click **OK**. You will see the following dialog box:

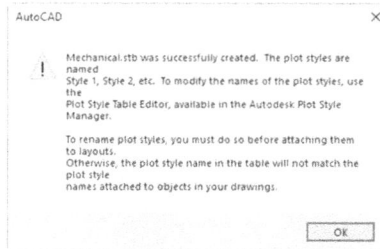

- Convert the drawing from Color-dependent Plot Style to Named Plot Style. At the command window type **convertpstyles** and you will see the following warning message:

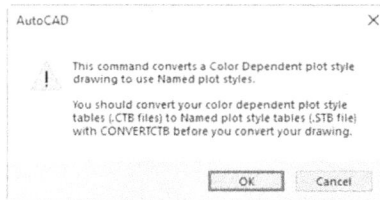

- Click **OK** and you will see the following dialog box:

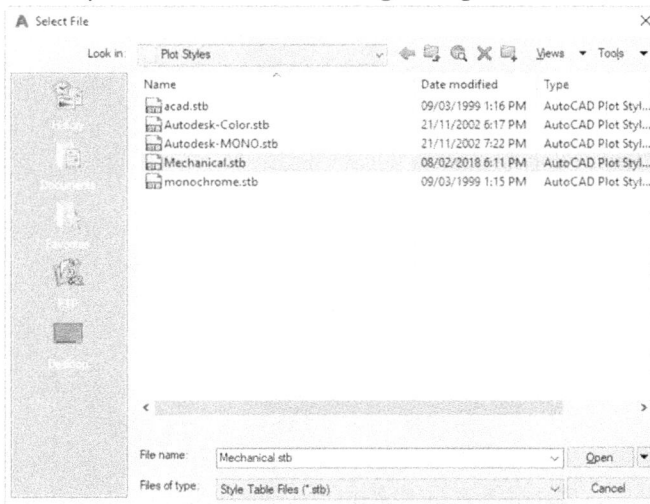

- Select the newly created Named Plot Style Table, then click **Open**, and you will see the following prompt:

```
Drawing converted from Color Dependent mode to Named
plot style mode.
```

- The second step: select the desired layout, start the **Page Setup Manager** at the upper right part of the dialog box, and under **Plot style table (pen assignment)** select the name of the newly created named plot style table. Click on the **Display plot styles** checkbox and end the Page Setup Manager command:

- Select the **Layer Properties Manager**, and for the desired layer(s) click the name of the plot style under the **VP Plot Style** column:

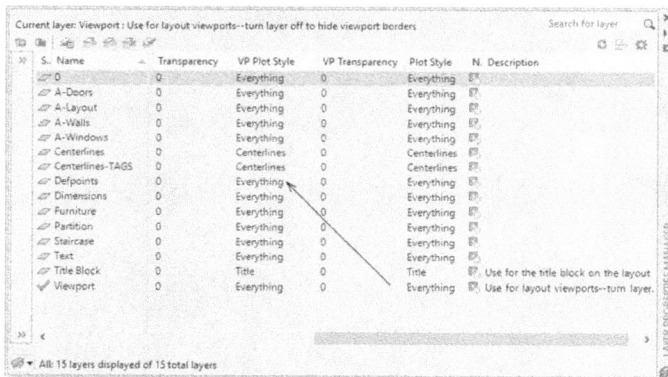

- You will see the following dialog box:

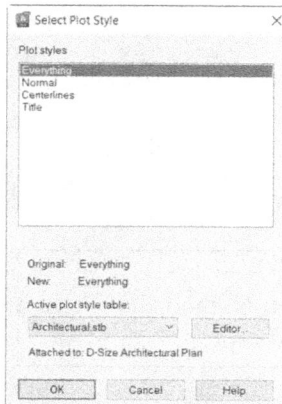

- Select the desired plot style. When done, click **OK**
- Repeat the same for other layers
- You may need to type **regenall** (means regenerate all viewports) at the command window in order to see the effect of what you did.

<u>NOTE</u> *If you create a new drawing using acad.dwt this means your drawing will accept only Color-dependent plot style tables. Use acad-Named Plot Styles. dwt to create a new drawing, which will accept only Named plot style tables.*

## PRACTICE 17-1   COLOR-DEPENDENT PLOT STYLE TABLE

**1.** Start AutoCAD 2025

**2.** Open **Practice 17-1.dwg** file

**3.** Create a new Color-dependent plot style table and call it **Architectural** using the following table:

| Color | 1 | 2 | 3 | 4 | 5 | 6 | 7 | 27 |
|---|---|---|---|---|---|---|---|---|
| Plot color | Black | Black | Use object color | Black | Black | Black | Black | Black |
| Lineweight | 0.3 | 0.3 | 0.3 | 0.3 | 0.7 | 0.3 | 0.3 | 0.3 |

**4.** Switch to D-Size Architectural Plan layout

**5.** Link this layout to Architectural plot style

**6.** Check that the Show/Hide Lineweight button at the status bar is turned on

**7.** Check the doors compared to the other objects; you will find them thicker, because the color (5) used for both the title block and doors uses a 0.7 lineweight

**8.** Save and close the file

## PRACTICE 17-2   NAMED PLOT STYLE TABLE

**1.** Start AutoCAD 2025

**2.** Open **Practice 17-2.dwg** file

**3.** Create a new Named plot style table and name it **Architectural** using the following table:

| Style name | Color | Lineweight |
|------------|-------|------------|
| Everything | Black | 0.3 |
| Centerlines | Green | 0.5 |
| Title | Black | 0.7 |

**4.** Link the newly created style table with the D-Size Architectural Plan layout

**5.** Using the Layer Properties Manager make the following:

**a.** Layer = Centerlines and Centerlines-TAGS linked to Centerlines

**b.** Title Block linked to Title

**c.** The rest of the layers linked to Everything

**6.** Zoom in to compare the Centerline lineweight to the other objects

**7.** Save and close

## 17.4 WHAT IS AN ANNOTATIVE FEATURE?

Since we will always print from layout, we need to use viewports. Moreover, since we will use viewports, we have to set the viewport scale for each viewport. The viewport scale will affect all objects in the Model space. Therefore, if you hatch, type text, insert dimensions, insert multileader, or insert a block containing text in Model space, all of these objects will be scaled. If they were scaled down, then the text and dimension would be unreadable, and the hatch would look like solid hatching.

What we need is a feature that will scale everything except the annotation objects (namely hatch, text, and dimension); this feature is **Annotative**.

The Annotative feature can be found in different places:

■ You will find it in Text style; under Size, you will see:

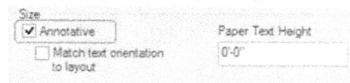

■ You will find it in dimension style; go to the Fit tab and you will see:

- You will find it in multileader style; go to Leader Structure tab and you will see:

- You will find it in the Hatch context tab; locate the Options panel and you will see:

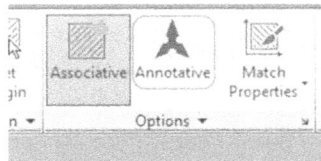

- You will find it in the Block creation command; under Behavior you will see:

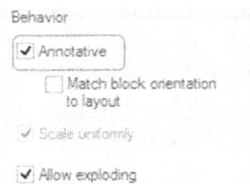

How you will know that you are dealing with a style supporting annotative feature? You will see at its name a special symbol; refer to the illustration below:

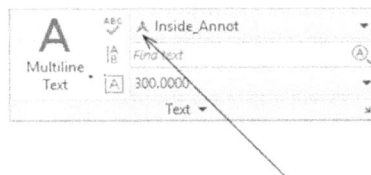

How you will know if an object was inserted using a style or a command supporting an annotative feature? Simply hover over it, and if you see the following picture, you know it was inserted with an annotative feature on:

In order to work with annotative objects, follow the simple steps listed below:

- Create your drawing in Model space without any annotation objects (namely hatch, text, dimensions, multileader, and blocks with text)

- Select the desired layout, add viewports, and then scale them. This scale will be for both the annotation objects and the viewport
- Double-click inside the viewport to make it active
- Insert all the annotation objects you need
- Once you insert annotation objects in the viewport, you can see them in this viewport and any other viewport holding the same scale value
- Changing the scale of the viewport will result in losing the annotation object
- To show the annotation object in more than one viewport holding different scale values, right-click on the **Annotation Visibility** button (a button at the right portion of the status bar) and you will see the following menu:

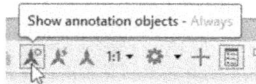

- Select either to Show Annotation Objects for Current Scale Only, or Show Annotation Objects for All Scales
- Control how to add scale values to the viewport: is it Automatic or manual? Right-click on the **Add Scale** button (a button at the right portion of the status bar) and you will see the following menu:

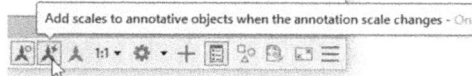

- If you used the zooming command inside the viewport, this will ruin the current viewport scale. To bring it back click the Synchronize button (a button at the right portion of the status bar) and you will see the following message (use the lock in the status bar to lock the viewport scale to avoid this problem):

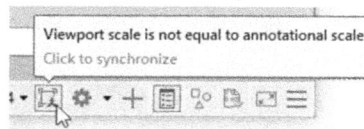

- Clicking this message will restore the viewport scale
- To make an annotation object appear in viewports holding different scales, select it, right-click, select the **Annotative Object Scale** option, and you will see the following sub-menu:

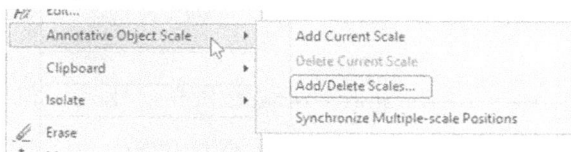

- Choose the **Add/Delete Scales** option and the following dialog box will appear:

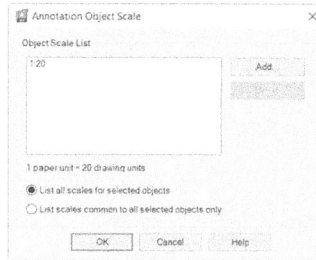

- Select the **Add** button and the following dialog box will appear:

- Select the desired scale value and click **OK** twice

## PRACTICE 17-3   ANNOTATIVE FEATURE

**1.** Start AutoCAD 2025

**2.** Open **Practice 17-3.dwg** file

**3.** Make layer Dimensions current

**4.** Switch to Details layout

**5.** There are three viewports, the first one at the left is scaled to be 1:20, and the other two at the right are scaled to be 1:10

**6.** Go to the Annotate tab, locate the Dimensions panel, and check the available dimension styles. You will see only one with a distinctive symbol at its left, which is mechanical. This dimension style is annotative

**7.** Double-click inside the big viewport

**8.** Start the Radius command and select the big magenta circle, then add the dimension block. Check that it did not appear in the upper right viewport

**9.** Add another Radius block to the small magenta circle

**10.** Add two linear dimensions for the total width and the total height

**11.** While you are still inside the same viewport, make layer Text current

**12.** Make text style "Annotative" current

**13.** Using Multiline text add beneath the shape the word "Bearing"

**14.** Make layer Dimensions current again

**15.** Click inside the upper right viewport to make it current, and add a Radius dimension to one of the small circles

**16.** Make layer Hatch current

**17.** Click inside the lower right viewport

**18.** Start the Hatch command, make sure that Annotative is on, set the scale = 10, and hatch the area between the two dashed lines, then finish Hatch command

**19.** Using the Add/Delete scales add scale 1:20 so the two hatches will appear in the big viewport

**20.** You should receive the following image:

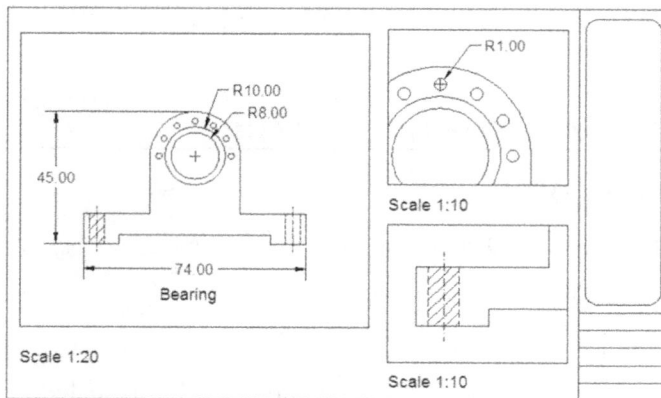

**21.** Save and close the file

## 17.5    EXPORTING DWG TO PDF: AN INTRODUCTION

You need to export your DWG to PDF for the following reasons:

- DWG files are vector files; hence, if you send them to anybody, they will be able to change them

- DWG files are always bulky (a file may reach to more than 50–100 MB depending on whether it is 2D or 3D), and very difficult to be sent by e-mail
- DWG files need AutoCAD to open it, and not everyone has AutoCAD installed on their machines

AutoCAD offer you the solution for this problem through exporting to PDF file. PDF files cannot be modified, are small in size, and can be viewed using Adobe Acrobat Reader (available from the Adobe web site)

## 17.6   EXPORTING TO A PDF

This command will export your current DWG file to a PDF. To start this command, go to the **Output** tab, locate the **Export to DWF/PDF** panel, then select **Export** button and click PDF:

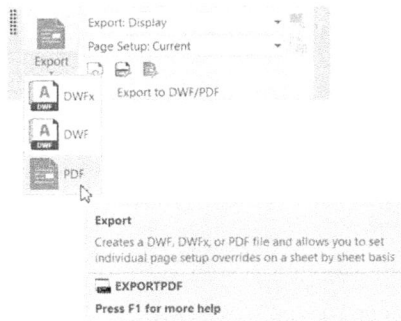

You will see the following dialog box:

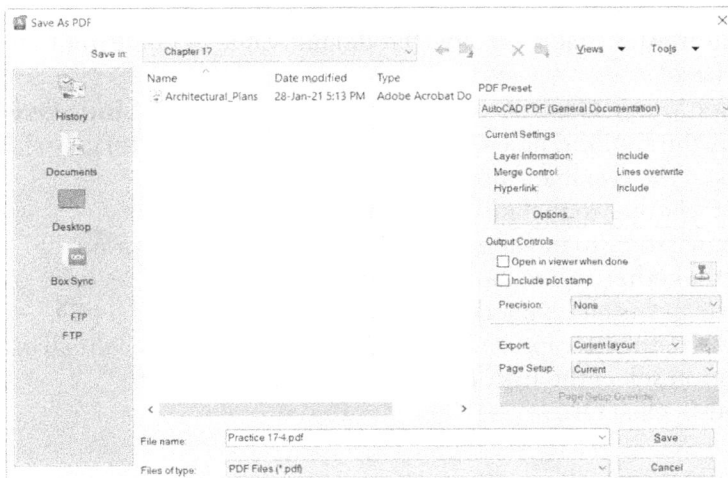

At the top right part of the dialog box, and under PDF Presets, AutoCAD lists the current settings, as in the image below:

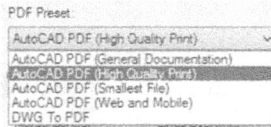

AutoCAD offers to export to PDF files with different quality which will affect the output file size.

You can edit the current settings by clicking the **Options** button and the following dialog box will appear:

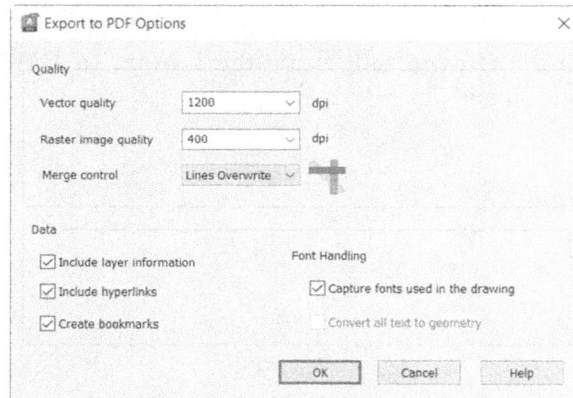

Change all or any of the following settings:

- Set the Vector quality (the default value depends on the quality preset you chose)
- Set the Raster image quality (the default value depends on the quality preset you chose)
- For Set Merge Control, you have two choices; either **Lines Overwrite** which means the top line hides the bottom line, or Lines **Merge**, which means the colors of the two lines mix together
- Select whether to include Layers in the PDF file or not
- Select whether to include Hyperlinks in the PDF file or not
- Select whether to create bookmarks or not (bookmarks here means creating links to sheets and named views in the PDF file panel)
- Select whether to capture fonts in the drawing, or to convert all text to geometry (once you select the first, the second will be turned off)
- To finish, click OK.

## 17.7   USING THE BATCH PLOT COMMAND

This command will allow the user to print to normal printer or produce PDF files containing multiple layouts from the current drawing and from other drawings. To issue this command go to the **Output** tab, locate the **Plot** panel, then select the **Batch Plot** button:

You will see the following dialog box:

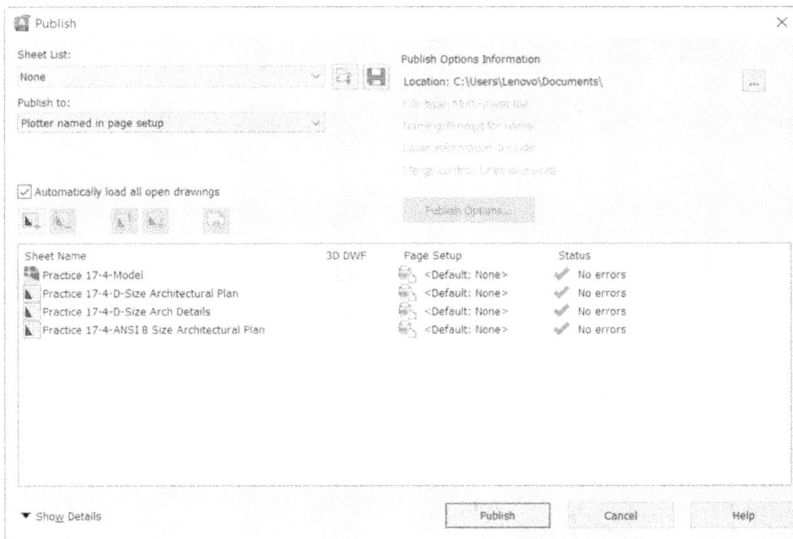

You will see a table contains Model Space and layouts, do all or any of the following:

- Indicate **Publish to**: it is either printer/plotter defined in the layout, or DWF, DWFx, PDF
- Indicate whether to load all open drawings automatically or not

- Using the five buttons above the table, you can add a sheet, remove a sheet, and move any sheet up or down to specify its order relative to the document sheets. Finally, you can preview the sheet:

- You can rename the sheets by clicking on any sheet name; you will see the name becomes editable
- Click the **Show Details** button at the lower left corner of the dialog box then set the following: Specify the number of copies; Select whether or not to include a Plot Stamp; Indicate whether or not you want to publish in the background; Indicate whether to open in viewer when done or not. Set the unit precision for the file

To finish, click the **Publish** button. A final message will come up as a bubble similar to the following:

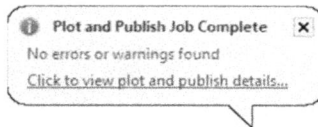

## PRACTICE 17-4    CREATING AND VIEWING A PDF FILE

1. Start AutoCAD 2025

2. Open **Practice 17-4.dwg** file

3. Produce a multisheet PDF file including all layouts except the Model space sheet. Make sure to include layers. Save the file under the name "Architectural Plans.pdf." Open it using Acrobat Reader

4. Save and close the file

## NOTES

## CHAPTER REVIEW

1. Color-dependent plot style is simulation to what we used to have before Auto-CAD 2000:

   **a.** True

   **b.** False

2. The best practice is to insert dimension, text, and hatch in _____, using _____ feature

3. _____ is better than Color-dependent plot style

4. There is no difference between DWF and DWFx:

   **a.** True

   **b.** False

5. In order to insert an annotative object, you have to be inside the viewport:

   **a.** True

   **b.** False

## CHAPTER REVIEW ANSWERS

**1.** a

**3.** Named plot style table

**5.** a

## CHAPTER 18

# HOW TO CREATE A TEMPLATE FILE AND INTERFACE CUSTOMIZATION

**In This Chapter**

- The contents of a template file and how to create one
- How to use CUI to customize the interface of AutoCAD

## 18.1  WHAT IS A TEMPLATE FILE AND HOW CAN WE CREATE IT?

Any company using AutoCAD should have a standard to follow in order to shorten production time. Templates can answer this issue in a simple way. Standardization includes using the same layer naming, colors, linetype, and lineweight. It also includes standard text and tables, standard dimension and leaders, and standard layouts. Needless to say, companies will use always the same shapes for the blocks.

So, to create a good template, you should include the following:

- Drawing units
- Drawing limits
- Grid and Snap settings
- Layers
- Linetypes
- Text Styles
- Table Styles
- Dimension Styles

- Multileader Styles
- Layouts (including Border blocks and Viewports)
- Page Setups
- Plot Style tables

Follow these steps to create a template file:

- Create a new file using template file **acad.dwt** or **acadiso.dwt**
- Prepare the above-mentioned settings
- Create inside this file all the above-mentioned settings
- From the Application menu, Save As/AutoCAD Drawing Template

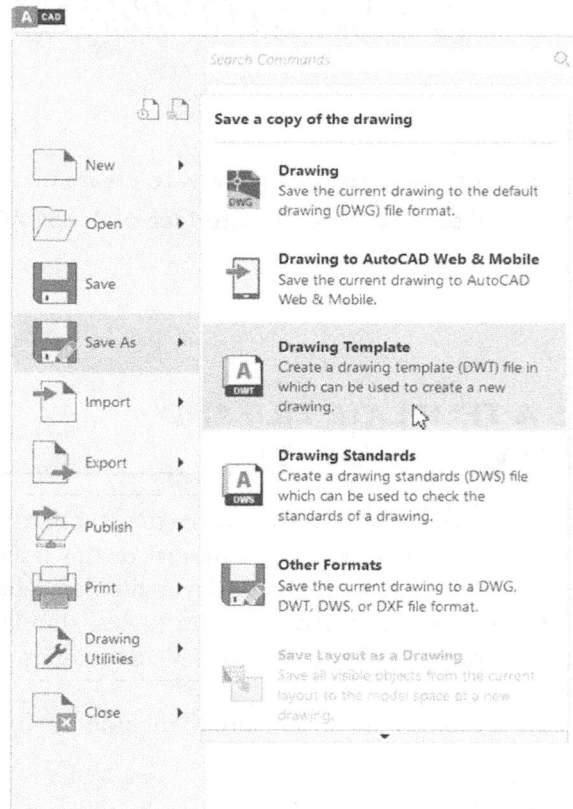

You will see the following dialog box:

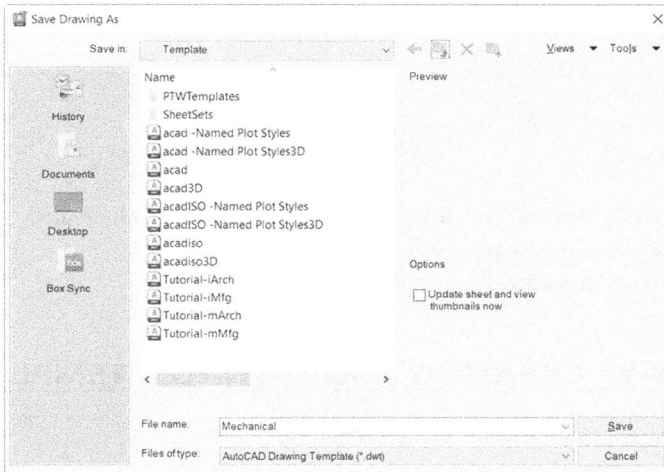

Input the name of the template file. By default, it will direct you to the same folder AutoCAD saved its default template files in. You can save your template there, or you can create your own folder, which we highly recommend. However, using this method will mean that you have to specify the new folder for AutoCAD. You can do this using the **Options** dialog box, as in the following illustration:

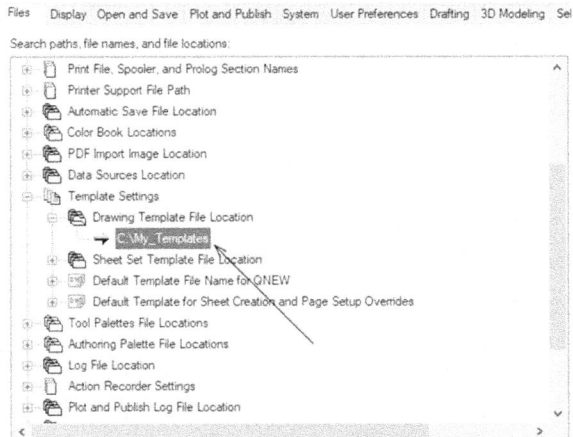

## 18.2  EDITING A TEMPLATE FILE

To edit an existing template, do the following steps:

- Use the normal **Open** file command

- You will see the following dialog box. Using **Files of type** pick **Drawing Template (*.dwt)**

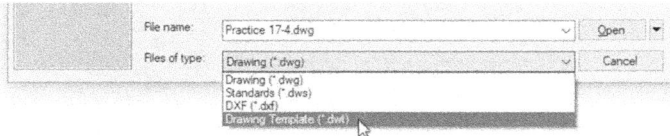

| File name: | Practice 17-4.dwg | Open |
|---|---|---|
| Files of type: | Drawing (*.dwg) | Cancel |
| | Drawing (*.dwg) | |
| | Standards (*.dws) | |
| | DXF (*.dxf) | |
| | Drawing Template (*.dwt) | |

- It will bring you to the default Template folder. Select and open the desired template then make your edits
- Save it using the same name or use a new name

## PRACTICE 18-1   CREATING AND EDITING A TEMPLATE FILE

1. Start AutoCAD 2025

2. Open **Practice 18-1.dwg** file

3. Go to Model and erase all objects

4. Delete D-Size Arch Details layout

5. Go to ANSI B Size Architectural Plan layout and delete the circular viewport

6. Rename ANSI B Size Architectural Plan to be ANSI B Size

7. Go to D-Size Architectural Plan layout and delete the viewport

8. Rename D-Size Architectural Plan to ANSI D Size

9. Go to the Annotate tab and check the existing dimension style and text style

10. Go to the Home tab and check the existing layers

11. Go to the Application menu and start the Units command; set the precision for length to be 0'-0 ½" and click OK to end the command

12. Save the file as a template file under the name My Company.dwt (it will be saved in the Template folder where AutoCAD will save all the templates). The user can save the template file in a different folder, but it should be specified in the Options dialog box in the Files tab

13. Close the file

14. Start a new file using the My Company.dwt template file; you will find that all of your settings are there in the new file, so you can start working without the need to set anything

15. Close the new file without saving

## 18.3   CUSTOMIZING THE INTERFACE – INTRODUCTION

Customizing the interface means changing the ribbons and panels to fit your own needs. The user can add a new workspace, a new tab, a new panel, and even new commands by using a single command: CUI (Customization User interface.

To issue this command, go to the **Manage** Panel, locate the **Customization** panel, then select **User Interface**:

You will see the following dialog box:

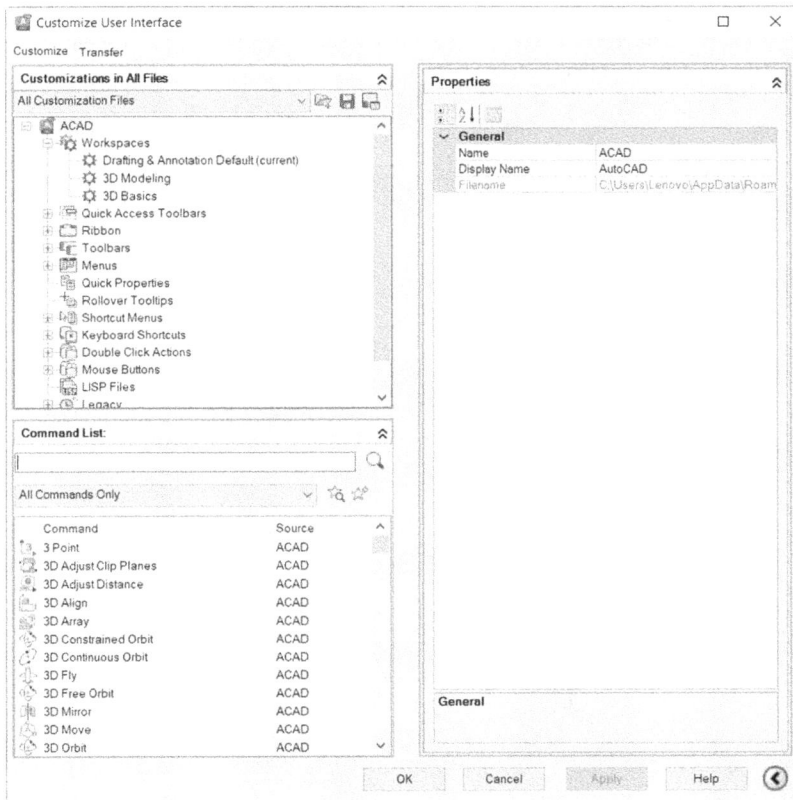

This dialog box is cut into the following parts, which are:

- The Customization part, in which you will specify how you want to customize the Workspace, Tab, Panel, Toolbar, etc. This part will allow you to create new, delete existing, rename, etc.
- The Command list part, which contains all AutoCAD commands
- Depending on what you choose from the left, the right part will change to Content and Properties

## 18.4   HOW TO CREATE A NEW PANEL

Panels are part of tabs; this step is the first step, which will be followed by creating tabs and then creating workspaces. To create a new panel, locate Ribbons at the upper left part of the CUI dialog box and expand it; beneath it you will find Panel. With or without expanding, right-click Panels as in the following, and select the **New Panel** option:

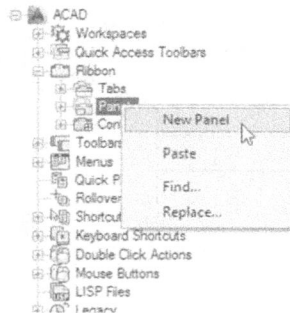

Type the name of the new panel and press [Enter], and you will see the following:

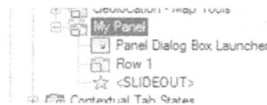

Meanwhile, at the right you will see Panel Preview with a small rectangle, as in the following:

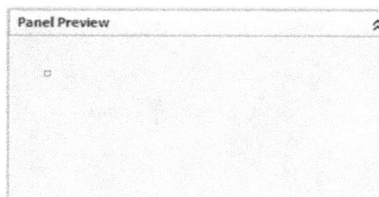

Beneath it, you will see the Properties part, which will look similar to the following:

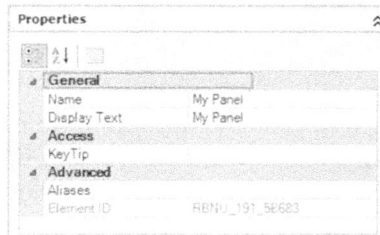

When you create a new panel, the following will automatically be added as the image below details:

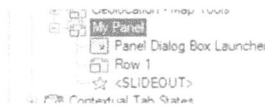

- Panel Dialog Box Launcher
- Row 1 (to add commands to this row or any other row, use the Command List part and drag-and-drop the desired command)
- SLIDEOUT

To understand these, see the following illustration:

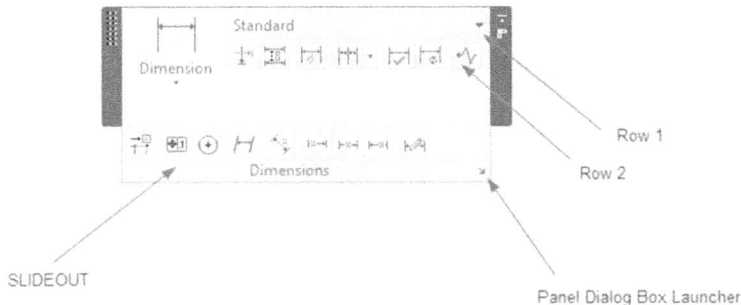

To create a panel, decide on the following:

- How many rows you will have
- How many sub-panels you will have
- What are the commands to be included and where (in rows or sub-panels)
- What command will be presented as the large icon at the left
- Will it have a Panel Dialog Box Launcher, and what is it (normally it is style or palette)
- Will it have Slideout or not and which commands will reside there?

Answer all six questions and write them on paper before you start working with AutoCAD CUI.

By default, all buttons will be small; in order to make a button large, do the following steps:

- Drag it to the panel
- From the Panel Preview, click the desired button
- At the Properties part, check Button Style and select Large (either with Text horizontal, vertical, or without text) as in the following:

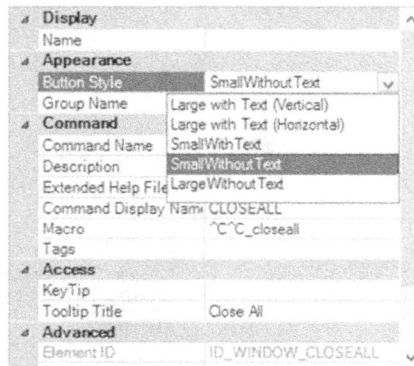

To add a new row in a panel, do the following steps:

- Select the desired panel
- Right-click the name and select the New Row option

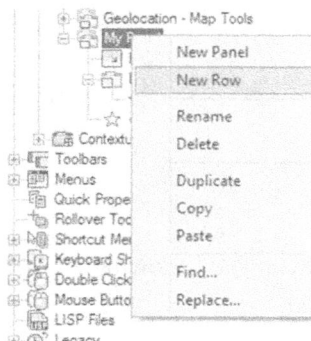

Sub-panels, as the name suggests, can cut the panel into smaller parts to control. Each sub-panel will have row1, row2, etc. To add a new sub-panel, do the following steps:

- Select the desired row

- Right-click and select New Sub-Panel:

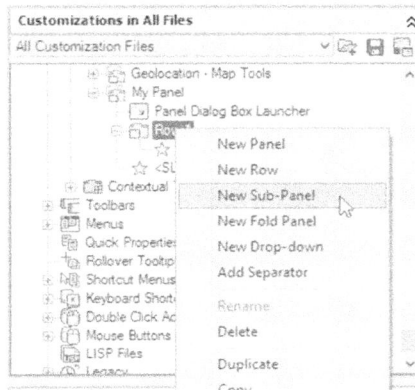

To add a vertical separator between buttons, do the following steps:

- Select the desired row
- Right-click and select the Add Separator option:

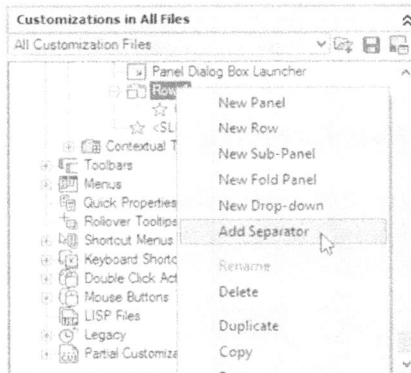

NOTE   *Any row that will be located below the slideout will be shown in the slideout.*

## 18.5   HOW TO CREATE A NEW TAB

A Tab consists of panels. The user can use panels that already accompany AutoCAD or can create their own panels or mix and match.

To create a new tab, start the CUI command, and at the upper left part expand Ribbon, right-click Tabs, and select the New Tab option:

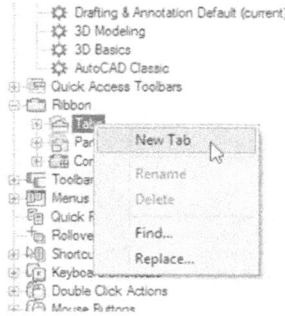

Type the name of the new tab.

You should fill the tab with panels; in order to do that use one of the following approaches:

- Drag-and-drop. This method is not practical, as we have a huge list of panels and tabs
- Go to the desired panel, right-click and select Copy, and then go to the desired tab, right-click and select Paste

## 18.6  HOW TO CREATE A QUICK ACCESS TOOLBAR

The Quick Access toolbar is a toolbar that appears at the top left part of the AutoCAD window to the right of the Application menu. The user can specify which commands should be included in this toolbar. To add a new Quick Access Toolbar, start the CUI command, at the upper left part locate the Quick Access Toolbar, and right-click and select the New Quick Access Toolbar option:

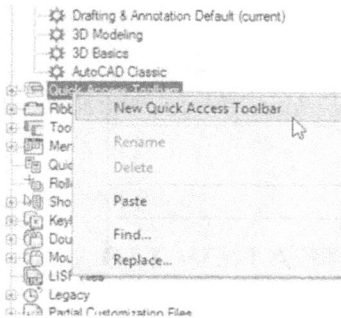

Type the name of the new Quick Access Toolbar. By default, six commands will be included, which are: New, Open, Save, Undo, Redo, and Plot. The user can drag-and-drop any new command from the Commands List.

## 18.7 HOW TO CREATE A NEW WORKSPACE

The Workspace is the set of tabs (such as panels) which will appear together along with the palettes, menus, toolbars, and Quick Access toolbars. To create a new workspace, start CUI command, at the upper left part locate Workspaces, right-click, and select the New Workspace option:

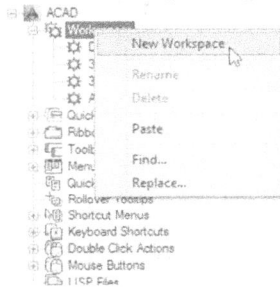

Type the name of the new workspace. Click the newly created workspace and look to the upper right part; you will see the following:

As you can see, there are five different things to be included inside a workspace, which are: Quick Access Toolbar, Toolbars, Menus, Palettes, and Ribbon Tabs. In order to add/remove the contents of the workspace, click the button at the top of the window, titled "Customize Workspace"; accordingly, the text will change to blue in this part, and at the upper left part you will see everything with a checkbox to its left, similar to the following:

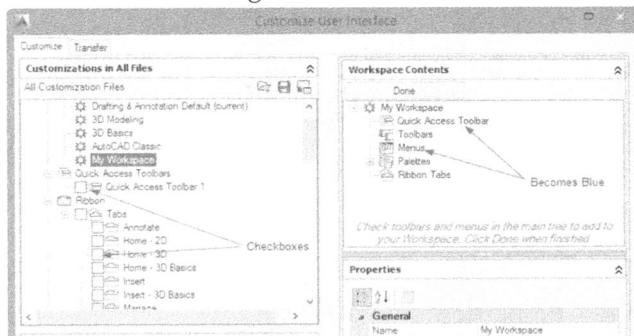

So, if you want to add anything to your workspace, simply click the checkbox on to copy it to your workspace. When you are through with your adding, click the "Done" button to finish the process of customizing your workspace.

## PRACTICE 18-2    CUSTOMIZING AN INTERFACE

**1.** Start AutoCAD 2025

**2.** Start a new file

**3.** Start the CUI command

**4.** Create a new workspace and call it My Workspace

**5.** Go to the Ribbon and expand it, locate Tabs

**6.** Create a new tab and call it My Tab, then collapse the list

**7.** Locate Panels, then create a new panel and call it My Panel. Do the necessary steps to create the following:

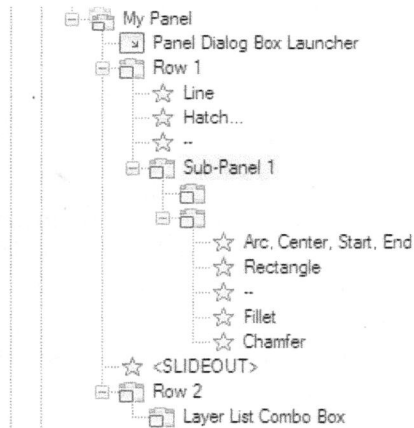

**8.** The Panel Dialog Box Launcher = Tool Palette command

**9.** Copy My Panel to My Tab

**10.** Add the following premade panels to My Tab:

    **a.** Annotate – Dimensions

    **b.** View – Viewports

    **c.** Home – Properties

**11.** Select My Workspace, then at the right click Customize Workspace button

**12.** At the left click My Tab to be the only tab included in this workspace (you can include other premade tabs if you like)

**13.** Collapse Tabs

**14.** Click Quick Access Toolbars, and click Quick Access Toolbar 1

**15.** Click OK to end CUI command

**16.** Go to the top of the screen and select My Workspace to make it current

**17.** Test the panel you created

**18.** Restore the Drafting and Annotation workspace

**19.** Close the file without saving

## 18.8   HOW TO CREATE YOUR OWN COMMAND

In this part, we will learn how to create a new command by creating a macro using AutoCAD commands with certain values, then assigning this macro to a button and put it in a panel, hence, in a tab.

We will use certain characters in the macro:

- ^C^C: which is equal to pressing [Esc] twice. Previously AutoCAD used [Ctrl] + C to cancel commands, then later on (AutoCAD R12) changed it to pressing [Esc] instead. We have to start each macro with ^C^C in order to make sure that the macro will cancel all running commands before it starts
- Use ";" to simulate [Enter]
- Use "\" to wait for the user input
- The macro cannot deal with commands that will initiate palettes or dialog boxes like LAYER or INSERT. Hence, we will use a command prompt version of these types by adding a hyphen before the command like -LAYER, and – INSERT

Before you start writing a macro, test the command prompts using the AutoCAD prompts, which appear at the command window. Try the following:

- Type "circle" at the command window, then press [Enter]
- In the prompts, what is the default option? The answer is "Center"
- The user must input the center (we will use \)
- The next prompt will be to input either the radius or diameter; with radius as the default, we will input value = 2

The macro will appear as the following:

$$^{\wedge}C^{\wedge}Ccircle;\backslash2;$$

To add a new command, do the following steps:

- Start the CUI command
- Using the lower left part (namely the Command List) click the Create a new command button:

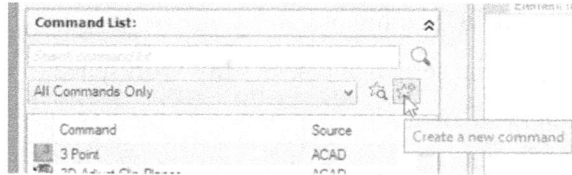

- A new command called command1 will be added
- A Properties window will open at the right part
- Type the new name of the command (in our case we will call it CircleR2)
- Type the macro as you produced it on the paper
- You will see the following:

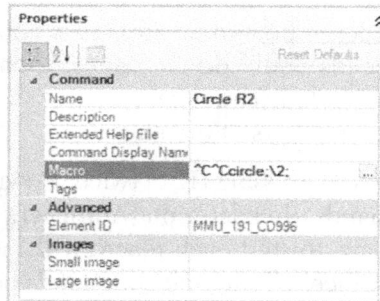

- At the top right, you will see part called Button Image. The user can select an existing image and use it as is or select it and make some modifications. If so, click the Edit button, and you will see the following dialog box:

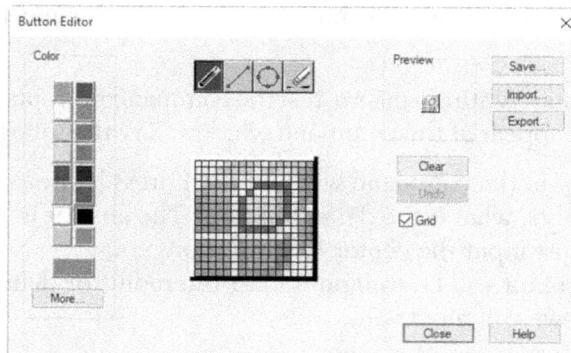

- All of the above are very familiar image-changing functions like adding a line or circle, or erasing. Use the Color part to set the current color to be used, and the Grid shows rows and columns to make it easier for you to draw
- When done, click Save, and give the new image a name
- Accordingly, the new image will be used as representation of the new macro you just created
- Copy it to an existing panel to be used

## PRACTICE 18-3   CREATING NEW COMMANDS

1. Start AutoCAD 2025

2. Start a new file using My Company.dwt

3. Start the CUI command

4. Create a new command with the following specifications:

    **a.** Name = Circle R2

    **b.** Description = This command will draw a circle with R = 2 units

    **c.** Macro = ^C^Ccircle;\2;

    **d.** The image should show a circle with number 2 inside it

5. Copy the command to My Panel, under Sub-Panel 1, under Row 1, beside the Rectangular Array button

6. Make another command with the following specifications:

    **a.** Name = CL 0

    **b.** Description = This command will make 0 the current layer

    **c.** Macro = ^C^C-layer;m;0;;

    **d.** The image should show a couple of layers and then 0 (zero)

7. Copy the command to My Panel, under Sub-Panel 1, under Row 2, beside the Chamfer button

8. Click OK to end the CUI command

9. Make My Workspace the current workspace and test the two new commands

10. Close the file without saving

# NOTES

## CHAPTER REVIEW

**1.** Creating a template file involves:

    **a.** Creating text styles

    **b.** Setting up units and limits

    **c.** Creating layers

    **d.** All of the above

**2.** _____ is the command to customize the interface

**3.** You cannot save your template file except in the Template folder designated by AutoCAD

    **a.** True

    **b.** Flase

**4.** While customizing the AutoCAD interface:

    **a.** You can create new a workspace

    **b.** You can create new a panel

    **c.** You can create new commands using macros

    **d.** All of the above

**5.** Macros cannot deal with commands that will show dialog boxes:

    **a.** True

    **b.** Flase

**6.** Which of the following simulate [Enter] in a macro:

    **a.** Backslash (\)

    **b.** Semicolon (;)

    **c.** ^C

    **d.** ^C^C

## CHAPTER REVIEW ANSWERS

**1.** d

**3.** b

**5.** a

## CHAPTER 19

# PARAMETRIC CONSTRAINTS

**In This Chapter**

- What are Parametric Constraints?
- Parametric constraints – Geometric
- Parametric constraints – Dimensional

## 19.1 WHAT ARE PARAMETRIC CONSTRAINTS?

Software such as Inventor has had the concept of parametric constraints for a long time since it is design software and not drafting software. Now AutoCAD includes this concept; hence, we can consider AutoCAD as a designing tool as well as a drafting tool.

The parametric constraints are: Geometric and Dimensional. The first type will set a relationship between different objects like parallel, horizontal, concentric, etc. Dimensional constraint will impose a certain dimension to a line, or a radius to an arc, or a circle, and can also set a relationship between two or more objects by writing a formula.

With this in hand, the designer will have the ability to express their own design intent without the fear that somebody will alter the design in an accidental or intentional way, as the design will hold by itself all the necessary measures to keep objects as they are.

This is a huge step forward for AutoCAD, which makes it competitive with other software in this specific category.

In the forthcoming pages, we will discuss the two types of the constraints and how can we apply them to objects. These are:

- Geometric constraints
- Dimensional constraints

## 19.2 USING GEOMETRIC CONSTRAINTS

Geometric constraints set rules for the objects. You can decide to make a line always horizontal, but perpendicular for another line. In addition, you can set two circles to always share the same center point, and so on. In order to reach to these type of constraints, select the **Parametric** tab and locate the **Geometric** panel:

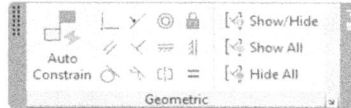

The panel shows twelve different types of constraints and other buttons like the Auto Constraint button and the Show and Hide buttons. We will discuss each one of them in the forthcoming pages.

### 19.2.1 Using Coincident Constraints

Locate the **Geometric** panel, then select the **Coincident** button:

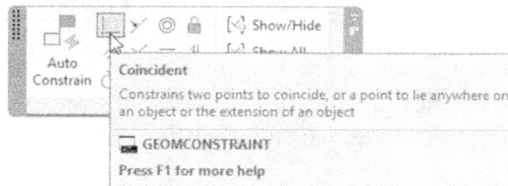

According to the AutoCAD definition this constraint will constrain two points to coincide, or a point to lie anywhere on an object or the extension of an object.
You will see the following prompts:

```
Select first point or [Object/Autoconstrain] <Object>:
Select second point or [Object] <Object>:
```

This command will require by its basic method to select two points on two existing objects. The first object will stay in its place yet the second will move to connect with the first object. A small blue square will appear at the point connecting the two objects:

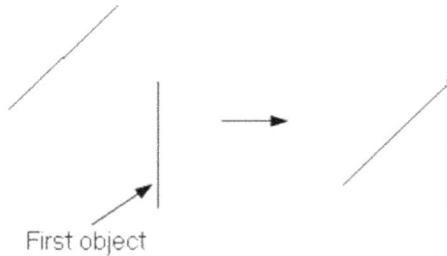

First object

Another variation of this command is to select an object rather than to select a point on an object. At the first prompt, right-click and select the Object option; you will see a prompt asking you to select the desired object, and once you will select it, the following prompt will be shown:

```
Select point or [Multiple]:
```

If you select a point on another object, the first object will stay in its place, and the other will link with the extension of the first object similar to the following:

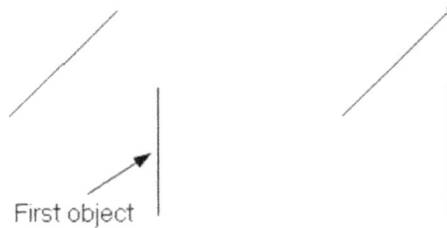

First object

If you use the Multiple option, you will see the following prompts:

```
Select Point:
Select Point:
```

You will be able to select multiple points to link them with the first object using the previous rule. The following example shows three objects coincident with the first object, connected using their endpoints (the user can select any other desired point):

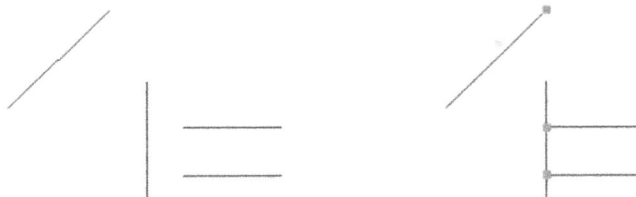

### 19.2.2   Using Collinear Constraints

Locate the **Geometric** panel, then select the **Collinear** button:

According to the AutoCAD definition, this constraint will constrain two lines to lie on the same infinite line.

You will see the following prompts:

```
Select first object or [Multiple]:
Select second object:
```

The simplest method is to select the first line (which will stay intact), then select the second line, which will move to be collinear. See the following illustration:

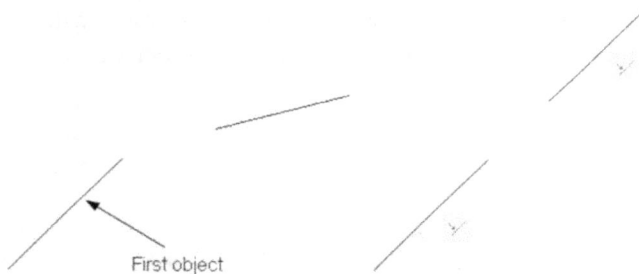

First object

If you select the **Multiple** option, this will allow you to set several lines to be collinear.

### 19.2.3   Using Concentric Constraints

Locate the **Geometric** panel, then select the **Concentric** button:

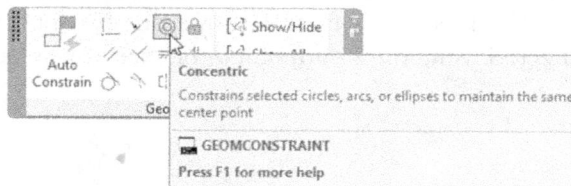

According to the AutoCAD definition, this constraint will constrain selected circles, arcs, or ellipses to have the same center point.

You will see the following prompts:

```
Select first object:
Select second object:
```

See the following illustration:

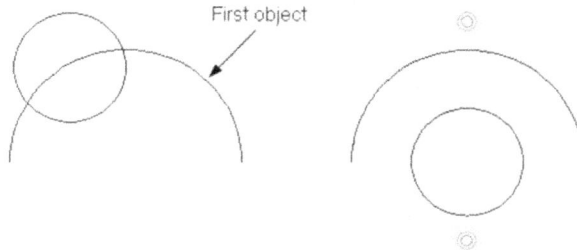

### 19.2.4   Using Fix Constraints

Locate the **Geometric** panel, then select the **Fix** button:

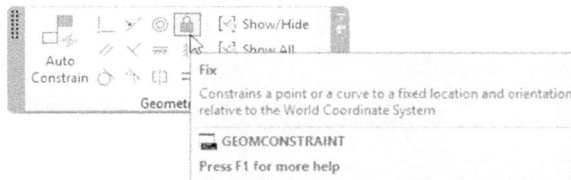

According to the AutoCAD definition, this constraint will constrain a point or a curve to a fixed location and orientation relative to the World Coordinate System. You will see the following prompt:

```
Select point or [Object] <Object>:
```

Either you will select a point on an object, or select then object option, to be allowed to select the desired object; see the following illustration, which will show the two cases:

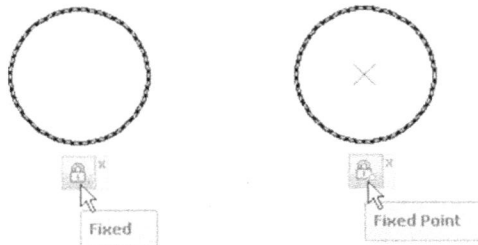

### 19.2.5   Using Parallel Constraints

Locate the **Geometric** panel, then select the **Parallel** button:

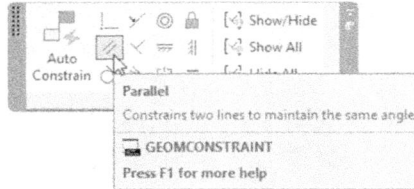

According to the AutoCAD definition, this constraint will constrain two lines to be parallel. You will see the following prompts:

```
Select first object:
Select second object:
```

The first object will preserve its current angle but the second object will rotate to have the same angle of the first. See the following illustration:

First object

### 19.2.6   Using Perpendicular Constraints

Locate the **Geometric** panel, then select the **Perpendicular** button:

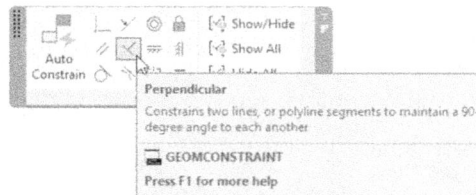

According to the AutoCAD definition, this constraint will constrain two lines or polyline segments to maintain a 90-degree angle to each other. You will see the following prompts:

```
Select first object:
Select second object:
```

The first object will keep its current angle; the second object will rotate to make it 90° relative to the first object. See the following illustration:

### 19.2.7 Using Horizontal Constraints

Locate the **Geometric** panel, and then select the **Horizontal** button:

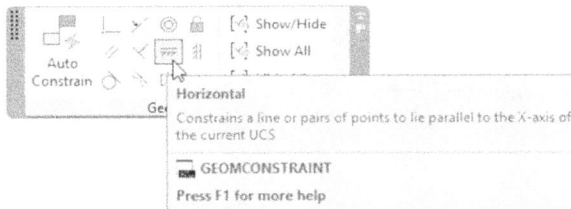

According to the AutoCAD definition, this constraint will constrain a line or pair of points to lie parallel to the X-axis of the current UCS. You will see the following prompts:

```
Select an object or [2Points] <2Points>:
```

If you select a line, it will become horizontal and the command will end. The endpoint nearest to the selection point will remain but the other end will move.

See the following illustration:

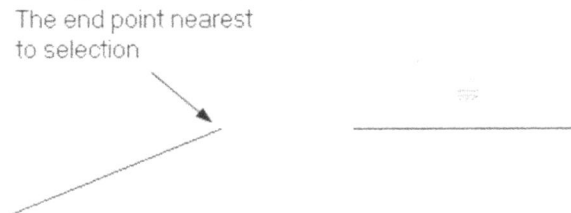

If you select the **2 Points** option, you will see the following prompts:

```
Select first point:
Select second point:
```

This option will make sure that the two points selected will form an imaginary horizontal line, as shown in the following example:

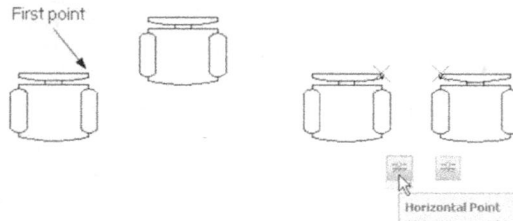

Note that although a polyline is considered a single object, if you use this constraint on it, AutoCAD will treat each object alone.

### 19.2.8 Using Vertical Constraints

Locate the **Geometric** panel, then select the **Vertical** button:

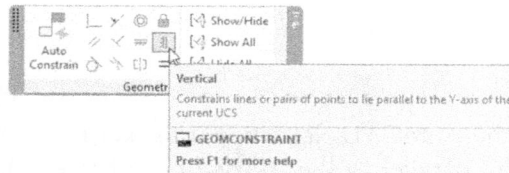

This command is identical to the Horizontal command allowing objects or points to be vertical.

### 19.2.9 Using Tangent Constraints

Locate the **Geometric** panel, then select the **Tangent** button:

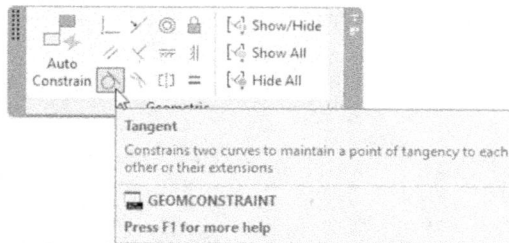

According to the AutoCAD definition, this constraint will constrain two curves to maintain a point of tangency to each other or their extension. You will see the following prompts:

```
Select first object:
Select second object:
```

The first object will stay in its current place, but the second object will move using the nearest tangent point close to its current position. See the following illustration:

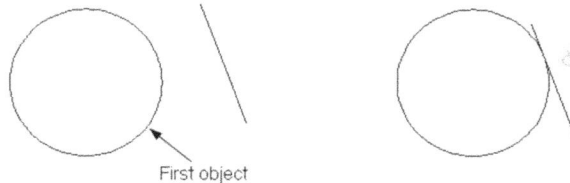

First object

### 19.2.10   Using Smooth Constraints

Locate the **Geometric** panel, then select the **Smooth** button:

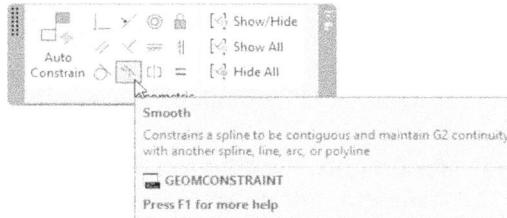

According to the AutoCAD definition, this constraint will constrain a spline to be contiguous and maintain G2 continuity with another spline, line, arc, or polyline. You will see the following prompts:

```
Select first spline curve:
Select second curve:
```

The first object should be a spline but the second object can be anything like: Spline, Line, Arc, or Polyline. The following illustration will demonstrate the continuity concept:

The user should be careful; although we are selecting objects, the point which will highlight during the selecting is very important to the end result.

### 19.2.11   Using Symmetric Constraints

Locate the **Geometric** panel, then select the **Symmetric** button:

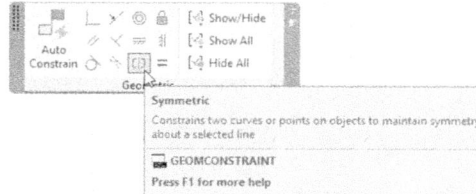

According to the AutoCAD definition, this constraint will constrain two curves or points on objects to maintain symmetry about a selected line. You will see the following prompts:

```
Select first object or [2Points] <2Points>:
Select second object:
Select symmetry line:
```

This is similar to the Horizontal command, because it uses the same prompts. You will select either two objects or two points. The first object will maintain its current angle, yet the second object will be rotating to make a mirror image of the first object around the symmetry line selected. See the following illustration:

### 19.2.12   Using Equal Constraints

Locate the **Geometric** panel, then select the **Equal** button:

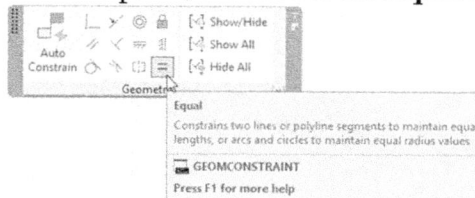

According to the AutoCAD definition, this constraint will constrain two lines or polyline segments to maintain equal lengths, arcs, or circles to maintain equal radius values. You will see the following prompts:

```
Select first object or [Multiple]:
Select second object:
```

The first object will maintain its length (radius for arcs and circles), the second object will change its length to match the first object. If you select the Multiple option, you can match the length of the first objects with multiple lines rather than with a single line. See the following illustration:

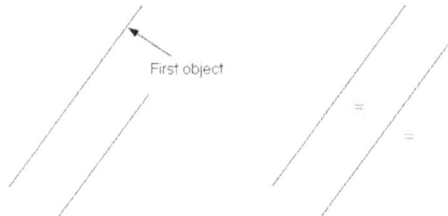

## 19.3 GEOMETRIC CONSTRAINTS SETTINGS

To control the display of the Geometric Constraints, the user should use the **Settings** dialog box. To issue this command locate **Geometric** panel, then click the arrow at the lower right corner of the panel:

You will see the following dialog box:

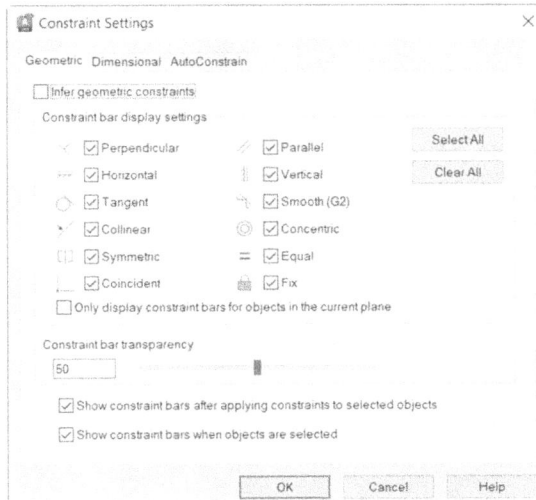

You will see that by default, all of the constraint will be displayed, and the user can select to hide any of the twelve constraints in the drawing. Also, control the following:

- Whether to display only constraint bar for objects in the current plane
- Constraint bar transparency value
- Whether or not to show constraint bar after applying constraint to selected objects
- Whether or not to show constraints bars when objects are selected

## 19.4 WHAT IS INFER CONSTRAINT?

In the previous dialog box, one checkbox was overlooked: the Infer geometric constraint checkbox at the top left of the dialog box. So, what is Infer Constraint? It will flip the process of constraining in AutoCAD, as it will set the constraints while drafting rather than after drafting. Also, commands like Fillet and Chamfer will be affected.

To activate this command, go to the Status bar and click the following button:

After you switch this checkbox on, AutoCAD will apply constraints to all objects as they are added to the drawing, using proper constraints depending on the object and the attached objects to it. Do not assume that this method is a replacement for the original method discussed at the beginning of this chapter, because you will be mistaken. This method will help you complete your work in fewer steps. Let's discuss the following example:

- Switch on the Infer Constraint checkbox
- Start Line command, and draw a vertical line. You will notice that AutoCAD applies a Vertical constraint
- Continue by drawing a horizontal line. You will notice that AutoCAD applies a Perpendicular constraint to the new horizontal line
- Continue drawing another vertical line and close the rectangle, and you will notice that AutoCAD applies Perpendicular constraint, along with Coincident, on all corner points
- Start the Fillet command and set the Radius to a suitable value, then fillet one of the corners. You will notice that AutoCAD applies Tangent constraint

between the two lines and the added arc, along with Coincident constraint at the two ends of the arc

- You will see the following:

If you right-click the button at the Status bar and select **Settings**, you will see the same dialog box we saw in the previous section, which will control which constraint is displayed on the screen and which is not.

## 19.5   WHAT IS AUTO CONSTRAIN?

The Auto Constrain command will apply multiple geometric constraints to the selected objects based on the current relationship between these objects, and the selected constraint to be applied in the Settings dialog box of Auto Constrain.

To issue this command, locate the **Geometric** panel, then select the **Auto Constrain** button:

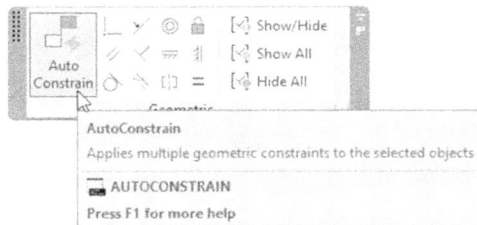

You will see the following prompt:

```
Select objects or [Settings]:
```

As a first step, select the **Settings** option and you will see the following dialog box:

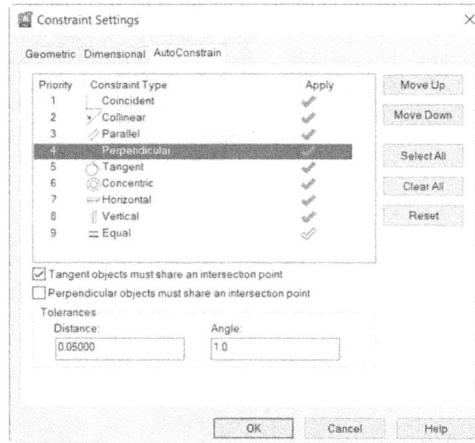

Using this dialog box you can do all or any of the following:

- Move the constraints up and down to set the priority for each constraint
- Turn off any unwanted constraint, by clicking the green (✓)
- Use Select All, Clear All, and Reset buttons
- Choose whether Tangent objects share an intersection point or not
- Choose whether Perpendicular objects share an intersection point or not

As you can see, three of the normal twelve constraints will not be used, they are: Symmetrical, Fix, and Smooth.

Let's assume the following example. You have the following shape:

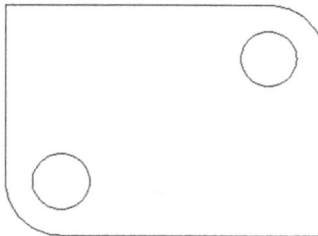

- Before you apply any constraint, select whole objects using grips and try to move one of the grips; take note of this movement
- Start the **Auto Constrain** command
- Select all objects, then press [Enter], the following message will appear:

```
16 constraint(s) applied to 8 object(s)
```

- The image will change to:

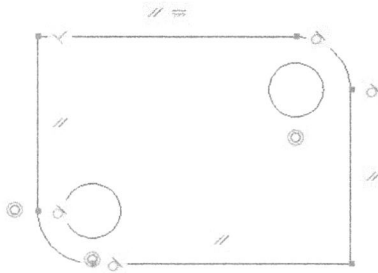

- Select the objects using grips and try to move one of the grips. You will feel that all the objects are moving together as a coherent set of objects that understands each other and keeps the relationships right at all times

As a final note for Auto Constrain, you should note that the first prompt of Coincident constraint contains Autoconstrain:

```
Select first point or [Object/Autoconstrain] <Object>:
```

Which means if you select this option, and then select objects, all of them will hold a coincident constraint.

## 19.6   CONSTRAINT BAR & SHOWING AND HIDING

### 19.6.1   Constraint Bar

The Constraint Bar is the small bar that appears beside the object after you apply a geometric constraint on it. Hovering over the small bar will highlight it along with the objects affected. See the following illustration:

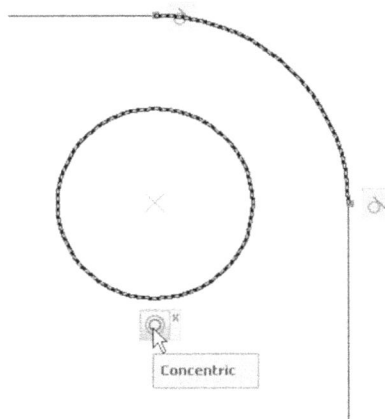

On the other hand, if you hover over an object with geometric constraint applied to it, Constraint Bars will highlight as well. Occasionally one object may hold more than one constraint, hence the bar will show more buttons.

Right-clicking the buttons in the Constraint Bar will show the following menu:

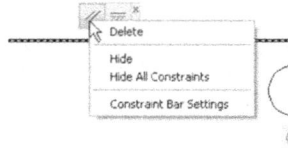

The following is a listing of the commands:

- The Delete command will delete the selected geometric constraint; you can also use the normal [Del] from the keyboard
- Hide the current Constraint Bar (even if it contains more than one button)
- Hide All Constraint in the current drawing
- Show Constraint Bar Settings, which was discussed previously

The User can move Constraint Bars from its default position to any other desired position; to hold the bar to any place, click and drag to the new position.

### 19.6.2 Showing and Hiding

The user can control the visibility of the Constraint Bar. The user can control whether to show it or not for all of the objects or for some objects.

To issue this set of commands, locate the **Geometric** panel to issue one of the following commands:

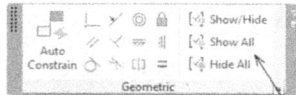

The commands are:

- **Show/Hide**: You can select some of the objects to show the constraints and some to hide
- **Show All**: Will show all constraints for all objects
- **Hide All**: Will hide all constraints for all objects

## 19.7 RELAXING AND OVER-CONSTRAINING OBJECTS

### 19.7.1 Relaxing Constraints

Relaxing an object means removing some of the constraints applied to it in order to fulfill a command, which wants to change the object's status. Assume you have a line with Parallel constraint to a vertical line, and you try to rotate it, how will AutoCAD respond to such an action? You will see the following message appear:

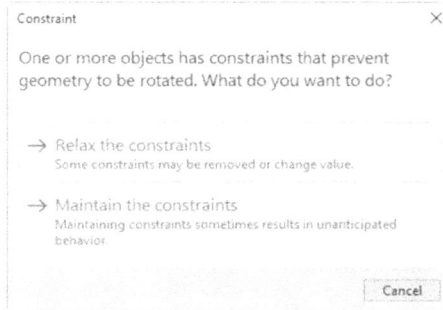

The message reads:

Constraint ✕

One or more objects has constraints that prevent geometry to be rotated. What do you want to do?

→ Relax the constraints
Some constraints may be removed or change value.

→ Maintain the constraints
Maintaining constraints sometimes results in unanticipated behavior.

Cancel

As you can see, AutoCAD will provide you with the choice either to relax the constraints or to maintain them. Selecting relaxing the constraint will mean removing only the associated constraint, which will allow the commands to be fulfilled.

### 19.7.2   Over-Constraining an Object

Let's assume you applied a Horizontal constraint to a line and then you apply Perpendicular constraint to an attached line. Applying Vertical to the second line will mean this object is over-constrained and AutoCAD will not allow this to happen. It will show you the following message:

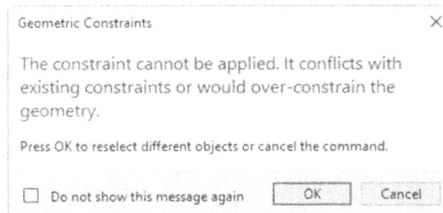

The message reads:

Geometric Constraints ✕

The constraint cannot be applied. It conflicts with existing constraints or would over-constrain the geometry.

Press OK to reselect different objects or cancel the command.

☐ Do not show this message again    OK    Cancel

The message indicates AutoCAD will not allow the action, hence, the user will click OK or Cancel to abort the process.

## PRACTICE 19-1   APPLYING GEOMETRIC CONSTRAINT

**1.** Start AutoCAD 2025

**2.** Open **Practice 19-1.dwg** file

**3.** Make sure that Infer Constraint is off at the Status bar

**4.** Start Coincident, select Autoconstrain, and select the whole shape

5. Start Concentric and select the arc, then the circle

6. Start Vertical and select the right vertical line

7. Start Perpendicular, select the right vertical line and the upper horizontal line

8. Repeat the same with the lower horizontal line

9. Start Parallel and select the right vertical line and the left vertical line

10. Start Horizontal and select the lower horizontal line; what is the message that AutoCAD produced? _____

11. Click on Infer Constraint at the Status bar

12. Using Fillet command, set Radius = 2 and fillet the upper right corner of the shape

13. Draw a circle with Radius = 1, using the center of the arc you just added using Fillet command

14. Select the whole shape using grips, click the midpoint of the right vertical line and move it, and see how the whole shape is responding to the movement

15. Switch off the Infer Constraint button

16. Pan to the shape at the right

17. Start the Auto Constrain command, select Settings, and click Select All button to select all constraints

18. Click OK and select the whole shape. How many constraints were added? _____ (18) on how many objects? _____ (8)

19. Check Equal constraints for both circles and arcs, delete

20. Select to Hide all constraints, then select to show them only for some objects

21. Save and close the file

## 19.8   USING DIMENSIONAL CONSTRAINTS

Geometric constraints are not sufficient; we need to set the Dimensional constraint to apply the constraint concept to its full. Dimensional constraints will specify a length for a line, an angle between two lines, a radius, or a diameter for an arc or a circle. It even will link the different dimensional constraints using formulas.

To reach all Dimensional constraint commands, select the **Parametric** tab, then locate the **Dimensional** panel:

The following is a discussion of all types of Dimensional constraints:

### 19.8.1    Using Linear, Horizontal, and Vertical Constraints

These are three different constraints that will deal with distances between points; if you click the button, you will see the following options:

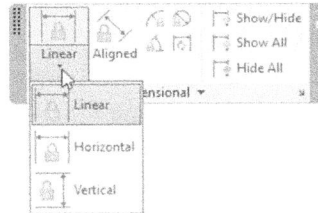

According to the AutoCAD definition, Linear will constrain either the horizontal or the vertical between two points (even of the two points form an angle other than 0, 90, 180, or 270). Horizontal command will constrain the X-axis distance between two points, and Vertical will constrain the Y-axis distance between two points. You will see the following prompt:

```
Specify first constraint point or [Object] <Object>:
Specify second constraint point:
Specify dimension line location:
Dimension text = 6.6615
```

Prompts for the three commands are the same. If you select the first point (a red cross with a circle will appear), then select the second point. The user will be asked to specify the dimension line location (just like we did in dimensioning), then whether to accept the real measured distance by pressing [Enter] or to input your own value. Of course the length will change according to the new value. If, at the first prompt, you selected the Object option, you will see the following prompts:

```
Select object:
Specify dimension line location:
Dimension text = 9.7586
```

The following illustration will show the shape of the dimensional constraint:

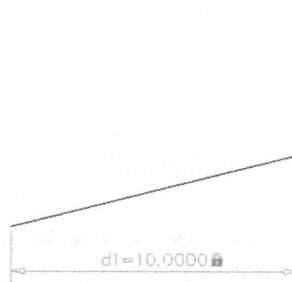

### 19.8.2 Using Aligned Constraints

To issue this command locate the **Dimensional** panel, then select the **Aligned** button:

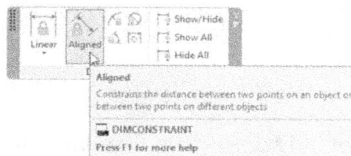

According to the AutoCAD definition, this constraint will constrain the distance between two points whether on the same objects or on different objects. You will see the following prompts:

```
Specify first constraint point or [Object/Point &
line/2Lines] <Object>:
```

If you specify the first constraint point, the user will be asked to specify the second point; if you use the Object option, the user will be asked to select an object. If you choose the Point & Line option, you will see the following prompts:

```
Specify constraint point or [Line] <Line>:
Select line:
Specify dimension line location:
Dimension text = 2.2466
```

The user will select a point, then a line (while selecting a line, make sure you are selecting the right point). If you select the 2 Lines option, you will see the following prompts:

```
Select first line:
Select second line to make parallel:
Specify dimension line location:
Dimension text = 2.9135
```

Select two line objects. The second line will be made parallel to the first line. Aligned constraint controls the distance between the two lines.

The following illustration will show Aligned constraint:

### 19.8.3 Using Radial and Diameter Constraints

To issue this command locate the **Dimensional** panel, then select the **Radial** or **Diameter** button:

According to AutoCAD definition, these two constraints will constrain the radius or the diameter of a circle or arc. You will see the following prompts:

```
Select arc or circle:
Dimension text = 3.4359
Specify dimension line location:
```

The following illustration will show the radial and diameter constraints:

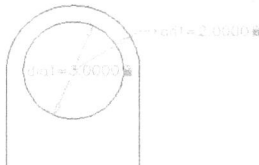

### 19.8.4 Using Angular Constraints

To issue this command locate the **Dimensional** panel, then select the **Angular** button:

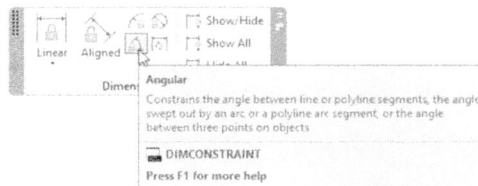

According to the AutoCAD definition, this Angular will constraint the angle between two lines or polyline objects, in an arc, or using three points. You will see the following prompts:

```
Select first line or arc or [3Point] <3Point>:
Select second line:
Specify dimension line location:
Dimension text = 45
```

The user will select two lines (or polyline segments) or an arc. If you select the 3 Points option, you will see the following prompts:

```
Specify angle vertex:
Specify first angle constraint point:
Specify second angle constraint point:
```

The user will select the vertex point first, and then the two other points to form an angle. The following illustration will show the angular constraint:

### 19.8.5 Using the Convert Command

To issue this command locate the **Dimensional** panel, then select the **Convert** button:

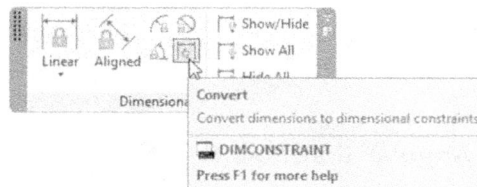

This command will convert normal dimensions to dimensional constraints. You will see the following prompts:

```
Select associative dimensions to convert:
Select associative dimensions to convert:
```

Simply select the desired dimensions to be converted to constraint. The following illustration will show the process:

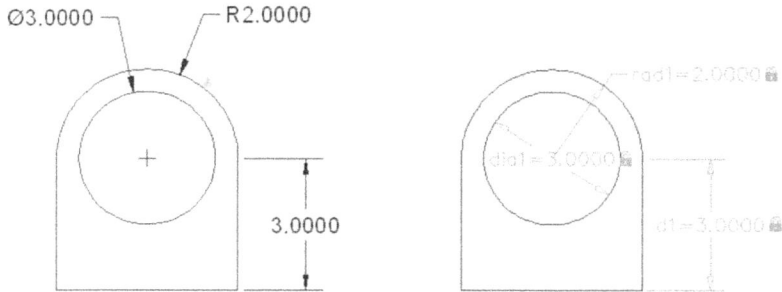

## 19.9 CONTROLLING DIMENSIONAL CONSTRAINT

In this section, we will discuss how to control Dimensional constraint, by:

- Using the Constraint Settings dialog box
- Deleting Dimensional constraint
- Showing and hiding Dimensional constraint

### 19.9.1 Constraint Settings Dialog Box

In this dialog box, you will control the appearance of the dimensional constraint on the screen. To issue this command select the **Parametric** tab, locate the **Dimensional** panel, then select the **Constraint Settings, Dimensional** button:

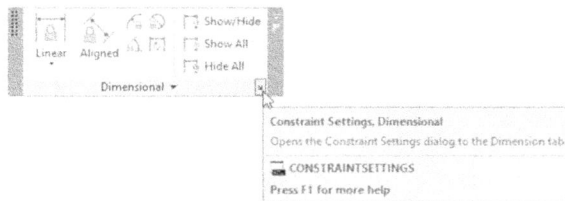

The following dialog box will appear:

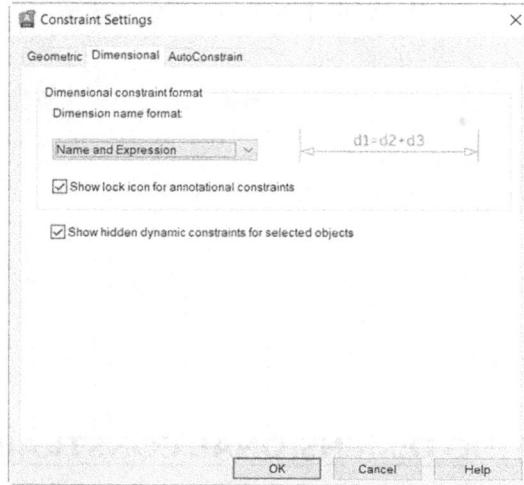

This dialog box will control what will appear at the dimensional constraint; there are three choices:

- Name
- Value
- Name and Expression

The user can select whether to show or to hide the lock symbol that appears near the measured value. Finally, the user can select whether to show the dimensional constraint for any hidden dimensional constraint when you select this object. See the following example:

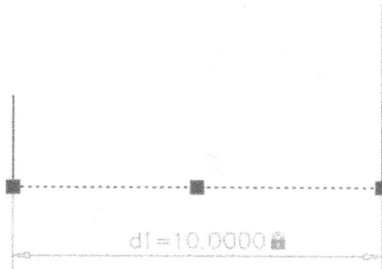

In the previous image, the Dimensional constraint was hidden, but when we selected the object, the Dimensional constraint appeared.

### 19.9.2 Deleting Constraints

The user can use the normal [Del] key on the keyboard to eliminate any Dimensional or Geometric constraints. Another way would be to locate the **Manage** panel and then select the **Delete Constraints** button:

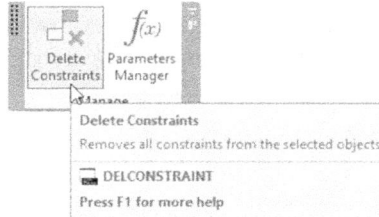

The following prompt will appear:

```
All constraints will be removed from selected objects…
Select objects:
```

Keep selecting the undesired constraint, then press [Enter] to end the command.

### 19.9.3 Showing and Hiding Dimensional Constraint

These commands are similar to what we learned in the Geometric constraint section. To issue this command locate the **Dimensional** panel, then select one of the following three buttons:

The three buttons are:

- The **Show/Hide** button will show a hidden constraint, or hide a visible constraint by selecting. You will see the following prompts:

```
Select objects:
Select objects:
Enter an option [Show/Hide]<Show>:
```

- The **Show All** button will show all dimensional constraints
- The **Hide All** button will hide all dimensional constraints

## 19.10    USING THE PARAMETERS MANAGER

The Parameters Manager is the place to create equations involving dimensional constraint, hence, linking dimensional constraint together. When using this method, any change in one of the dimensional constraints will affect the others as well. In order for us to create the right equations, AutoCAD allows us to create user-defined parameters. To issue this command, locate the **Manage** panel, then select the **Parameters Manager** button:

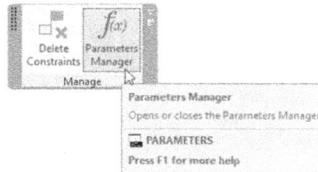

You will see the following palette:

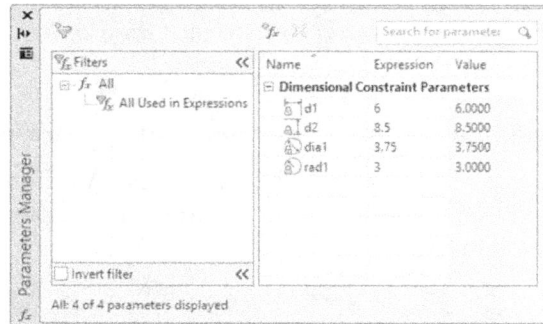

The Parameters Manager palette will be displayed showing all the existing Dimensional Constraint Parameters. As you can see, AutoCAD is using the default names, but the user has the ability to rename these parameters. Simply click the name once and you will be able to input your own names:

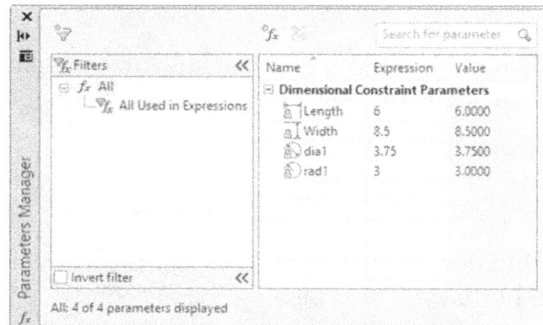

The user can create an equation including one or more of the parameters in the following example:

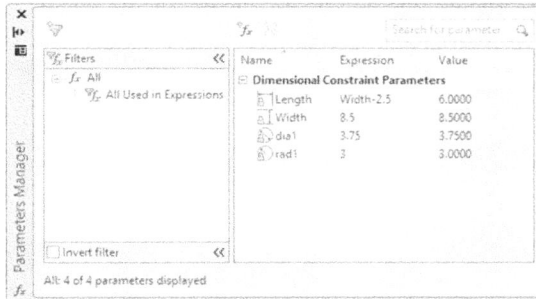

The previous means that whenever you change Width, Length will change as well. The user can create their own parameters by creating User-Defined parameters using the following button:

Input the name of the user-defined parameters, an expression (if valid); otherwise, input the current value. Later on, you can include an expression involving dimensional constraint parameters. Refer to the example below:

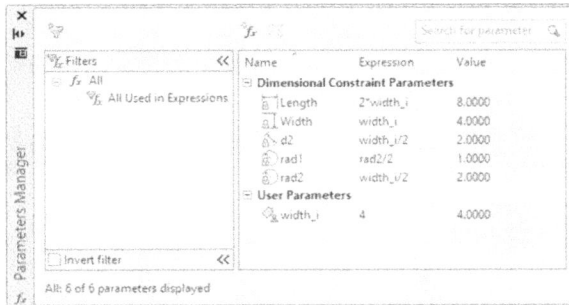

The previous example shows that both Length and Width are now linked with the user parameter called width_i.

AutoCAD provides filters to categorize your parameters in categories of your making. To create a new filter, click the following button:

Once you click this button, a new filter will be added; simply type the name of the filter, then press [Enter]. To fill it with parameters, drag parameters from the right, and drop them at the name of the filter.

You will receive the following:

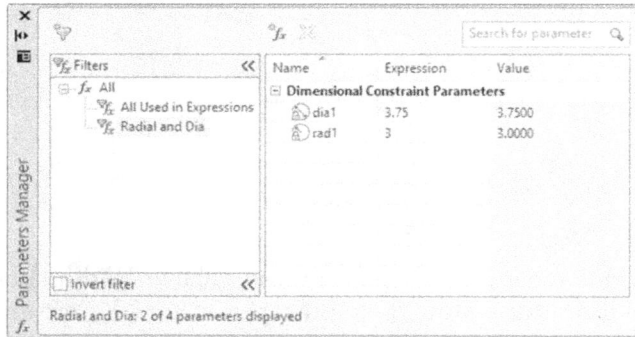

The user can easily differentiate between normal parameters and equations: equations starts with **fx:**. The user can overwrite the equation by double-clicking it and inputting their own value.

## 19.11  WHAT IS ANNOTATIONAL CONSTRAINT MODE?

The blocks displayed after you add a dimensional constraint will not be printed, and zoom in and zoom out commands will not affect its displayed size. But this means you will do two tasks instead of doing only one; the user will add normal dimensions and will add dimensional constraints. AutoCAD can help you to limit the workload so it needs to be done only once. How? The user has the choice to convert all of the dimensional constraint to Annotational constraints. Annotational constraints will be printed and will use the current dimension style.

There are two methods to accomplish that:

- Before you start, the user can change the mode to be Annotational constraint. With this method, the user will combine the two actions into a single action
- For the dimensional constraints already placed, the user can convert them using the Properties palette

### 19.11.1 Annotational Constraint Mode

Before you start placing dimensional constraints, locate the **Dimensional** panel and click the arrow at the bottom to show more buttons, to ensure **Annotational Constraint Mode** is selected:

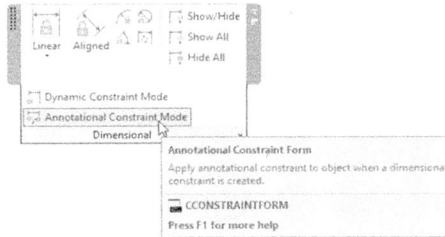

Now adding a dimensional constraint using the normal ways we discussed previously, you will see the following:

Using the Dimensional Constraint Settings dialog box, the user can choose to show/hide the lock symbol along with the showing the name.

### 19.11.2 Converting Dimensional Constraints to Annotational

If you already placed dimensional constraints, you can choose to convert them to be Annotational constraints using the Properties palette. Simply select the desired dimensional constraint to be converted, right-click, and select Properties; you will see the following palette. Click the **Constraint Form** and select **Annotational** instead of **Dynamic**:

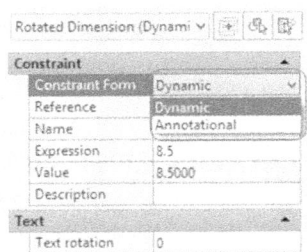

## 19.12   USING DIMENSIONAL GRIPS

The user can manipulate dimensional constraints using grips. The following image includes three types of dimensional constraints, namely: Linear, Aligned, and Angular:

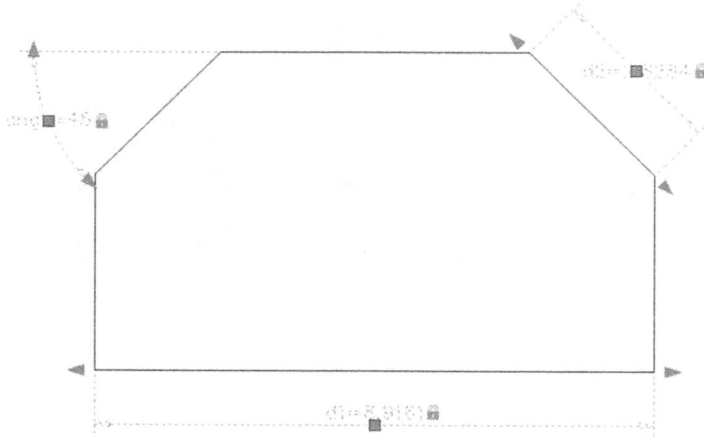

As you can see there are grips at the middle of each dimensional constraint that will move the whole block back and forth. The other two arrows will stretch the dimensional constraint and, hence, stretch the object by itself. Dimensional parameters have their own grips depending on the type.

In the example below, you will see a radius and diameter constraint. The grip can relocate the dimensional constraint block. While the radius has one arrow to change the value of the radius, the diameter has two arrows to change the value in either direction. Arcs and circles will be affected by this change:

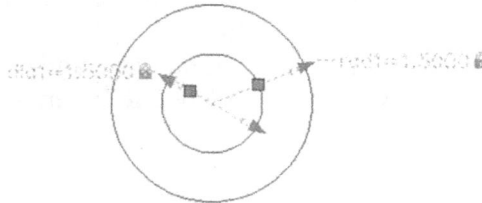

## PRACTICE 19-2   APPLYING DIMENSIONAL CONSTRAINT

**1.** Start AutoCAD 2025

**2.** Open **Practice 19-2.dwg** file

**3.** Using different types of dimensional constraints, add the following constraints using the same names:

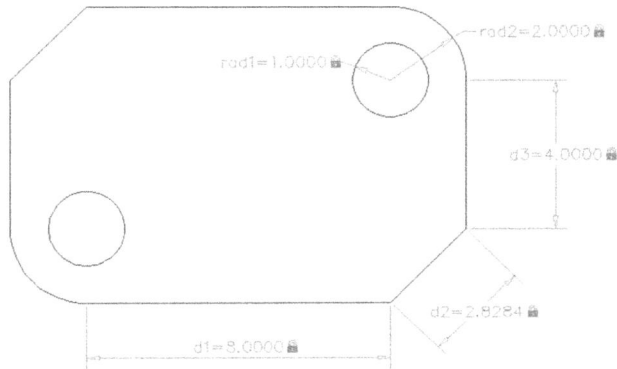

**4.** Using the Parameters Manager do the following:

   **a.** Rename d1 to become Length

   **b.** Rename d3 to become Width

   **c.** Create a user-defined parameter width_i with a default value of 4

   **d.** Change Width = width_i

   **e.** Change Length = 2 * width_i

   **f.** rad2 = width_i / 2

   **g.** d2 = width_i / 2

   **h.** rad1 = rad2/2

**5.** Using the Parameters Manager change the value of width_i to the following values: 2, 3, 5, 6. How did the other objects react to the change? _____
_____
_____

**6.** Change width_i to 4, and close the Parameters Manager

**7.** Change one of the dimensional constraints to be Annotational, using Zoom in and Zoom out; how did this block respond compared to the other blocks?

**8.** Using the Dimensional Constraint Settings, choose to show only the values without the lock symbol

**9.** The lock symbol is still there because all the constraints are equations

**10.** Save and close the file

# NOTES

## CHAPTER REVIEW

1. One of the following is not related to dimensional constraints:

    **a.** It will not be affected by zooming

    **b.** Linear

    **c.** Equal

    **d.** Angular

2. Parallel, Fix, and Coincident are all _____ constraints

3. Infer Constraints can be applied while:

    **a.** Adding line command

    **b.** Filleting two lines

    **c.** Stretching objects

    **d.** Chamfering two lines

4. You can add a single block to work as a normal dimension and as a constraint:

    **a.** True

    **b.** False

5. There is an option called Autoconstraint in Collinear command:

    **a.** True

    **b.** False

6. You have to select a spline as a first object for _____ constraint

7. In _____ you can create equations linking different dimensional constraints together

## CHAPTER REVIEW ANSWERS

**1.** c

**3.** c

**5.** b

**7.** Parameters Manager

## CHAPTER 20

# DYNAMIC BLOCKS

## In This Chapter

- What are Dynamic Blocks?
- Parameters and Actions
- Dynamic Blocks and Constraints

## 20.1 INTRODUCTION TO DYNAMIC BLOCKS

All we know until now is that AutoCAD can provide the facility to create and insert a block with one shape, and a dimension that can be scaled to be bigger or smaller. AutoCAD, however, provides more than this with concern to blocks; it provides the possibility to create a block with multiple shapes, multiple sizes, and multiple views. This type of block is called a Dynamic Block and can be created in the Block Editor.

Dynamic blocks will use Parameters and Actions. Actions are very similar to AutoCAD modifying commands, like Stretch, Scale, and Array. Each action needs a parameter to work on. Another method is to create dynamic blocks using the Geometric and Dimensional constraints discussed in Chapter 19.

In order to differentiate between the normal blocks and the dynamic blocks, dynamic blocks will always have a lightning symbol to the right, whether you are at the Insert command dialog box or tool palettes. The following are some examples:

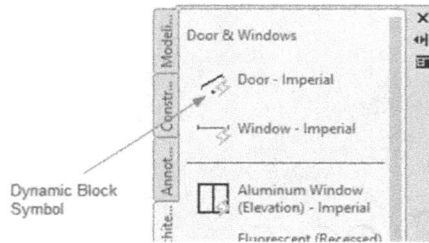

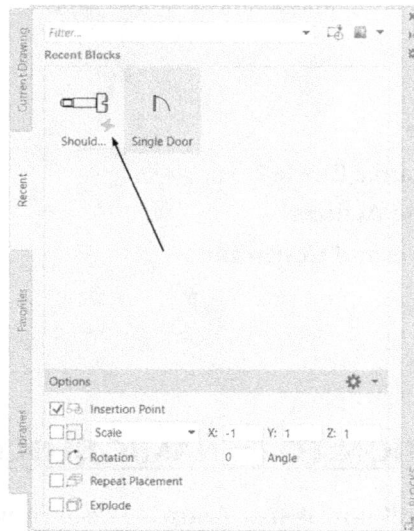

Dynamic blocks are very useful tools to help you to finish drawing production faster and with less mistakes, reducing the number of blocks to work with.

## 20.2   METHODS TO CREATE A DYNAMIC BLOCK

Dynamic blocks are made in Block Editor, so how can we reach the Block Editor? There are multiple ways to do this:

- By using the **Open in block editor** checkbox in the Block Definition dialog box
- By double-clicking an existing block, then selecting the name of the block
- By issuing the **Block Editor** command

### 20.2.1   Using the Block Definition Dialog Box

In general, when you create a normal block, you will use the Block Definition dialog box as in the following; simply click on the Open in block Editor checkbox, and then click OK. AutoCAD will bring you directly to the block editor where you can add some dynamic features to your normal block:

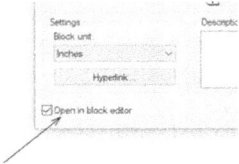

### 20.2.2   Double-Clicking an Existing Block

You can convert an existing normal block to be dynamic by double-clicking it, and you will see the following dialog box:

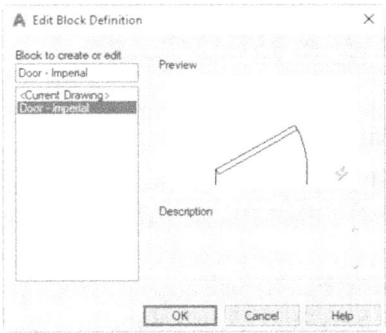

AutoCAD will show all the blocks in the current drawing; select the block by double-clicking, and simply click OK to go directly to the block editor to assess the desired features.

### 20.2.3   Block Editor Command

To issue this command go to the **Insert** tab and **Block Definition** panel, then select the **Block Editor** button:

The following dialog box will appear:

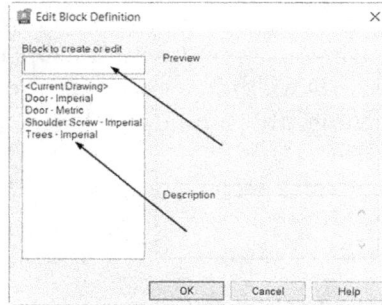

You can do the following in this dialog box:

- Type a name of a new block
- Select the name of one of the existing blocks to convert (or edit)

## 20.3 INSIDE THE BLOCK EDITOR

Using one of the methods mentioned previously, you will be inside the block editor. The following will occur, which are:

- The background will change to a different color (most likely it will be dark gray)
- A new contextual tab called **Block Editor** will appear, which includes lots of panels
- The **Block Authoring** Palette will appear, which contains four tabs: Parameters, Actions, Parameter Sets, and Constraints, just as shown in the following:

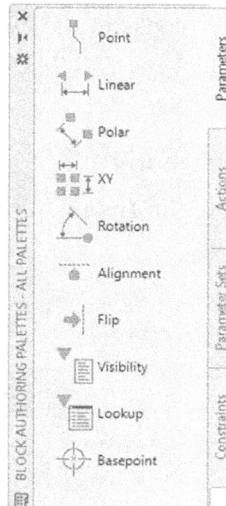

We will use both the context tab "Block Editor" panels and the Block Authoring palettes in our pursuit to add dynamic features to our block.

## 20.4   WHAT ARE PARAMETERS AND ACTIONS?

There is a very simple rule in creating a dynamic block; first add a parameter to your block, and then assign an action to work on it. Parameters are geometric features of the objects forming the block, like linear, polar, visibility, lookup, etc. Actions are a simulation of AutoCAD modifying commands like Move, Stretch, Scale, etc.

You will find that there are some parameters that do not need any actions, and some actions do not need parameters to work on. The rule of thumb, however, is we need them in this order: parameter first, then action second. AutoCAD, in order to help us expedite our work, offers Parameter Sets, which include a set of parameters with action in one shot.

Both parameters and actions have properties, which will be very important to us in order to control the outcome of the dynamic block.

In Chapter 19, we discussed the geometric and dimensional constraints and how they will affect normal objects. AutoCAD allows us the ability to enhance these constraints inside a block, which will make it a very interesting thing to watch; how can you utilize the constraints in order to create a dynamic block?

The following is the list of the available parameters:

- Point parameter will add a point definition so the likes of Stretch and Move actions will use it
- Linear parameter will measure a distance, which can have any angle. But once applied, the user cannot change it. The following actions can act on this parameter: Move, Stretch, Scale, and Array
- Polar parameter, similar to Linear parameter, has the ability to change the angle. The following actions can act on this parameter: Move, Stretch, Polar Stretch, Scale, and Array
- The XY parameter will help you define two associated dimensions in X and Y directions. The following actions can act on this parameter: Move, Stretch, Scale, and Array
- Rotate parameter will help you add a rotation parameter to the block definition. Rotate action can act on this parameter only
- Alignment parameter will help you add the ability for the block to align itself with an existing object. This parameter does not need an action
- Flip parameter will flip the existing block to its mirror image using the mirror line. Flip action can act on this parameter only

- Visibility parameter will help you create a visibility state for some of the objects to be visible and others to be invisible. This parameter does not need an action
- Lookup parameter will help you create table of possible values of existing parameters. Lookup action can act on this parameter only
- Basepoint parameter will define a new base point for the block. This parameter does not need an action

The following is a list of the available actions:

- Move action will move the selected objects from one position to another; you will see the following prompts:

```
Select parameter:
Specify selection set for action
Select objects:
```

- Scale action will scale the selected objects; you will see the following prompts:

```
Select parameter:
Specify selection set for action
Select objects:
```

- Stretch action will stretch selected objects using the Crossing concept (just like in a normal Stretch command). You will see the following prompts:

```
Select parameter:
Specify parameter point to associate with action or
enter [sTart point/Second point] <Second>:
Specify first corner of stretch frame or [CPolygon]:
Specify objects to stretch Select objects:
```

- Polar Stretch action is equal to Stretch action, except it has an angle option
- Rotate action will rotate objects using Rotate parameter, you will see the following prompts:

```
Select parameter:
Specify selection set for action
Select objects:
```

- Flip action will work on the Flip parameter only and will create a mirror image using the mirror line created in the Flip parameter. You will see the following prompts:

```
Select parameter:
Specify selection set for action
Select objects:
```

- Array action simulates the rectangular array command (the old array prior to the 2012 version). You will see the following prompts:

```
Select parameter:
Specify selection set for action
Select objects:
Enter the distance between columns (||||):
```

- Lookup action will work only on a Lookup parameter in order to create a table of existing parameters; accordingly, it will show them as a list for you to pick one of the available choices. You will see a dialog box, which will be discussed later on
- Block Properties Table, which will be discussed later

Parameter Sets are a one-step job. AutoCAD wants to help us reduce the steps to complete our work and, hence, provided us with Parameters Sets that include both parameters and actions working on them in one shot. There are two exceptions for that:

- The concept of Pairs, which means there are two actions acting on a single parameter. There are several pairs like Linear Stretch Pair, etc.
- The concept of the Box Set, which means there are four actions on a single parameter. There are multiple box sets, they are XY Stretch Box Set, XY Array box set, etc.

The user will find the same set of parameters and actions using the context tab Block Editor, locating **Action Parameters** panel, which appears as the following:

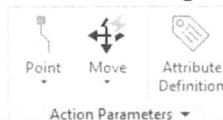

After you finish adding parameters and actions you will see the following;

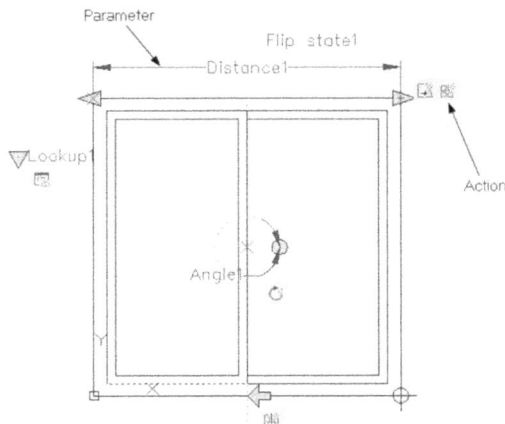

If you hover over one of the actions, all of the parameters acting on it will be highlighted, along with the objects selected:

If you right-click any action, you will see the following menu:

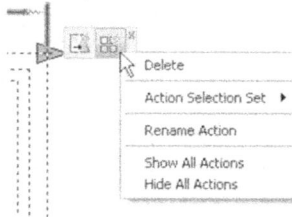

Using this menu, the user can delete the action, change the selection set, rename the actions, and show/hide all actions in the current file. This menu may show different options for different actions.

## 20.5   CONTROLLING PARAMETER PROPERTIES

The user can control the parameter properties to control the behavior of the block once the user starts to use it. To see the properties palette, select the parameter, right-click, then select the Properties option, and the following will appear:

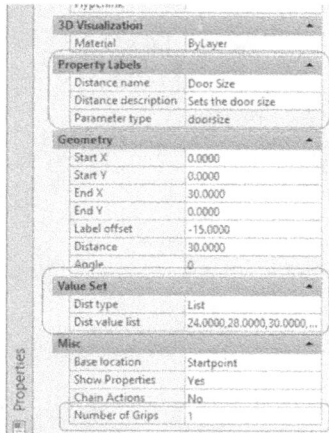

When you add a new parameter, AutoCAD will give it a temporary name, but as a user, you can change it to a proper name.

To do that locate **Property Labels**, and change the **Distance name** (in this example for Linear parameter), similar to the following:

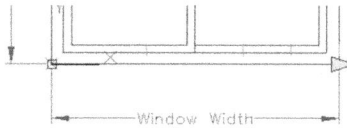

Another thing users can control is size change using **Value Set**. There are three different types:

- **None type**: This type means the user can input two values, the Dist minimum and Dist maximum. These two values can be different and can be equal:

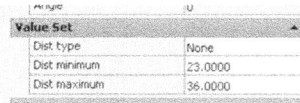

- **Increment type**: This type means the value will change using a Dist Increment. The user should input this as well the Dist minimum and Dist maximum:

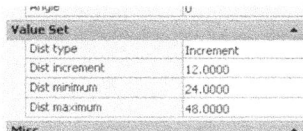

- **List type**: This type means the value will change using a list of values which will be input using the small button at the right with the three dots inside it; when clicked, you will see the next dialog box:

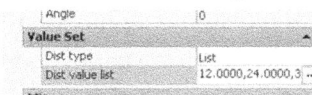

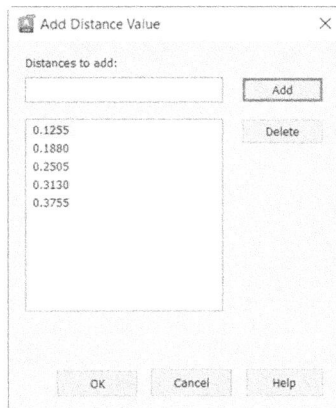

Finally, the user can control the number of grips to appear in the parameter. The default number of parameters in linear is two, and in XY is four. In order to prevent you from scaling or stretching your block from points you do not want, simply reduce or remove the grips. Refer to the following illustration:

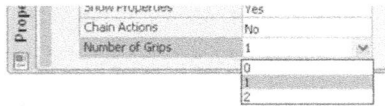

## 20.6 CONTROLLING THE VISIBILITY PARAMETER

One of the features a user can add to a dynamic block is visibility. The technique is very simple: you have several objects (blocks) and you create a state that will show some of these objects and hide others. The user will create several states. To use this block, you will see a list of states to select from beside the block. You need to follow this procedure:

- Add a visibility parameter (this parameter does not need an action)
- Once you add it, the **Visibility** panel in the Block Editor context tab will come to life, similar to the following:

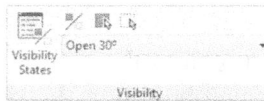

- As you can see, AutoCAD created a new visibility state and gave it a temporary name, VisibilityState0. Click the Visibility States button to rename the current visibility state; you will see the following dialog box:

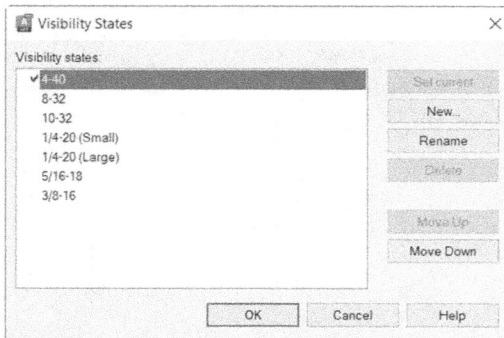

- Click the Rename button, input a new name, then click OK.
- Select object(s) you want to make visible, then click the **Make Visible** button at the Visibility panel. Then select object(s) you want to make invisible, then

click the **Make Invisible** button at the Visibility panel. See the following two buttons:

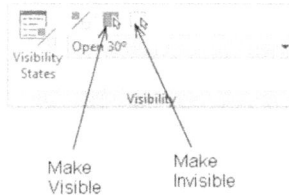

- The default is to hide the invisible objects 100%. However, you can choose to show them with transparency degrees by using the **Visibility Mode** button:

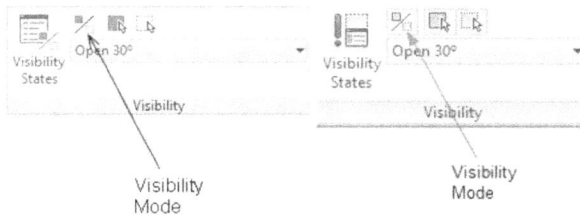

- Create another Visibility State by clicking on the Visibility State button, then clicking on the New button, and you will see the following dialog box:

- Type the name of the new state, and then select one of the choices available. You can select to Hide/Show all of the objects or keep objects with their current visibility settings. Click OK, then change the visibility of the objects as you wish
- Create another state and so on ...
- Test your settings by using the list in the panel, as shown in the following:

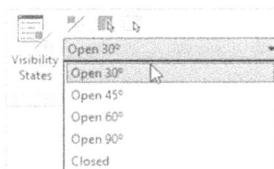

## 20.7    USING LOOKUP PARAMETER AND ACTION

Lookup parameter and action are used to display all different sizes of a block in a simple list. With this approach, you will fully control the output of the block, because you will limit the choices for the user to the list. You should add the parameters and set the grip and value conditions; the last thing would be adding Lookup parameter, and Lookup action, or you can add both in one shot, using Parameter Sets. After adding the parameter and action, do the following steps:

- You will see the following dialog box:

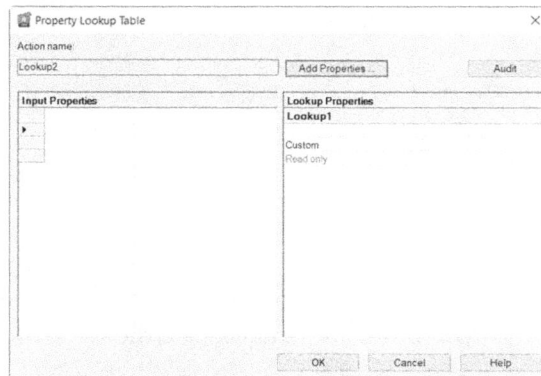

- To add the parameters you want to show in the list click **Add Properties** button, to see the following dialog box:

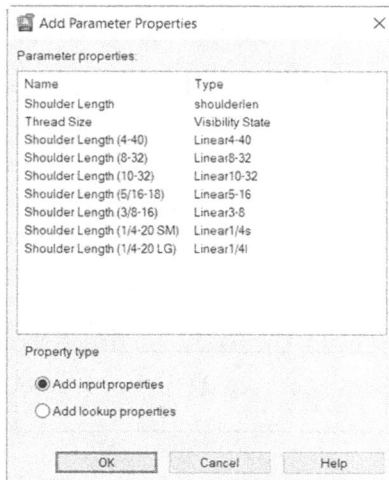

- Select the desired parameters (you can use [Ctrl] key to select multiple parameters), then click OK to end the adding process

- The left part of the dialog box will change to the following:

| Input Properties | | | | |
|---|---|---|---|---|
| Door Size | Wall Thickness | Hinge | Swing | Opening Angle |
| ▶ 600.0000 | 100.0000 | Left | Inside | Open 60° |
| <Unmatched> | | | | |

- Start selecting the different values for the Door. At the right table, input the text that should appear to the user, "Door Size 1," "Door Size 2," etc. This is what you will have at the end:

| Input Properties | | | | | Lookup Properties |
|---|---|---|---|---|---|
| Door Size | Wall Thickness | Hinge | Swing | Opening Angle | Lookup1 |
| 600.0000 | 100.0000 | Left | Inside | Open 60° | Door Size 1 |
| 700.0000 | 100.0000 | Right | Outside | Open 45° | Door Size 2 |
| 700.0000 | 150.0000 | Left | Inside | Open 45° | Door Size 3 |

- The final product will look similar to the following:

```
        Door Size 1
        Door Size 2
        Door Size 3
  ✓     Custom
```

## 20.8  FINAL STEPS

Once you are finished with adding your features, you have to go to the outmost panel at the left, then the outmost panel at the right. The one at the left is the Open / Save panel which looks similar to the following:

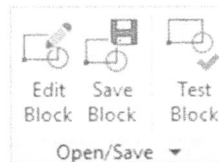

```
   Edit    Save    Test
   Block   Block   Block
        Open/Save  ▾
```

This panel will:

- Save the block with the current addition
- Test the block without going out of the Block Editor
- Edit another block
- If you expand the panel, you will see a button to Save As the block under another name

The second panel is at the rightmost panel which looks similar to the following:

```
        ✓
      Close
   Block Editor
      Close
```

To finish the Block Editor command, the user should click the only button in the panel which is the **Close Block Editor** button. If user did not save changes, AutoCAD will show a warning message to save or discard changes you made.

## PRACTICE 20-1  DYNAMIC BLOCKS – CREATING A CHEST OF DRAWERS

1. Start AutoCAD 2025

2. Open **Practice 20-1.dwg** file

3. Create a new block calling it **Chest of Drawers**, setting up the upper right corner to be the base point, setting up the block units to be Centimeter, and making sure that the Open in block editor checkbox is on

4. Add a Linear parameter for the left vertical line (pick the lower point first then the upper point second). Using the Properties palette change the name to Height, and set the number of grips to one

5. Using Properties change Dist Type to Increment, with Dist minimum = 35, Dist maximum = 130, and Dist Increment = 25

6. Add a Stretch action using the following information:

   **a.** Select Height parameter

   **b.** Select the upper left corner as your parameter point

   **c.** Set the stretch frame (and objects to select) as the picture below, then press [Enter]:

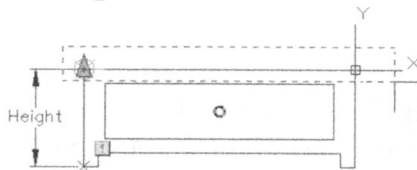

7. Add an Array action selecting the Height parameter, setting the distance to be 25

8. Without leaving the block editor, test the block. If everything is correct, save it and test it normally

9. Save and close the file

## PRACTICE 20-2   DYNAMIC BLOCKS – DOOR CONTROL

1. Start AutoCAD 2025

2. Open **Practice 20-2.dwg** file

3. There will be a block representing a door. Double-click the block, a dialog box will appear, select Interior Door (it is already selected), and click OK. You are in block editor

4. Add a Flip parameter selecting the midpoint of the two jambs as the reflecting line, specifying the label location beneath the line

5. Add a Flip action selecting the parameter and all objects representing the door

6. Test the block

7. Add a Flip parameter. To specify reflecting line use [Shift] + right-click, and select Mid between Two Points option, select the lower left corner of the left jamb, and the lower right corner of the right jamb. AutoCAD will select the point between these two points, then using Polar Tracking, select any point upwards or downwards. Select label location to be at the middle of the door opening

8. Add a Flip action selecting the parameter and all objects representing the door

9. Test the block

10. Draw a rectangle (using the Rectangle command or Line command) to represent the door is closed, as shown in the following:

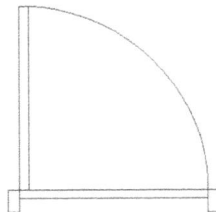

11. Add a Visibility parameter, selecting the upper left part of the door to be the location of the parameter. Note how the Visibility panel was turned on

12. Using the Visibility panel create a new state and call it Open, by renaming the existing state

13. Select the rectangle (or lines) you just added, then click the Make Invisible button

14. Click the Visibility Mode button to see the Invisible objects (they will be transparent)

15. Create a new state calling it Closed

16. Select the arc and the vertical lines and click the Make Invisible button. Select the rectangle (or lines) you just added, and click the Make Visible button

17. Test the block

18. Save the block. Test it outside block editor

19. Save and close the file

## PRACTICE 20-3    DYNAMIC BLOCKS – WIDE FLANGE BEAMS

1. Start AutoCAD 2025

2. Open **Practice 20-3.dwg** file

3. Create a block and name it Wide Flange Beam, select the lower left corner to be the base point, make sure block unit is Inch and that Open in block editor is on

4. Add four Linear parameters and name them as following:

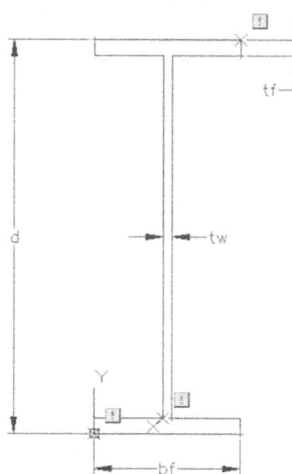

5. Select all linear parameters and set Grips to 0

**6.** Add a Scale action and select the **d** parameter and all objects

**7.** Using the following table set the Value Set to List and input the following values (the first value is already set, so no need to re-input it):

| d (in) | tw (in) | bf (in) | tf (in) |
|--------|---------|---------|---------|
| 44.02  | 1.02    | 15.95   | 1.77    |
| 34.21  | 1.63    | 15.865  | 2.95    |
| 27.63  | 0.61    | 10.01   | 1.10    |

**8.** Going to the Parameter Sets tab, select Lookup Set and add to the drawing, select a proper location for it

**9.** Right-click the Lookup action icon and select Display Lookup Table option. A dialog box will open, click the Add Properties button, and you will see a list of all parameters added previously. Select them all and click OK

**10.** Using the table above select all the matched values together, and name the first W 44×335, the second W 30×477, and the third W 27×129

**11.** Test the block

**12.** Save and test the block outside block editor

**13.** Save and close the file

## 20.9   USING CONSTRAINTS IN DYNAMIC BLOCKS

In block editor, the user can replace parameters and actions with geometric and dimensional constraints. This will make things much easier for the user as it will simplify the steps and cut the time to produce dynamic blocks. Using geometric and dimensional constraints in creating dynamic blocks is the same as using it in ordinary objects; yet, geometric and dimensional constraints are meant to be for blocks rather than for ordinary objects. To be clear, it is not being said here that the user should abandon parameters and actions. The user will likely see more power in using and controlling geometric and dimensional constraints.

It is necessary to mention that there are some parameters like Visibility, and some actions like Array, that cannot be replaced in geometric and dimensional constraints.

Even the interface is similar to what we learned in Chapter 19, with only three new buttons. These are:

- In the Dimensional panel in block editor there is a button called **Block Table**

- In the Manage panel in block editor there is a button called **Construction**
- In the Manage panel as well in block editor there is a button called **Constraint Status**

### 20.9.1  Block Table Button

This button is a replica of the Lookup parameter and action, and it works with dimensional constraints. Why was this feature added? The answer is simple; the user cannot use the Lookup parameter and action on dimensional constraints, so we need this feature to produce a pop-up list for the user to select sizes, dimensions, etc. To use this feature, while you are in block editor, locate the **Dimensional** panel and click the **Block Table** button:

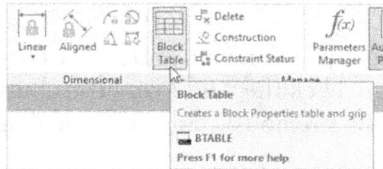

The following prompts will appear:

```
Specify parameter location or [Palette]:
Enter number of grips [0/1] <1>:
```

The user will be asked to specify any proper location for the list to appear, then to specify the number of grips to be included.

The following dialog box will appear:

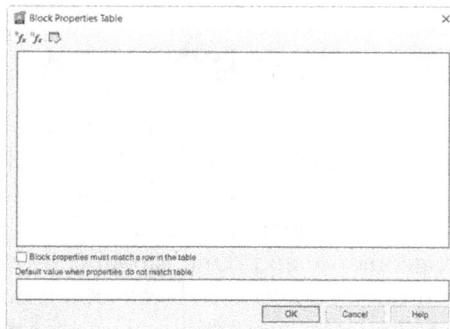

To add dimensional constraints to the list, click the following button:

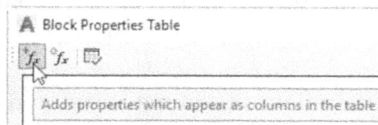

The following dialog box will appear:

Select the desired dimensional constraints and then click OK. You will see the following:

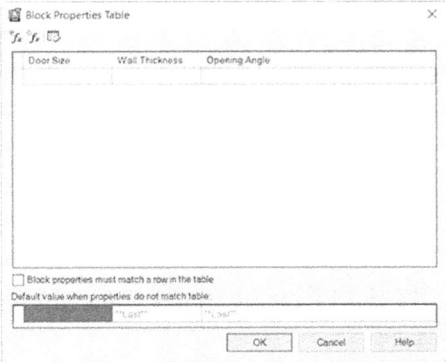

Pick the first field in the first row and select one of the available values:

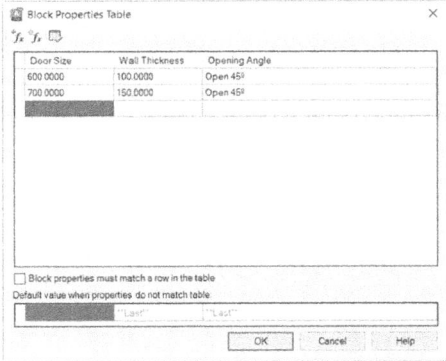

Keep inputting values until you receive the following:

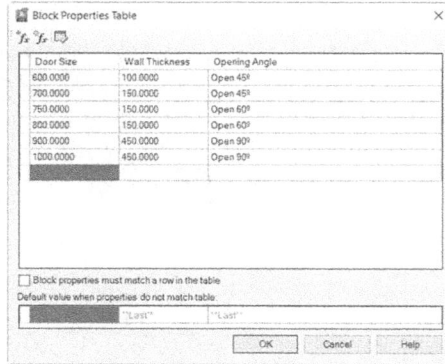

In order for these values to appear here, the user should go to the Properties palette of each dimensional constraint and add a list or increment.

The other two buttons at the top left part of the dialog box are to create a user parameter; you will see the following dialog box:

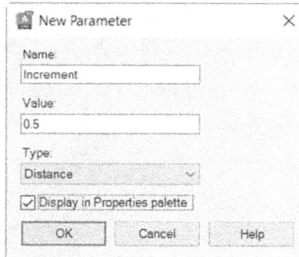

The third is to audit the block table; if it does not contain any errors, you will see the following message:

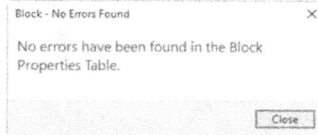

If AutoCAD finds any mistake (such as one of the fields being empty), you will see the following message:

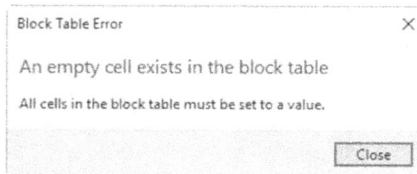

### 20.9.2    Construction Button

A Construction object is an object that you will use to define your dynamic block and you do not want others to see. Inside block editor, locate the **Manage** panel, and click the **Construction** button:

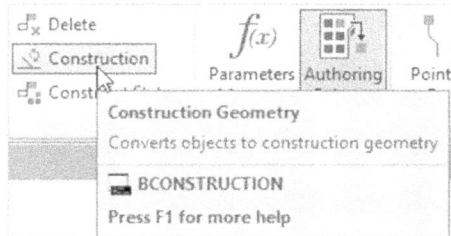

You will see the following prompt:

```
Select objects or [Show all/Hide all]:
```

Select the desired object(s).

### 20.9.3    Constraint Status Button

This button will allow AutoCAD to check your current geometric constraint status. There are two options: either blue which means partially constraint, or magenta which means fully constraint. This is an on/off button. It is preferable to keep it switched on to monitor your block. Most likely if you add all of your constraints and you still have a blue color, you will need to add Fix constraint to one of the points, which is part of AutoCAD requirements. To switch on/off this button, while you are inside the block editor, locate the **Manage** panel, then select the **Constraint Status** button:

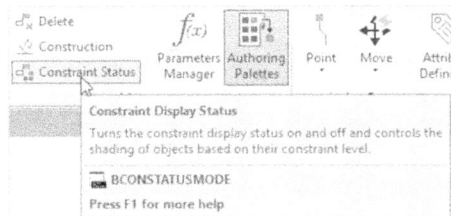

## PRACTICE 20-4    DYNAMIC BLOCKS USING CONSTRAINTS – WIDE FLANGE BEAMS

1. Start AutoCAD 2025

2. Open **Practice 20-4.dwg** file

3. Redo Practice 20-3 using geometric and dimensional constraint instead of parameters and actions

4. Save and close the file

## PRACTICE 20-5    DYNAMIC BLOCKS USING CONSTRAINTS – CREATING WINDOWS

1. Start AutoCAD 2025

2. Open **Practice 20-5.dwg** file

3. Create a new block from the existing shape naming it **Window Elevation**, using the lower left corner as base point, and Units to be Inch, making sure that Open in block editor is on

4. Locate the Geometric panel, select the Auto Constrain button, and select all objects to define the geometric constraints, then hide all of them

5. Using dimensional constraints set the following linear constraints, keeping the naming used below:

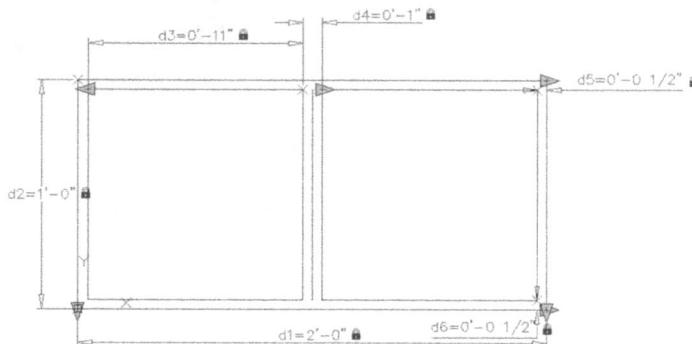

6. Using the **Parameters Manager**, make the following changes to the names: d1=Width, d2=Height, and d3=Glazing

**7.** Set the following equations:

    **a.** Width = Height * 2

    **b.** Glazing = (Width/2)-d4

**8.** Select all dimensional constraints except Height, and set Grips = 0

**9.** Set the Value Set for Height to be Increment, set the min = 1', max = 2', and increment = 6"

**10.** Set the line between the two glazing to be Construction object

**11.** Test the block

**12.** Save the block, and test it outside the block editor

**13.** Save and close the file

## NOTES

## CHAPTER REVIEW

**1.** You can add an action to a dimensional constraint:

   **a.** True

   **b.** False

**2.** The name of the palette in the block editor is _____

**3.** One of the following parameters does not need an action:

   **a.** Linear

   **b.** Visibility

   **c.** Lookup

   **d.** Rotation

**4.** Rotate action works only with the Rotation parameter:

   **a.** True

   **b.** False

**5.** While you are at the Properties palette of a linear parameter you can set the value set to:

   **a.** None, Increment

   **b.** None, List

   **c.** Increment, List

   **d.** None, increment, and List

**6.** All actions need parameters:

   **a.** True

   **b.** False

**7.** You can test your block inside the block editor before you save the changes you made:

   **a.** True

   **b.** False

## CHAPTER REVIEW ANSWERS

**1.** b

**3.** b

**5.** d

**7.** a

## CHAPTER 21

# BLOCK ATTRIBUTES

### In This Chapter

- What are Block Attributes?
- How to create a block attribute
- How to edit block attributes and value
- How to extract attribute values from drawings

## 21.1  WHAT ARE BLOCK ATTRIBUTES?

We discussed in the previous chapter how to add dynamic features to the block to be more versatile. Still the block does not include any data of any sort. Block Attributes is the way of including information inside a block. When we want to discuss this feature in AutoCAD, we have to ask ourselves, who are we? Are we the creator of the block? Or the people who will use the block? Depending on the answer, we will understand the role of each one:

- If you are the creator of the block, which contains attributes, then you will define the attribute definitions, which are the questions. Examples would be: Color, Cost, Height, and Width
- If you are the user, who will insert the block, you will be the one who inputs values for these questions. Examples would be: Red, 12.99, 3'-6", and 2'-6"

The procedure is very simple:

- Draw the graphical shape of the block
- Using the attribute definition command define the attributes you want to collect data for
- Using the Block command (the normal command to create blocks) define the block which contains the attributes
- Insert the block, and fill in the values
- Set the visibility of the attributes
- Edit the attributes either as values or definitions
- Extract data to a table inside the drawing or as Excel, or database file

## 21.2   HOW TO DEFINE ATTRIBUTES

This will be the first step, which is creating the attribute definitions. To issue this command, go to the **Insert** tab, locate the **Block Definition** panel, then select the **Define Attributes** button:

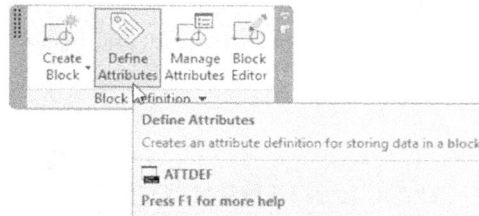

The following dialog box will appear:

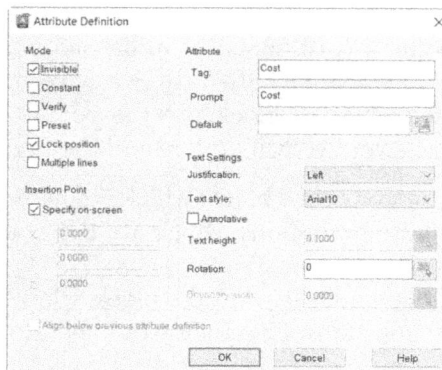

There are four parts in this dialog box, they are:

### 21.2.1 Attribute Part

This part contains three different things to control, which are:

- **Tag**: which is the attribute name, you can select any desired name
- **Prompt**: which represents the message that appears to you asking for a piece of information
- **Default**: if there is a value that will be repeated more than once, this is the place to input it. If this value is connected to a field, then click the button at the right, and input this field

### 21.2.2 Mode Part

There are six modes for the attribute definitions, these are:

- **Invisible**: will create an invisible attribute, existing but doesn't appear in the drawing
- **Constant**: will create a constant value for all of the insertions
- **Preset**: will create an attribute which will be equalized to the default value
- **Verify**: the user will be asked to input the attribute twice to verify the value (applies only for command window input)
- **Lock position**: this mode will lock the position of the attribute definition
- **Multiple lines**: to allow you to input a multiline text rather than a single line text

### 21.2.3 Text Settings Part

The following settings will control the appearance of the attribute definition, these are:

- **Justification**: will allow you to specify text justification
- **Text style**: specify the used text style (recommended to create text style prior to this step) whether the text will be annotative or not
- **Text height**: input the text height to be used if the text style height is 0
- **Rotation**: input the rotation angle of the attribute definition
- **Boundary width**: on only when Multiple lines option is selected. It will be the line length for the multiple lines

### 21.2.4 Insertion Point Part

This part is to specify the position of the attribute definition either by typing the coordinates or by selecting the **Specify on-screen** option. If you are inserting the second attribute, you will have the choice of aligning the new attribute with the previously inserted attribute. Be certain the following checkbox is on:

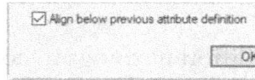

After defining the attributes, start Block command, and create a block from the graphics and the attributes. While selecting the attributes, select them in the same order you prefer them to appear.

## 21.3   INSERTING BLOCKS WITH ATTRIBUTES

You should control how the attribute questions will appear to you. System variable ATTDIA will do the job. There are two values to choose from, they are:

- If **ATTDIA = 0**, questions will be shown at the command window
- If **ATTDIA = 1**, questions will appear in a dialog box, as shown in the following:

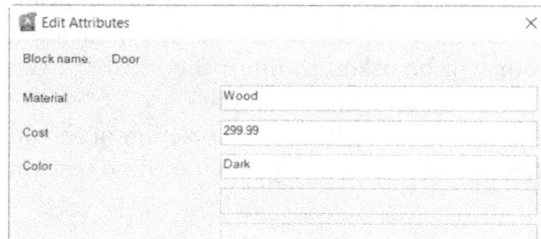

## 21.4   HOW TO CONTROL ATTRIBUTE VISIBILITY

Attributes are either visible or invisible. This part will discuss how to change this status temporarily. In order to control attribute visibility go to the **Insert** tab and locate the **Block** panel. Extend the panel and click the list:

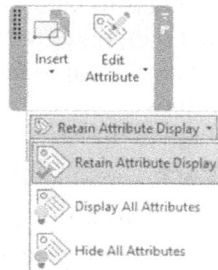

As displayed above, user has three choices to choose from:

- **Retain Attribute Display**, which means AutoCAD will show the visible and hide the invisible
- **Display All Attributes**, which means AutoCAD will show all attributes
- **Hide All Attributes**, which means AutoCAD will hide all attributes

## PRACTICE 21-1   DEFINING AND INSERTING BLOCKS WITH ATTRIBUTES

1. Start AutoCAD 2025

2. Open **Practice 21-1.dwg** file

3. Zoom to the shape at the left

4. Define the following attributes, inserting them below the door shape:

| Tag | Material | Cost | Color |
|---|---|---|---|
| Prompt | Material | Cost | Color |
| Default | Wood | 299.99 | Dark |
| Invisible | ✓ | ✓ | ✓ |
| Preset | ✗ | ✓ | ✗ |
| Justification | Left | Align | |
| Text Style | Arial10 | Align | |

5. Define a new block, naming it "Door," selecting the lower left corner of the door as the insertion point, selecting all objects except attributes, then selecting attributes one by one from top to bottom; to finish click OK

6. Make layer A-Door current

7. Using ATTDIA command, make sure the value is 1

8. Insert the new block into the three openings available in the architectural plan, using the following information:

   **a.** The one at the left, leave the default values

   **b.** The one at the middle, change the material to be Aluminum

   **c.** The one at the right, change the cost to 399.99, and the color to Light

9. Change the visibility mode to show all attributes

10. Save and close the file

## 21.5  HOW TO EDIT INDIVIDUAL ATTRIBUTE VALUES

This part will discuss your ability to edit individual block attributes. The user can edit not only the values, but many other features. There are two ways to reach this command:

- Double-click an existing block with attributes
- Go to the **Insert** tab, locate the **Block** panel, click **Edit Attribute**, then select the **Single** button

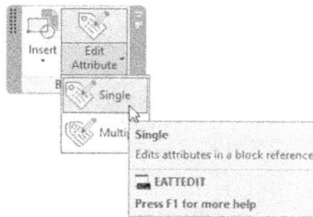

AutoCAD will ask you to select a block, then you will see the following:

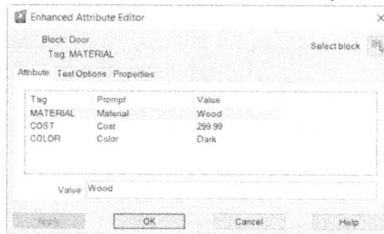

There are three tabs, Attribute, Text Options, and Properties. You will see the Attribute tab when you start editing. In this tab, you can change the value of the attribute for this specific block. Click the **Text Options** tab to see the following:

As you can see, you can change everything related to text style for this specific block and attribute. Select the **Properties** tab to see the following:

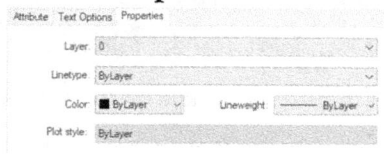

The user can change anything related to properties like layer, linetype, color, and lineweight.

## 21.6   HOW TO EDIT ATTRIBUTE VALUES GLOBALLY

What if we have so many incidences of a block and we want to change one specific value for all incidences? The above method will be lengthy and tedious. AutoCAD offers a simple command that everybody knows, the **Find/Replace** command. To issue this command, go to the **Annotate** tab and locate the **Text** panel; you will find a field with ***Find text*** inside it:

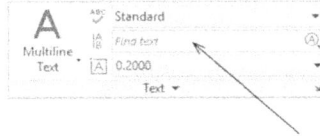

Click inside the field, and type the value of attribute you want to change globally, and then click the small button at the right. You will see the following dialog box:

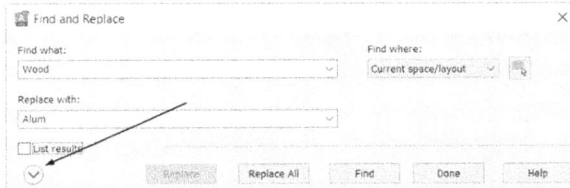

Extend the dialog box by clicking the small arrow at the lower left corner of the dialog box to see it at full. You will see the following:

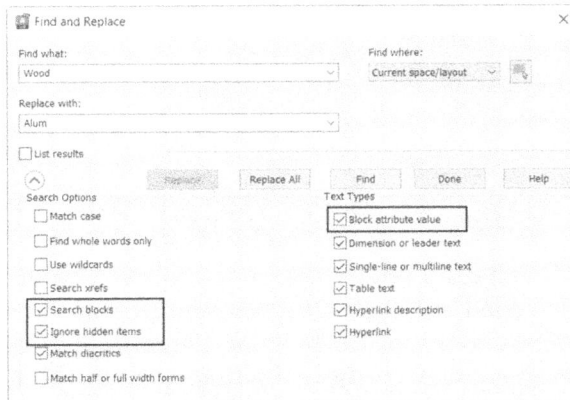

Be certain of the following:

- Under Text Types, make sure Block attribute value is on
- Under Search Options, make sure that Search blocks is on
- Under Search Options, make sure that Ignore hidden items is off

Finally click one of the three execution commands, Find, Replace, or Replace All.

## PRACTICE 21-2 EDITING ATTRIBUTE VALUES INDIVIDUALLY AND GLOBALLY

1. Start AutoCAD 2025

2. Open **Practice 21-2.dwg** file

3. There have been some mistakes in inputting the attribute values, and we need to unify them. Change the following:

   **a.** All doors are wood, none is aluminum

   **b.** All doors are 299.99, none is 399.99

   **c.** All doors are dark, none is light

4. Double-click any two doors and make sure that changes took place

5. Show all attributes

6. Double-click the main door, then select Cost attribute

7. Go to the Text tab, and set the Text style to be Standard, and set the Height = 0.15

8. Go to the Properties tab, and select Text layer

9. Click OK, and compare this attribute to the others

10. Retain Attribute Display

11. Save and close the file

## 21.7 HOW TO REDEFINE AND SYNC ATTRIBUTE DEFINITIONS

What if you found that there is a missing attribute in your block? Or, you want to remove an unneeded attribute? Or, you want to edit the modes of an existing attribute? This means you need to redefine the attributes, using Block Attribute Manager. To issue this command, go to the **Insert** tab, locate the **Block Definition** panel, then select the **Manage Attributes** button:

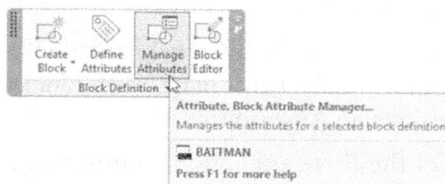

You will see the following dialog box:

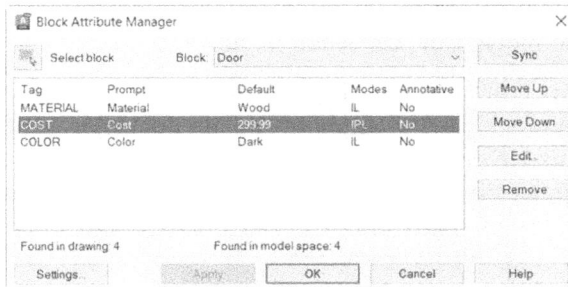

In this dialog box, you will be able to edit Tag, Prompt, Default value, Mode, and all other text options and properties as well

There are two ways to start working:

- Click the button at the top right to select a block
- Click the list to the right of the button to select the desired block

Based on the selected block you will see a list of attributes for this block. You can do four things with these attributes:

- Changing the position of the attribute in relation to other attributes, by moving the selected attribute one position up
- Changing the position of the attribute in relation to other attributes, by moving the selected attribute one position down
- Removing (deleting) the selected attribute
- Edit the definition of the selected attribute, you will see the following dialog box:

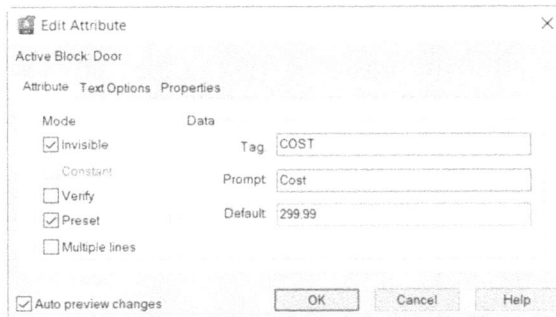

In this dialog box, you will be able to edit Tag, Prompt, Default value, Mode, and all other text options and properties as well

What about adding a new attribute to an existing block? To answer this question, follow these steps:

- Insert the desired block in your drawing
- Explode it

- The existing attributes will appear
- Define the new attribute
- Redefine the block using the same name

Whether, we edit, delete, or add new attributes, this will affect the new insertions of the block. But what about the already inserted incidences? The user should synchronize in order to see the changes. In order to issue this command go to the **Insert** tab, locate the **Block Definition** panel, then select the **Synchronize** button:

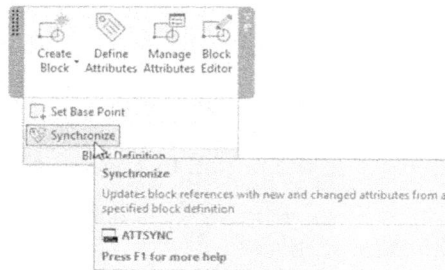

You will see the following prompt:

```
Enter an option [?/Name/Select] <Select>:
```

Type the name of the block, or choose Select option to select it, and you will see the following prompt:

```
ATTSYNC block Window? [Yes/No] <Yes>:
```

If you input Yes (default), AutoCAD will carry on the synchronization process. The newly added attributes will be empty, so the user is invited to fill them by double-clicking the desired incidence of the block.

## PRACTICE 21-3    REDEFINING ATTRIBUTE DEFINITIONS

1. Start AutoCAD 2025

2. Open **Practice 21-3.dwg** file

3. Using the Block Attribute Manager, do the following:

   **a.** Remove Color attribute

   **b.** Make Cost attribute modes only Invisible

   **c.** Change the cost to appear first in the list of questions

**4.** Insert block Door in an empty space in the drawing, then explode it

**5.** Make layer 0 current

**6.** Add a new attribute with the following data:

    **a.** Tag: SUP

    **b.** Prompt: Supplier

    **c.** Default: MYNE Wood

    **d.** Modes: Invisible

    **e.** Insert it below the existing two attributes

**7.** Redefine the block under the same name, using the same base point (very important)

**8.** Double-click any of the blocks, you will see that Supplier attribute is not added, since the sync command was not issued

**9.** Issue the Synchronize command, and select one of the doors

**10.** Now double-click one of the blocks again, do you see the Supplier attribute?

**11.** Save and close the file

## 21.8   HOW TO EXTRACT ATTRIBUTES FROM FILES

The final step will be extracting the data from a drawing and putting it in a table inside the same drawing or another drawing, or creating an Excel sheet or an Access database. Extracting data in AutoCAD will be done through a wizard which will produce the desired table without any hassle. To access this command, go to the **Insert** tab, locate the **Linking & Extraction** panel, then select the **Extract Data** button:

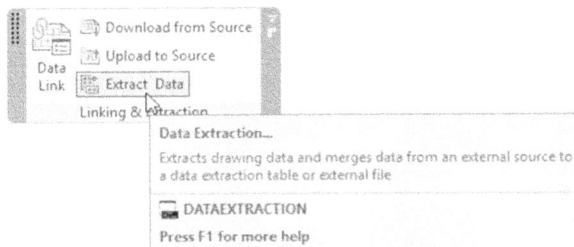

The following dialog box will appear:

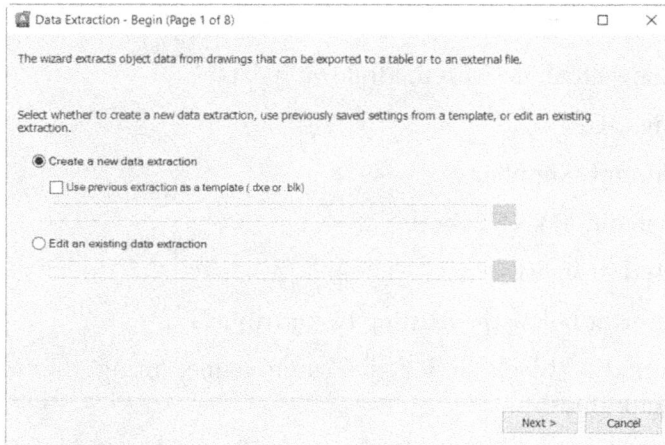

User can select one of the three available options:

- Create a new data extraction
- Use previous extraction as a template
- Edit an existing data extraction

Let's assume we want to create a new extraction. We will select the first option, and click **Next** button. AutoCAD will ask you to save this extraction (file format is *.dxe) for future use, so the following dialog box will appear:

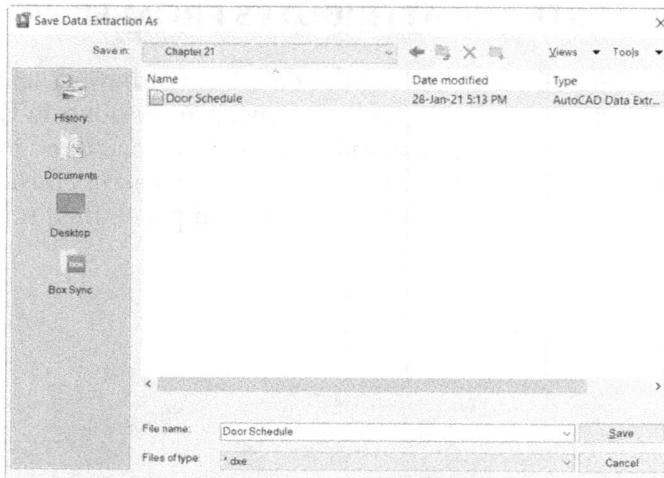

After saving the dxe file, you will see the following dialog box:

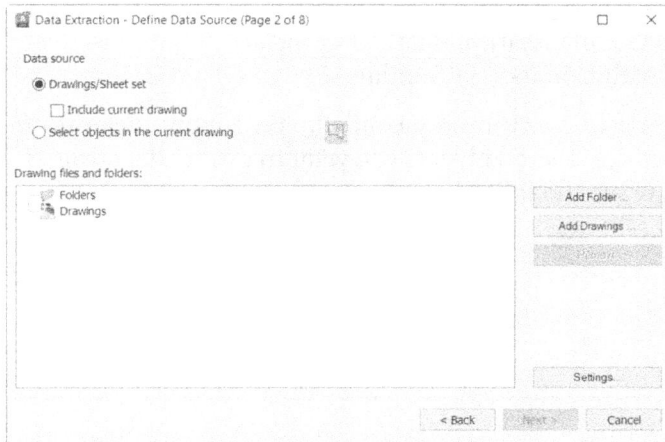

AutoCAD needs to know where it will find the files to extract data from; you have two choices to select from:

- Certain files (or sheet sets), and whether to include or to exclude the current file
- Select objects in the current drawing

Selecting the first choice means you have to specify the folders and drawing files using the two buttons at the right. Clicking the **Add Folder** button, you will see the following dialog box:

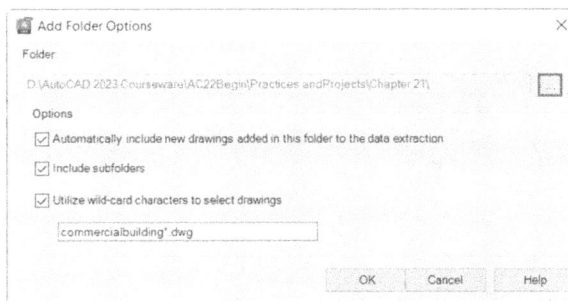

There are four steps to control in this dialog box, they are:

- Click the small button with three dots to specify the folder that contains the desired files
- Select whether or not you want to "Automatically include new drawings added in this folder to the data extraction
- Select whether or not you want to include all subfolders of the selected folder

- Select whether or not you want to "Utilize wild-card characters to select drawings." This is similar to what we can do in Windows Search tool. For instance CommercialBuilding *.dwg means you want to select all drawing files starting with CommercialBuilding

If you want to go with the second choice, simply click the small button at the right to select the desired objects you want to extract data from. If you make a wild card, you may see the following:

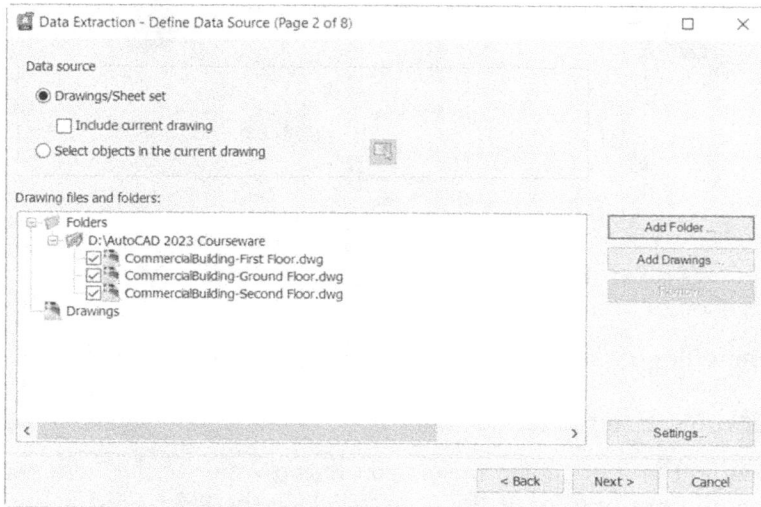

When done click the **Next** button and you will see the following dialog box:

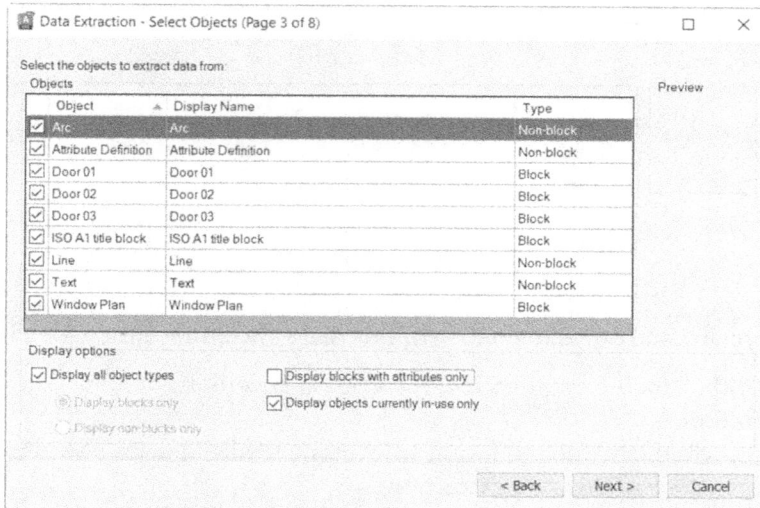

The list is showing objects and blocks. In order to refine this list, use the checkboxes beneath. You have three choices:

- Select whether or not to Display all object types. If you turn this off, you have to select whether to Display blocks only, or to Display non-blocks only
- Select whether or not to Display blocks with attributes only
- Select whether or not to Display objects currently in-use only

The below example will show only blocks with attributes, then we deselected one of the blocks that we do not want to include in the extraction process:

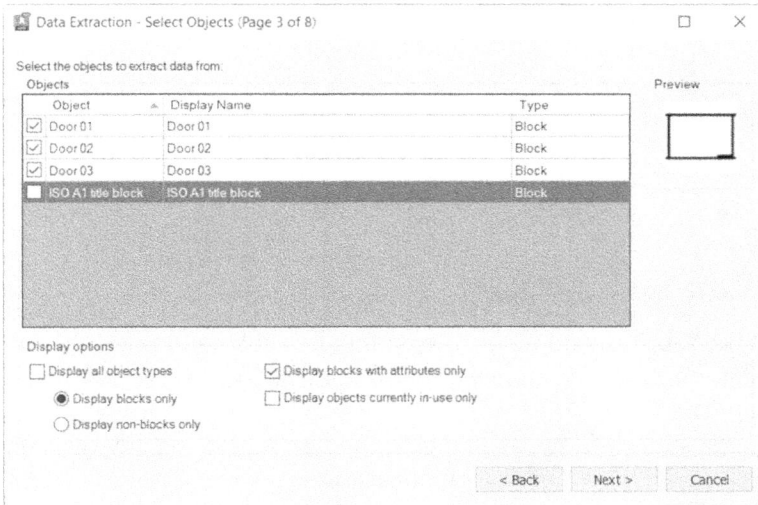

When done, click **Next** button, and you will see the following dialog box:

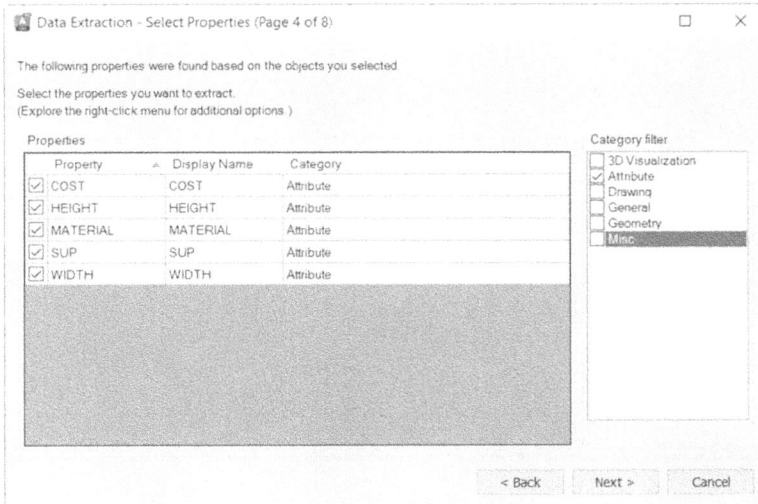

At the right and under Category filter, deselect all choices except Attribute. Also, you can further deselect the attributes that you do not want to include in the extraction process. When done click **Next** to continue to the fifth step in the extraction wizard. You will see the following dialog box:

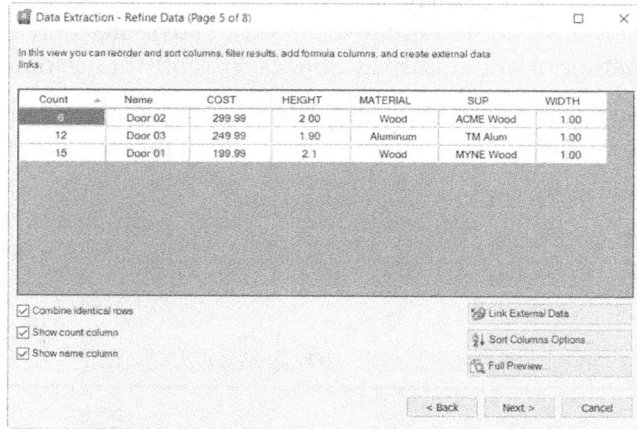

There are some basic functions you can do with this page:

- Sort the table ascending or descending by clicking the desired column header
- Change the location of any column, by clicking the header and dragging and dropping
- Change column width by dragging the line separating any two columns
- You can select to Combine identical rows
- Show count column or not
- Show name column or not

If you right-click any header, you will see the following menu:

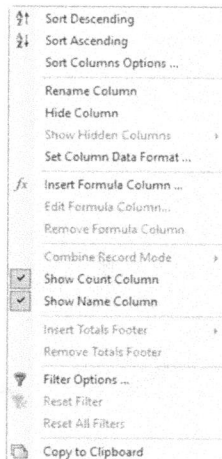

Use this menu to do all or any of the following:

- Rename a column
- Hide and Show columns
- Set Column Data Format, you will see the following dialog box:

- Insert a new formula column; you will see the following dialog box. In this dialog box type the name of the new column, then drag-and-drop the names of the attributes that you want to use, using the arithmetic function between them

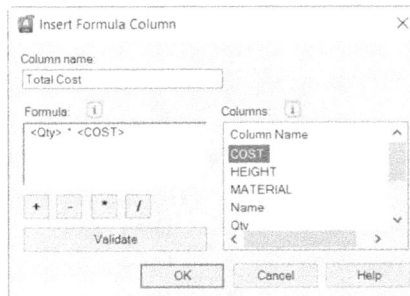

- Edit and Remove Formula Column
- Insert Totals Footer
- Create a filter for the selected column

If you want to establish a link between this table and an Excel file, click the Link External Data button at the right.

If you want to set advanced sorting criteria then click the Sort Columns Options button at the right.

If you want to establish a data link between these extracted data and Excel sheet.

Whereas the last button (Full Preview) will show a preview of the table.

When done, click Next to the next step, you will see the following dialog box:

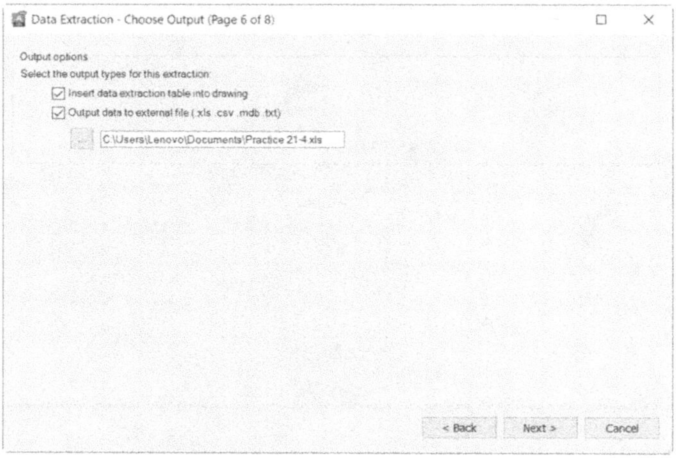

The user in this page can decide to insert the data extraction as a table in the current file, and produce an external file such as an Excel sheet file and an Access database file.

When done click **Next** to go to the seventh step, the following dialog box will appear:

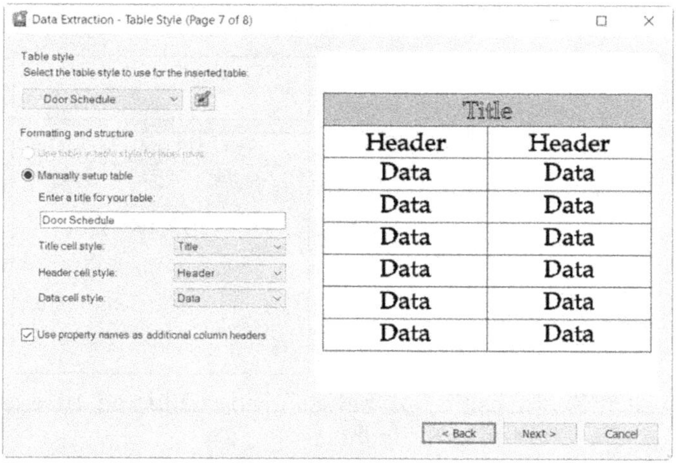

Select the table style you want to use. When done click **Next** to go to the last step, the following dialog box will appear:

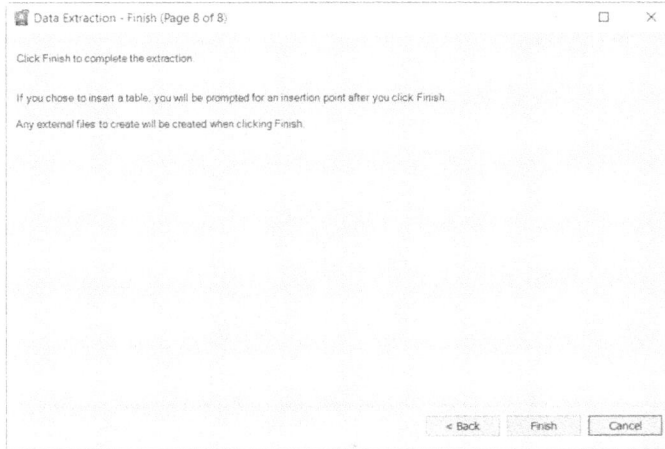

Click **Finish** to wrap up the extraction process.

## PRACTICE 21-4   EXTRACTING ATTRIBUTES

**1.** Start AutoCAD 2025

**2.** Open **Practice 21-4.dwg** file

**3.** Extract data from the following three files which exist in your Chapter 21 folder of practices:

   **a.** CommercialBuilding-Ground Floor.dwg

   **b.** CommercialBuilding-First Floor.dwg

   **c.** CommercialBuilding-Second Floor.dwg

**4.** This is the final product of the extraction, do whatever it needs to produce an identical replica:

| Door Schedule | | | | | | | |
|---|---|---|---|---|---|---|---|
| Name | HEIGHT | WIDTH | MATERIAL | Supplier | Qty | COST | Total Cost |
| Door 01 | 2.1 | 1.00 | Wood | MYNE Wood | 15 | 199.99 | $2,999.85 |
| Door 02 | 2.00 | 1.00 | Wood | ACME Wood | 6 | 299.99 | $1,799.94 |
| Door 03 | 1.90 | 1.00 | Aluminum | TM Alum | 12 | 249.99 | $2,999.88 |
| | | | | | 33 | | $7,799.67 |

**5.** Save and close the file

## NOTES

## CHAPTER REVIEW

1. While you are defining a new attribute you can make the default value equal to a field:

   **a.** True

   **b.** False

2. One of the following statements is not true about the extracting process:

   **a.** While you are extracting you can add a formula column

   **b.** You can add more than one formula column

   **c.** Formula columns will use attributes and fields

   **d.** You can add a total footer

3. You can edit attribute values by double-clicking the desired block:

   **a.** True

   **b.** False

4. After inserting a block containing attributes in my drawing you can do all but:

   **a.** Edit the values one by one

   **b.** Edit the attribute definitions

   **c.** Allow AutoCAD to synchronize the values automatically

   **d.** Add a new attribute to the block

5. _____ mode will create an attribute which will be equalized to the default value

6. Invisible and constant are _____

7. _____ is a system variable which will control to show or not show a dialog box while inputting values for attributes

## CHAPTER REVIEW ANSWERS

**1.** a

**3.** a

**5.** Preset

**7.** ATTDIA

## CHAPTER 22

# EXTERNAL REFERENCING (XREF)

## In This Chapter

- What is External Reference (XREF)?
- Inserting External Reference
- External Reference Layers
- Editing External Reference Editing
- External Reference Functions
- External Reference Clipping
- External Reference Special Function
- eTransmit and External Reference

## 22.1 INTRODUCTION TO EXTERNAL REFERENCES

We are working these days in a collaborative work environment. Each user needs his/her coworkers' drawings to work as an underlay. External Reference (or xref as many people like to call it) will help the engineers/draftsmen to collaborate and coordinate their drawings with each other by allowing them to bring in the others' drawing to your drawing with minimal add-on size and segregating the content of each drawing. This is an old technique in AutoCAD, but in the last two or three versions of AutoCAD, it was allowed to bring in other file formats like DWF, DGN, PDF, and any image file format, which makes it a very interesting subject to discuss.

External Referenced files will establish a link with the original file, so any change will take place in the original file, and you will be notified of any updates. When a user opens their file, which uses xrefed files, the latest version of these files will be loaded into the drawing.

External Reference is a better method than using the Insert command to bring in contents from DWG files, as this method will add the whole file into your file, it cannot be updated, and will mix all coming layers and blocks with your current layers and blocks.

## 22.2 INSERTING EXTERNAL REFERENCES INTO DIFFERENT FILE FORMATS

There is a single command that can insert all the five types of files in the current drawing file. To reach this command go to the **Insert** tab, locate the **References** panel, then select the small arrow at the lower right:

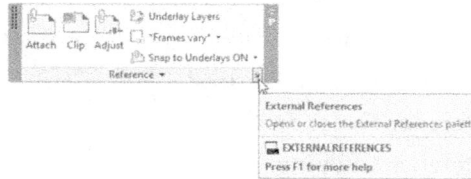

The **External Reference** palette will appear, showing only the name of the current file:

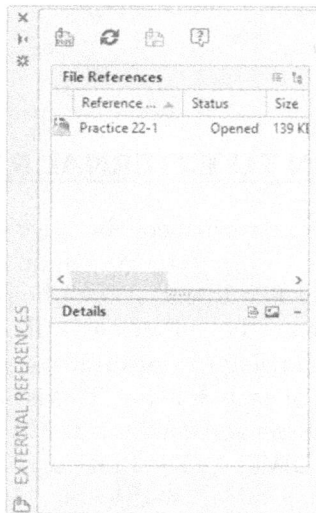

To attach any file format, simply go to the top left part of the palette, where a small list of button exists; click it, and you will see the following:

As shown above, you can select one of the following file formats to bring into your current file:

- Attach DWG file
- Attach Image file
- Attach DWF file
- Attach DGN file (Microstation V8 file)
- Attach PDF file
- Attach Point Cloud
- Attach Coordination Model (Naviswork Document File NWD)

### 22.2.1 Attaching a DWG File

You will use this command to attach AutoCAD drawing in the current drawing. The normal Open dialog box will appear, and once you click OK, you will see the following dialog box:

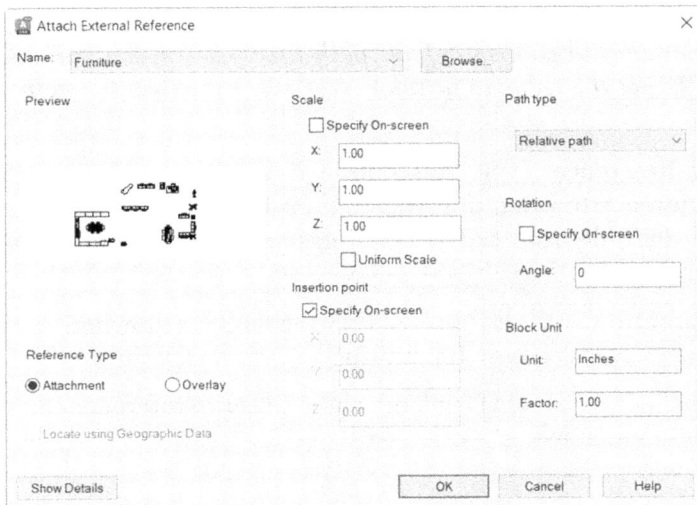

As shown, four parts of this dialog box are identical to the Insert block command:

- Scale
- Insertion point
- Rotation
- Block unit

The two different things are: Path type and Reference type.

As for Path type, we already said that AutoCAD keeps a link to the original file to keep track of the changes that will occur. In order to do that AutoCAD saves the current path of the DWG file attached. It will ask you the type of the path; there are three possible choices:

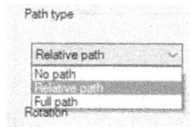

- **Full path**: AutoCAD will save the exact path of the attached DWG file, remembering the drive and folder(s), which means if the file is moved to a different place, AutoCAD will report an error
- **Relative path**: The default option. AutoCAD will only save the name of the folder containing the file, ignoring the other superior folders and drives. If your folder is moved to another place, AutoCAD will be able to locate it
- **No path**: AutoCAD will not save the path for the DWG file; as an alternative, it will search the current folder, then Search paths mentioned in the **Options** dialog box and **Files** tab, both paths mentioned under Project and Support

**NOTE** *If you want to control default path type, use system variable **REFPATH-TYPE**. Set the value to be 0 for No Path, 1 for Relative Path, and 2 for Full Path.*

As for Reference Type, this option is dealing with multiple people using external reference to bring in DWG to their files. Let's assume that B brings in a DWG to his/her file from A. Then C brings in B's file. The following is what C will see in their file:

- **Attachment**: C will see both A's drawing, and B's drawing
- **Overlay**: C will see only B's drawing

As a final note, AutoCAD will not allow a user to reference the same DWG file twice in the same drawing.

### 22.2.2 Attaching an Image File

You will use this command to attach an image file in the current drawing. You will see a normal Open dialog box to select your desired file, and once you click OK, you will see the following dialog box:

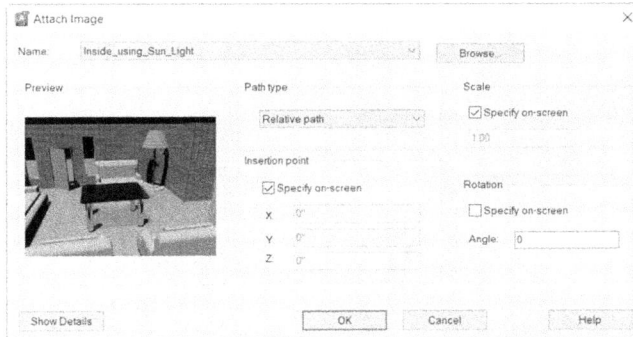

We already discussed all the parts that appear in this dialog box. AutoCAD permits reference to the same file more than once in the same drawing.

### 22.2.3 Attaching a DWF File

You will use this command to attach a DWF file in the current drawing. You will see a normal Open dialog box to select your desired file, and once you click OK, you will see the following dialog box:

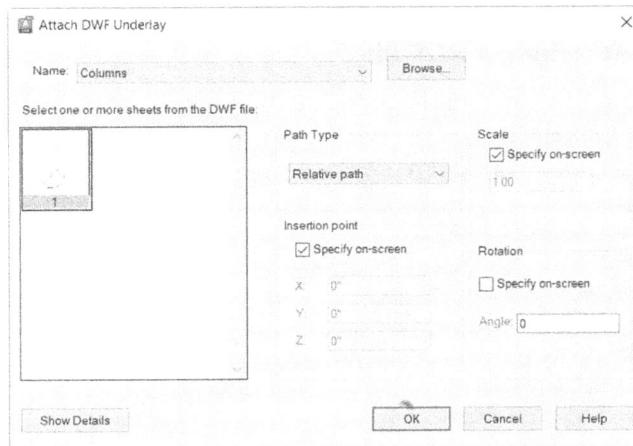

This dialog box is identical to the attaching DWG file dialog box except for the part for selecting which sheet of the DWF file you want to attach, that is, if you have a multiple sheet DWF file. AutoCAD permits to reference the same file more than once in the same drawing.

### 22.2.4  Attaching a DGN File

You will use this command to attach a DGN (Microstation) file in the current drawing. You will see a normal Open dialog box to select your desired file, and once you click OK, you will see the following dialog box:

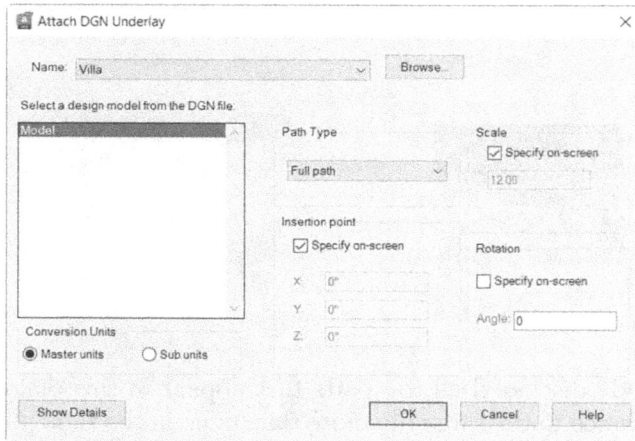

This dialog box is identical to DWG file except for one thing: the part about Conversion Units. If you know that MicroStation files deal with two units, Master and Sub, you have to tell AutoCAD whether the current AutoCAD unit is Master or Sub. AutoCAD permits reference to the same file more than once in the same drawing.

### 22.2.5  Attaching a PDF File

You will use this command to attach a PDF file in the current drawing. You will see a normal Open dialog box to select your desired file, and once you click OK, you will see the following dialog box:

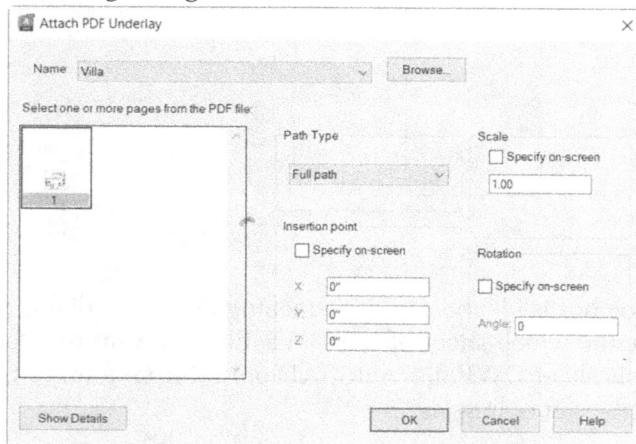

This dialog box is identical to the attaching DWG files dialog box except the part about selecting which sheet of the PDF file you want to attach, that is, if you have a multiple sheet PDF file. AutoCAD permits reference to the same file twice in the same drawing.

Note the following:

- The current layer will be always the carrier of the reference file, based on that, the user is advised to remember the name of the current layer when they attache the reference file, so as not to freeze it in the future
- You can use as well use the system variable XREFLAYER. If you type this system variable in the command window, you will see the following prompt:

```
Enter new value for XREFLAYER, or . for use current
<"use current">:
```

- The default is Current layer, but if you typed a new name, this will be where the XREF will reside, even if this layer does not exist in the current drawing
- The first time you insert a reference file of any type, you will notice at the right part of the Status bar an icon called **Manage Xrefs**; clicking it once will show the External Reference palette:

## 22.3 EXTERNAL REFERENCE PALETTE CONTENTS

Mastering the different parts of the palette is very helpful to obtain the desired information. We will discuss each part of the palette and what it means for you. The upper part has two views:

- **List view**: it shows all the files attached right now to your file. Each attachment has a different icon to its left to distinguish it from the others. Also, you can see to the right of the name lots of information about the file, such as its status, size, path, etc.:

| Reference Name | Status | Size | Type |
|---|---|---|---|
| Practice 22-1 Solved | Opened | 120 KB | Current |
| Columns - Columns-Model | Loaded | 7.35 KB | DWF |
| Furniture | Loaded | 138 KB | Attach |
| Inside_using_Sun_Light | Loaded | 77.9 KB | JPG |

- **Tree view**: it will show the same files but in tree view. This is very handy if your referenced DWG file has reference files by itself. This view has no details

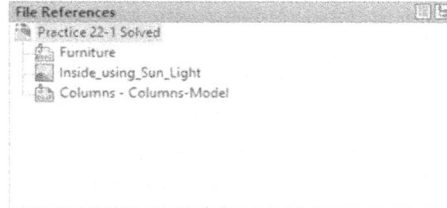

As for the lower part, it has two views:

- Details, which will show the following. You can see that the Saved Path is editable and can be changed from this part:

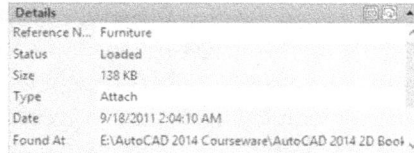

- Or, you will see a preview of the attached file, similar to the following:

## 22.4    USING THE ATTACH COMMAND

AutoCAD offers another way to attach all types of files without using a palette, which is the **Attach** command. To access this command go to the **Insert** tab, locate the **Reference** panel, then select the **Attach** button:

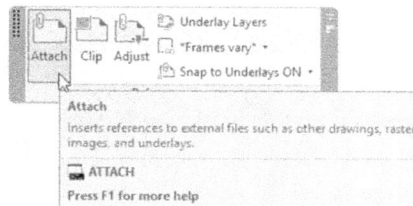

The following dialog box will appear to select first the type of the file (use the list shown below), then to select the desired file:

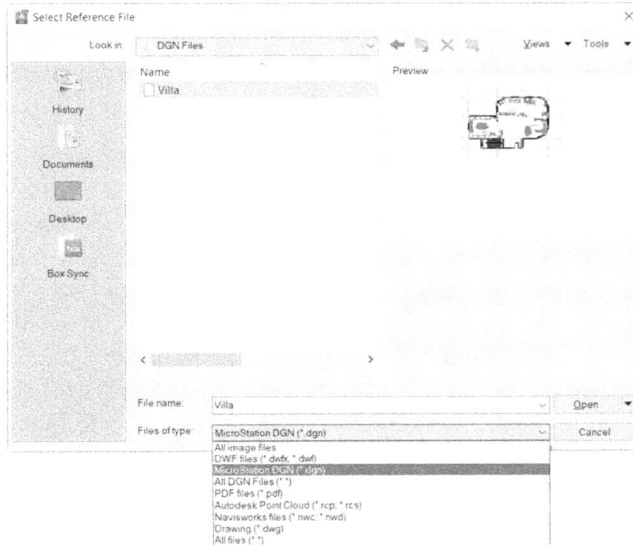

## 22.5    REFERENCE FILES AND LAYERS

External Reference – as we mentioned previously – segregates coming layers from reference DWG files and the original layers. To check this out, you should visit the Layer Properties Manager palette after the attachment, and you will see the following:

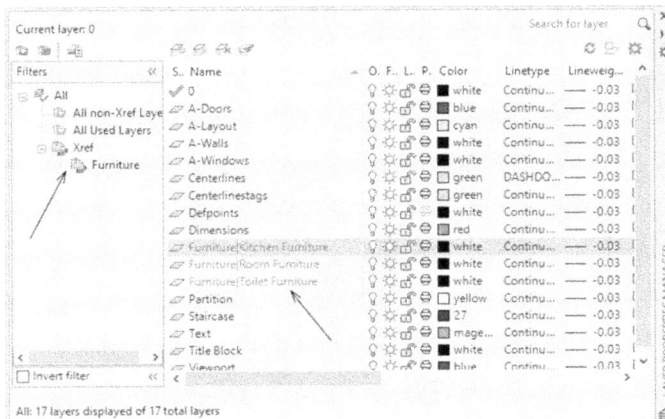

Seeing the above palette, the user can easily distinguish the original layers of the current file, and layers coming from a DWG reference file. At the left pane, you can see an automatic filter called Xref.

Layers coming from a DWG reference file always start with the name of the file, then a pipe (|), and then the name of the layer: *filename|layername.*

We have to mention two big facts about layers coming from a reference:

- AutoCAD disallows you to draw on reference layers, so you cannot make any of them current
- AutoCAD allows you to change the properties of referenced layers (color, linetype, lineweight, etc.) and to change the status of a layer (on/off, freeze/Thaw, etc.) without affecting the original file. Use and set
- You can retain the changes you made to the XREF layer, or you can select to ignore these changes once the file is saved and closed. Using Layer Properties Manager, click the Settings button at the top right part of the palette. At the top right part under Xref Layer Settings, you will see the following:

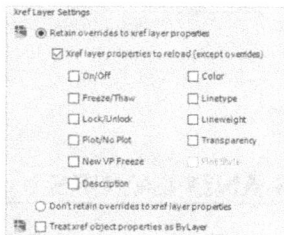

- Clicking the top radio button, you are selecting to keep all changes you made to the Xref layers once the file is saved and closed. You can pick and choose which one of these changes you want to retain
- Clicking the bottom radio button, you are telling AutoCAD whatever changes you made to the Xref layers should not be saved once the file is saved and closed
- If you made any change to Xref layer, a new filter called Xref Overrides will be created which will hold only modified layers from Xref layers. Check the following picture:

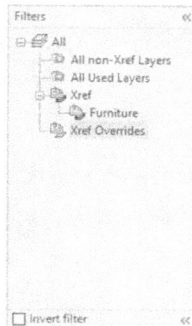

## 22.6 CONTROLLING THE FADING OF A REFERENCE FILE

When you attach a DWG file to your current file, you will notice it is faded. This is true because the default setting for any DWG reference file is 70% as it shows in the Reference panel.

You can do two things regarding this issue:

- Control the fading percentage by increasing it or decreasing it using the Reference panel:

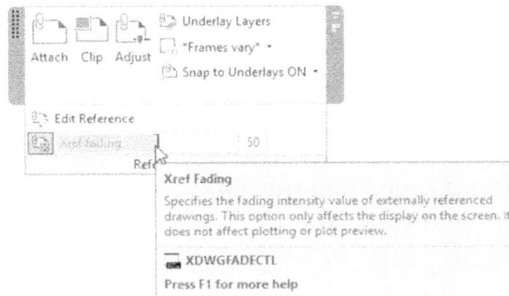

- Turning on/off the fading using the Reference panel:

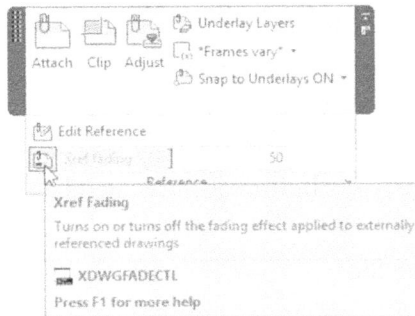

## PRACTICE 22-1 ATTACHING AND CONTROLLING REFERENCE FILES

**1.** Start AutoCAD 2025

**2.** Open **Practice 22-1.dwg** file

**3.** Make sure that layer 0 is the current layer

**4.** Using External Reference, attach Furniture.dwg file using 0,0

**5.** Using External Reference, attach Columns.dwf using 0,0

**6.** Go to D-Size Architectural Plan layout

**7.** Using External Reference, attach Inside_using_Sun_Light.jpg and using the user base point and scale, go to the frame and insert it in the upper part of the legend using proper scale to fit within

**8.** Close the External Reference palette

**9.** Start Layer Properties Manager

**10.** How many layers are added with the Furniture.dwg? _____

**11.** What are their names? _____, _____, _____

**12.** Change the color of Toilet Furniture to Red

**13.** Freeze layer Kitchen Furniture

**14.** Click Settings in Layer Properties Manager, and click Don't retain overrides to xref layer properties option

**15.** Save the file and close it

**16.** Reopen the file, what happens to the reference file layers? _____

_____

Why? _____

_____

**17.** Turn off the fading of reference DWG file, then turn it on to see the difference

**18.** Click the Manage Xrefs button at the Status bar to show the External Reference palette, then examine the four parts of the palette

**19.** Save and close the file

## 22.7 EDITING AN EXTERNAL REFERENCE DWG FILE

AutoCAD allows you to edit any external referenced DWG file using two methods, they are:

- Using Edit Reference command
- Using Open command

Ethically, the user should not edit another person's file without their permission. What we will discuss here is the possibility of doing so.

### 22.7.1 Using the Edit Reference Command

This command will edit the external referenced DWG file in its place. To issue this command, go to the **Insert** tab, locate the **References** panel, then select the **Edit Reference** button:

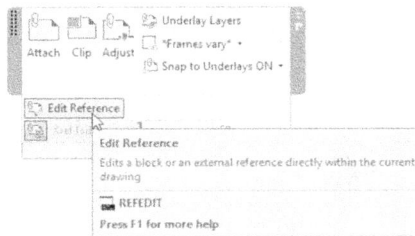

You will see the following prompt:

```
Select reference:
```

Click on the desired DWG to edit, you will see the following dialog box:

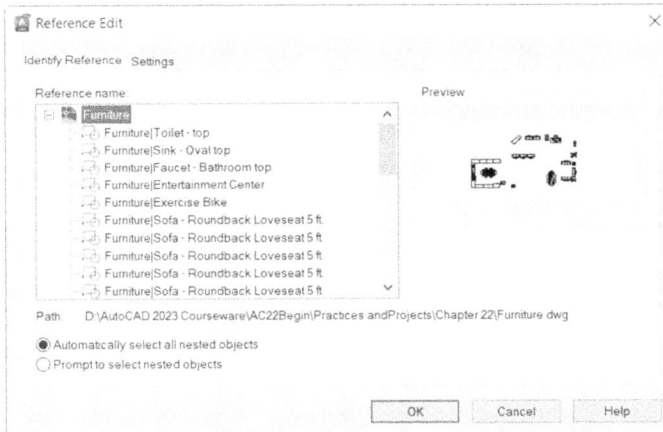

As you can see, the filename of the selected file appears with all of the blocks coming with it, which proves that this command can edit not only external referenced files, but blocks as well. If this is the file you want to edit, click the OK button. Once you do, two things will occur:

- All the drawing will be dimmed except the selected file
- A new context tab titled Edit Reference will appear showing four buttons

Meanwhile, the user can use all the Modifying commands in order to alter the file in its place.

The context tab will contain four buttons, they are:

- **Add to Working Set**: to allow you to add more objects to the current set of objects to edit
- **Remove from Working Set**: to allow you to remove objects from the current set of objects to edit
- **Save Changes**: this is the final step after you finish editing
- **Discard Changes**: this is the final step after you finish editing and decide you no more need these changes to take place

### 22.7.2 Using the Open Command

This command will open the external referenced DWG file for editing, then save the file with the new amendments. To do that, click any object in the external referenced DWG file, then right-click; a long menu will appear, select Open Xref option:

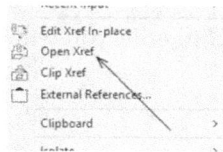

The file will open, ready for editing. Make the changes you need, then save the file and close it.

There are two things worth noting while discussing this issue:

- There is no way you can prevent anybody who referenced your file from opening it and making changes. But you can prevent them from doing editing in place. To do that go to the **Options** dialog box, then select the **Open and Save** tab, and click off "Allow other users to Refedit current drawing" option.

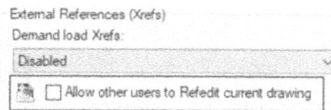

- Regardless of the method used, if any changes were saved to the external referenced file, you will see a bubble stating that changes took place to the file, and you need to reloaded it, as shown in the following:

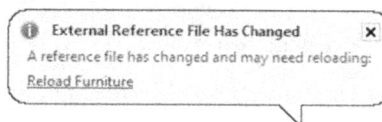

## 22.8  EXTERNAL REFERENCE FILE-RELATED FUNCTIONS

When you are dealing with the External Reference palette using the top part, you can right-click the name of the file to receive the following menu:

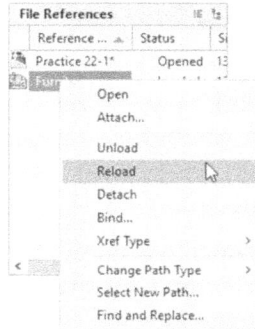

As shown, there are several commands:

- Open (already discussed)
- Attach (will show you the Attachment dialog box to edit the settings of attachment)
- Unload
- Reload
- Detach
- Bind
- Xref Type (Attach or Overlay)
- Change Path Type (Make Absolute, Make Relative, Remove Path)
- Select New Path
- Find and Replace

We will discuss the new commands:

### 22.8.1  Unload Command

Unloading an attached file means the file will disappear from the display, but the link will stay. This is (in principle) similar to the freeze function in layers.

The unloaded file will appear as the following in the list:

### 22.8.2 Reload Command

Reload command can be used to reload an unloaded file. Or you can use it to refresh all of your current links if needed.

### 22.8.3 Detach Command

This command will make the link disappear just like the Unload command, except the user will lose the link as well. Therefore, you will not see the file in the list. If you need the file again, you have to attach it again.

### 22.8.4 Bind Command

If for any reason the user opts to keep the latest copy of an external referenced DWG file in his/her current file and cut the link between the two files, then the user should use the Bind command. Bind command will bind the contents of the desired file inside your current drawing and will remove the link. Bind command offers two methods for binding, as you will see in the following dialog box:

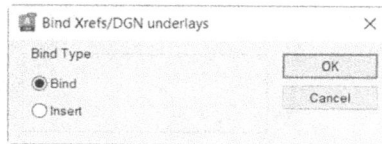

As shown you have two choices:

- Bind
- Insert

The difference between the two methods is the naming of layers, styles, and blocks after the binding process end. Assume we have a layer called:

*Furniture|Frame*

If you use bind the name will become: Furniture$0$Frame (the name of the layer that the file comes from is still there, and "|" was converted to $0$).

On the other hand, if we use Insert, it will become Frame.

That applies to styles (text style, dimension style, etc.) and blocks.

### 22.8.5 Xref Type

This option will change the xref type from Attach to Overlay and vice versa.

### 22.8.6 Change Path Type

Upon the attachment of xref, the user should specify whether they want to make the path Absolute, Relative, or No Path. With this option, the user can change this setting to any of the other options.

### 22.8.7   Select New Path

This option will open Select new path dialog box for you to select new location for a lost reference file, and then provides you the ability to apply the same new location for other missing references.

### 22.8.8   Find and Replace

Assume you want to change the path of a group of reference files, this is the answer for you. This option will find all the references that use a specified path from all the references that you selected and replace them with a new path.

## PRACTICE 22-2   EXTERNAL REFERENCE EDITING

1. Start AutoCAD 2025

2. Open **Practice 22-2.dwg** file

3. Click on any object in Furniture.dwg, then right-click and select the Open Xref option

4. Erase the table and chairs in the kitchen and save the file

5. A bubble will come to reload the file, reload the file to get the latest version of it

6. Using the External Reference palette, unload Furniture.dwg

7. Reload Furniture.dwg

8. Open the Layer Properties Manager palette and take a look at the three layers coming from Furniture.dwg

9. Using the External Reference palette, bind Furniture.dwg using the Bind option

10. List the things took place after binding:

    a. _____

    b._____

    c._____

11. Save and close the file

## 22.9   CLIPPING AN EXTERNAL REFERENCE FILE

AutoCAD allows you to clip an external referenced file (of any type) to show portions of the file. To reach this command go to the **Insert** tab, locate the **Reference** panel, then select the **Clip** button:

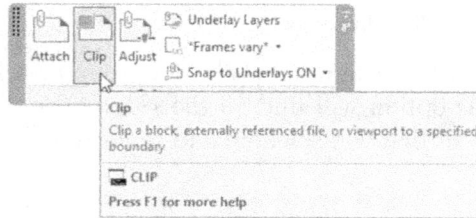

The following prompt will appear:

```
Select Object to clip:
```

You can select any type of external referenced file. For simplicity, we will show the DWG file prompts. The prompts you will see are:

```
Enter clipping option [ON/OFF/Clipdepth/Delete/
generate Polyline/New boundary] <New>:
```

The default option is to create a new boundary to clip the existing referenced file. So if you press [Enter] the second prompt will appear:

```
[Select polyline/Polygonal/Rectangular/Invert clip]
<Rectangular>:
```

Using this prompt you can:

- Select a drawn polyline to act as a boundary
- Define a new boundary using polygon option which will draw any irregular shape
- Define a new boundary using rectangular shape
- Invert an existing boundary

Based on the option you will select, define the desired boundary; hence you will now see some of the referenced file rather than all of it.

The other options are:

- Turn the clipping on/off while keeping it
- Deleting the clipping boundary
- Clip depth (that has to do with the 3D)

Note that by default the user can't see the clipping boundary, due to the hidden nature of it. To show it and accordingly manipulate it go to the **Insert** tab, locate the **Reference** panel, and check the second list at the right; you will see:

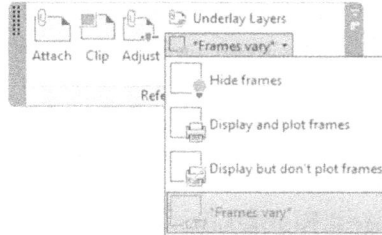

There are four choices to select from, they are:

- Hide frames
- Display and plot frames
- Display but don't plot frame
- Frames vary

While the frame is shown, the user can click the frame itself to show the following:

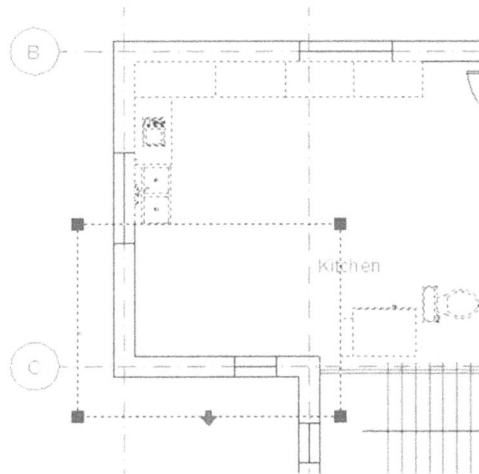

With grips are displayed you can:

- Resize the frame
- Invert the frame
- Delete the frame (deleting the frame using [Del] button will mean deleting the external reference file—be careful)

## 22.10   CLICKING AND RIGHT-CLICKING A REFERENCE FILE

There are two actions that you can make while you are dealing with an external referenced file:

- When you click it, a context tab will appear with related panels for editing, clipping, and other options
- When you click, then right-click, similar specific functions will appear with the menu relating to the type of referenced file selected

### 22.10.1   Clicking and Right-Clicking a DWG File

Clicking (one click) a DWG referenced file will lead to the External Reference context tab, which includes three panels, as shown in the following:

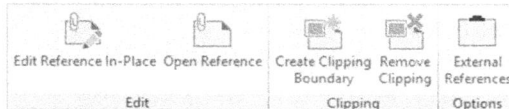

The Edit panel contains the Edit Reference In-Place and Open Reference buttons (both discussed), the Clipping panel contains the Create Clipping Boundary and Remove Clipping buttons (both discussed), and finally the Options panel contains an External References button which will show the External Reference palette.

Right-clicking a DWG referenced file will show the following menu:

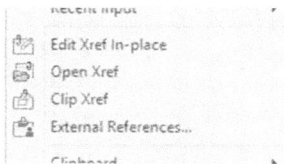

They are identical to the buttons in the context tab.

### 22.10.2   Clicking and Right-Clicking an Image File

Clicking (one click) an image referenced file will lead to the Image context tab, which includes three panels, as shown in the following:

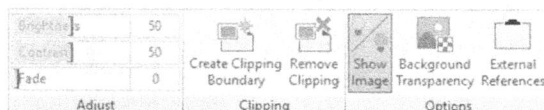

The Adjust panel contains Brightness, Contrast, and Fade slider, to control the image colors; the Clipping panel contains Create Clipping Boundary, and Remove

Clipping buttons; finally the Options panel contains Show Image, Background Transparency (whether to show objects behind the image), and External References buttons.

Right-clicking an image referenced file will show the following menu:

They are identical to the buttons in the context tab.

### 22.10.3   Clicking and Right-Clicking a DWF File

Clicking (one click) a DWF referenced file will lead to show the DWF Underlay context tab, which includes three panels, as shown in the following:

The Adjust panel contains Contrast, Fade sliders, and Display in the Monochrome button to control the image colors; the Clipping panel contains Create Clipping Boundary and Remove Clipping buttons; the Options panel contains Show Underlay, Enable Snap, and External References buttons; and finally the DWF layers panel includes Edit Layers, which will present the following dialog box:

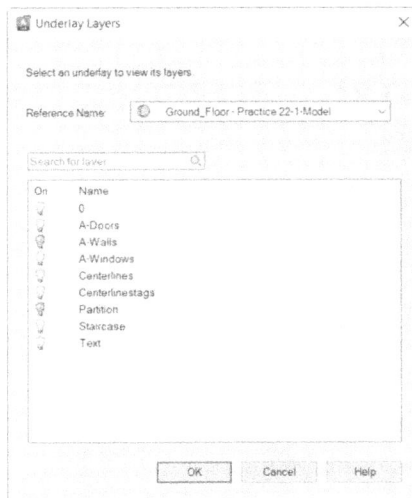

Right-clicking a DWF referenced file, will show the following menu:

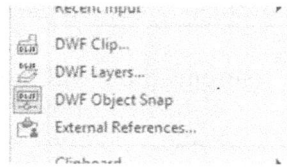

They are identical to the buttons in the context tab.

DGN and PDF tabs are identical to DWF, so there is no need to repeat them.

To wrap up this subject, you will notice that there are two buttons in the Reference panel that we did not explore; they are:

- **Underlay Layers button**: which will show a dialog box of Edit Layers mentioned above for all DWF, DGN, and PDF attached to your current drawing
- **Snap to Underlays button**: which will acquire a point on DWF, DGN, and PDF files, you will see the following list:

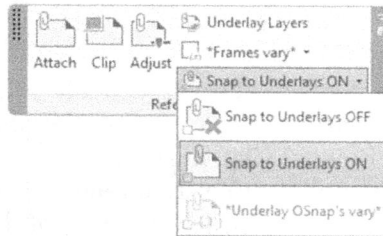

# PRACTICE 22-3   EXTERNAL REFERENCE FILE CLIPPING AND CONTROLLING

1. Start AutoCAD 2025

2. Open **Practice 22-3.dwg** file

3. Using the Insert tab and Reference panel, click Clip, click one of the furniture, select rectangular shape, and create a rectangle around the kitchen and toilet

4. What happened to the other furniture? _____

5. Select to show the frame of the clip

6. Click the frame and shrink it to include only the kitchen

7. Invert the clipping frame

**8.** Using the clip command delete the clipping frame

**9.** Go to D-Size Architectural Plan layout, and zoom to the image at the top right of the title block

**10.** Click the frame of the image, what happened? _____

**11.** Change the Fade % to 30%

**12.** Using the context tab, clip the image to a smaller size

**13.** Go back to Model space, select one of the columns, then right-click, and set DWF snap to off

**14.** Hide frames of clipping

**15.** Save and close the file

## 22.11 USING THE eTRANSMIT COMMAND WITH EXTERNAL REFERENCE FILES

If you attached to your current file several DWG files, along with image and DWF files, then your boss told you that we need to send your file to somebody via e-mail. If you simply write your e-mail and attach your current file, you are making a big mistake, because your file is now dependent on other files to look complete; hence AutoCAD will show the recipient an error message of the sort "Can't locate xxxx.dwg attached" or "Can't locate xxxx.jpg attached."

So we definitely need other methods to dispatch a DWG file with the attached files, even if you are not using reference files, because DWG files use font, hatch, linetype, and shape, files which may not exist in the recipient's computer. This method is eTransmit command, which will (using a simple click) collect all related files to your current file, and then zip them, and prepare them to be sent either using e-mail or other means.

Using the Application Menu, select **Publish/eTransmit**. If you did not save your file, you will see the following message:

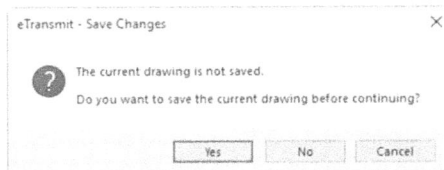

Save your file by clicking Yes, and then the following dialog box will appear:

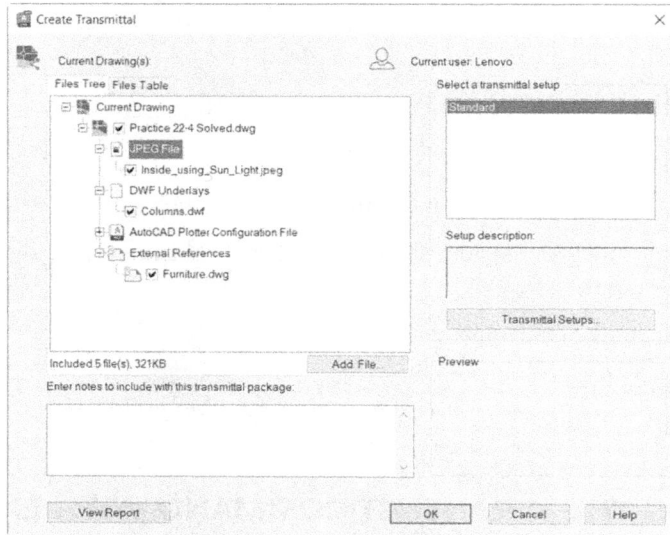

At the top left you can see two tabs; the default is the Files Tree, and another view is Files Table as shown in the following:

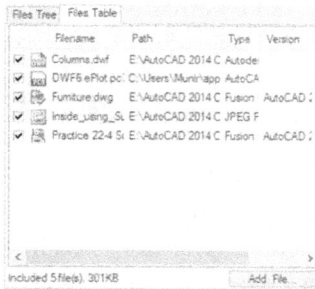

You can include or exclude any undesired file. Using the button labeled **Add File,** you can add other files to the existing list.

Use the Transmittal Setups button to create your own parameters. Click the Transmittal Setups button, then click New button, and you will see the following dialog box:

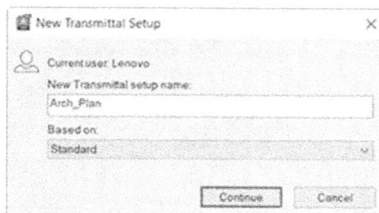

Type the name of the new setup, then click Continue button to see the following dialog box:

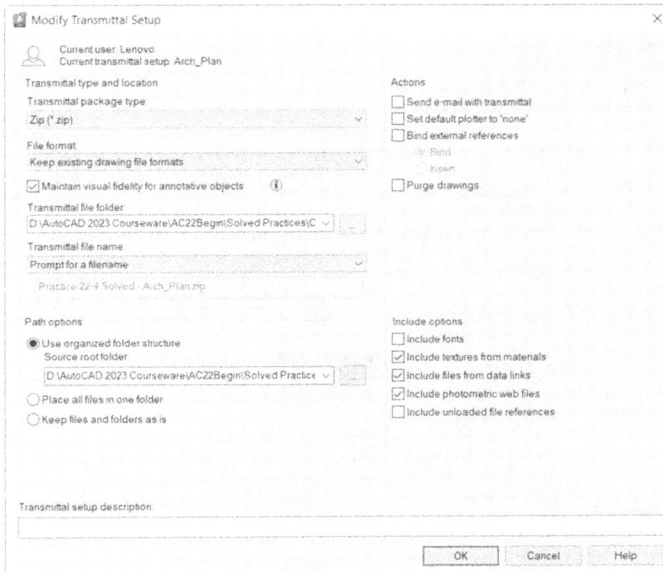

All the options in this dialog box are very simple so that any common computer user will change them effectively.

Make the necessary changes, then click OK to end this dialog box.

As a final step, click the button at the lower left labeled View Report to decide if everything is OK before the final go ahead. You will see the following:

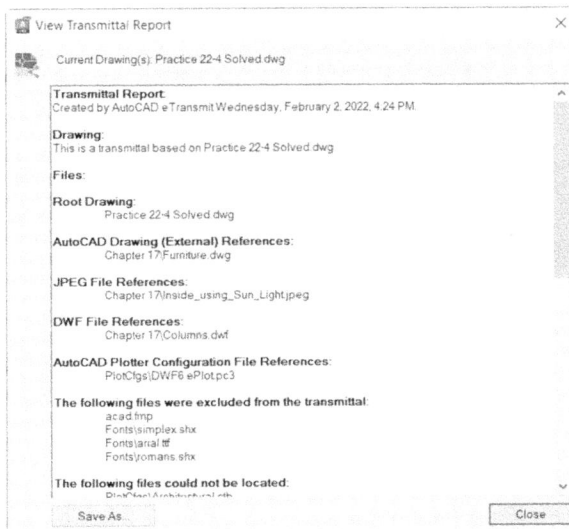

## PRACTICE 22-4    USING THE ETRANSMIT COMMAND

1. Start AutoCAD 2025

2. Open **Practice 22-4.dwg** file

3. Start command eTransmit

4. See the Files Tree and Files Table. Is the AutoCAD Plotter Configuration File listed? _____ If yes, what is it? _____

5. Create a new Transmittal Setup and name it Architectural Drawing. Make sure of the following:

    a. Package type = Zip file

    b. File format = AutoCAD 2013

    c. Include fonts = yes

6. As you noticed, font files are added

7. View the report

8. Save the file as Architectural Drawing.zip

9. Save and close the file

# NOTES

## CHAPTER REVIEW

1. You can attach DOC and XLS files to my AutoCAD drawing using the External Reference palette:

    **a.** True

    **b.** False

2. There is only one reference type: "Overlay"

    **a.** True

    **b.** False

3. If you attached a DWF file, all is correct except one:

    **a.** You can control DWF layers

    **b.** You can Unload and Reload

    **c.** You can Bind

    **d.** You can Detach

4. You can attach the same image file more than once in the current file:

    **a.** True

    **b.** False

5. One of the following statements is not true:

    **a.** Once you attach one file, the icon will appear at the right part of the Status bar

    **b.** The current layer will be always the carrier of the reference file

    **c.** Layer 0 will be always the carrier of the reference file

    **d.** Relative path is better than Full path

6. If we have a layer in a reference DWG file named Furniture|Chair, it will be after using Bind command and Bind option _____

7. _____ will collect all related files to your current file and then zip them, and prepare them to be sent either using e-mail or other means

## CHAPTER REVIEW ANSWERS

**1.** b

**3.** c

**5.** c

**7.** eTransmit

## CHAPTER 23

# SHEET SETS

## In This Chapter

- What are sheet sets?
- Types of sheet sets
- Understanding the Sheet Set Manager
- Example sheet set creation steps
- Existing drawing sheet set creation steps
- eTransmit, Archive, and Publish sheet sets
- Label blocks & Callout blocks

## 23.1   INTRODUCTION TO SHEET SETS

The ultimate goal for any company, especially Architectural Engineering Construction firms, is to print out the final design in a sheet set. Printing individual drawings then gathering them into a set, labeling them, and putting them in a certain sequence proved to be burdensome. International bodies like US National CAD Standard (NCS—*http://www.nationalcadstandard.org*) produced a CAD standard, which addresses the sheet sets, specifying discipline designator, sheet type designator, and sheet sequence number. Accordingly, different companies who want to adopt US NCS wants AutoCAD to help them do their print job quickly and accurately. AutoCAD responded, and since AutoCAD 2005 has provided Sheet Set creation and manipulation.

AutoCAD will create a sheet set then save it as a template to be used several times as needed, which will help you to avoid repeating the work.

There are two methods to create a sheet set, they are:

- **Example sheet set**: Example Sheet set is a premade sheet set to provide the organizational structure of the sheet set depending on the discipline. There will be no sheets attached to the sheet set, hence you should create sheets, add views, and scale them
- **Existing drawings**: this method will use the layouts in the drawings as sheets, and folders as subsets

There are pros and cons for each method. The first method advocates will tell you: "This is a lengthy method, yes! But it will give you full control over all the elements." The advocates for the second method will tell you: "While we are masters at creating layouts, why do the work twice, once for the layout, and the again for sheet sets?"

## 23.2  DEALING WITH THE SHEET SET MANAGER PALETTE

In this part, we will discuss:

- How to open and close an existing sheet set
- How to deal with the Sheet Set Manager palette

### 23.2.1  How to Open and Close an Existing Sheet Set

To show the Sheet Set Manager palette go to the **View** tab, locate the Palettes panel, then select the **Sheet Set Manager** button:

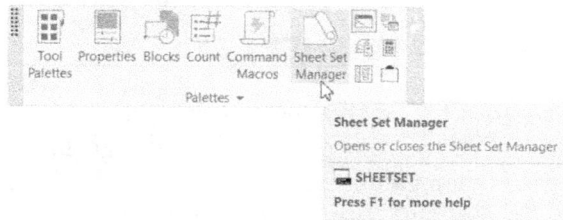

Or, if there are no opened files, you can click the same icon in the Quick Access menu. Either way, you will see the following palette appearing:

From the list at the top select Open:

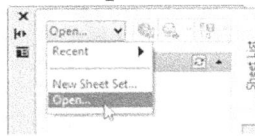

The normal Open dialog box will appear; select your desired *.DST* file to open it. There are several sample DST files in the following path: \Autodesk\AutoCAD 2025\Sample\Sheet Sets\, and you will find three folders: Architectural, Civil, and Manufacturing.

The following will appear:

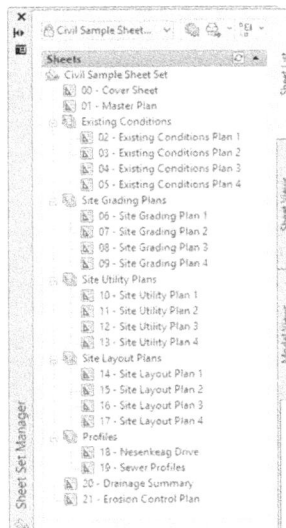

An existing sheet set can be closed by clicking the name of the sheet set in the Sheet Set Manager, then right-clicking to bring up a menu; select the first option, **Close Sheet Set**:

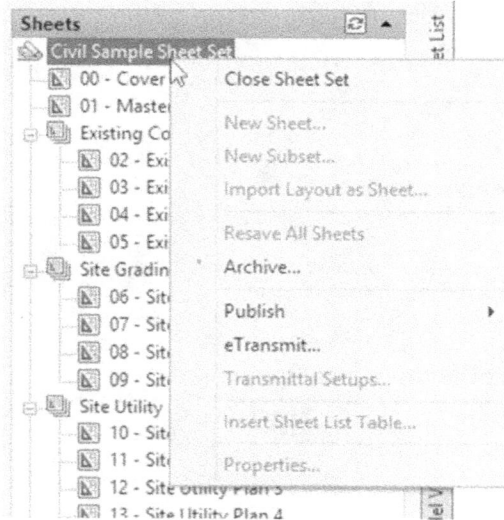

### 23.2.2 How to Deal with the Sheet Set Manager Palette

As shown above, the Sheet Set manager palette has three tabs:

- Sheet List tab
- Sheet Views tab
- Model View tab

While you are in the Sheet List tab (default tab) you will see the following:

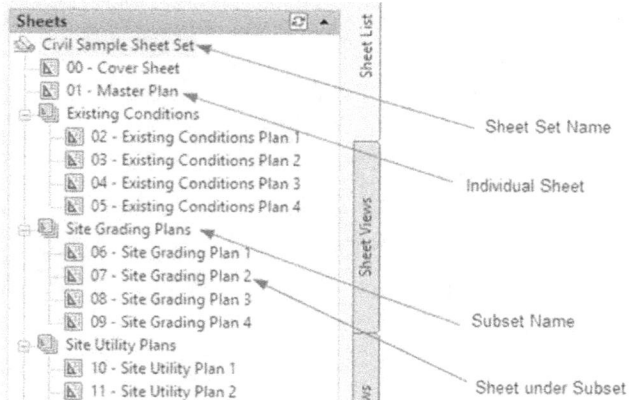

In the Sheet List tab, you will see the following:

- Sheet Set name
- Individual sheets (not listed beneath Subsets)
- Subsets (method to group sheets)
- Sheets under Subsets
- You will see the sheets sorted in a way from top to bottom
- For each sheet, there is a number and a title (for instance 01 is the number for the Master Plan sheet)

We will cover the Model View tab first, then we will tackle the Sheet Views tab. When you click you will see the following:

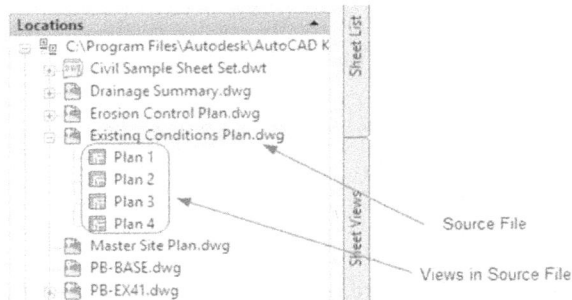

The Model View tab contains the AutoCAD files named source files, which contain the views which will be used in the first method of sheet sets (the user will be using the drag-and-drop method to insert these views as an external reference to the current file).

Clicking on the Sheet Views, the following will be shown:

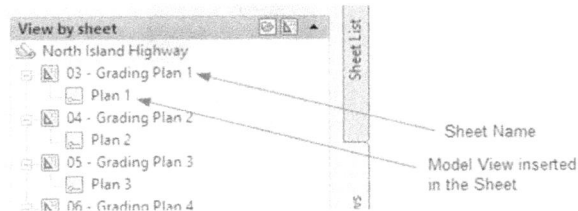

Using the Model Views tab, if you dragged and dropped a model view, then the Sheet View tab will show the name of the sheet and the name of the view inserted in it.

While you are in any of the three tabs, you will always be able to see a preview of the file, sheet, and view; see the following two examples:

- In the Sheet List tab, if you point (without clicking) on any of the sheets, you will see the following:

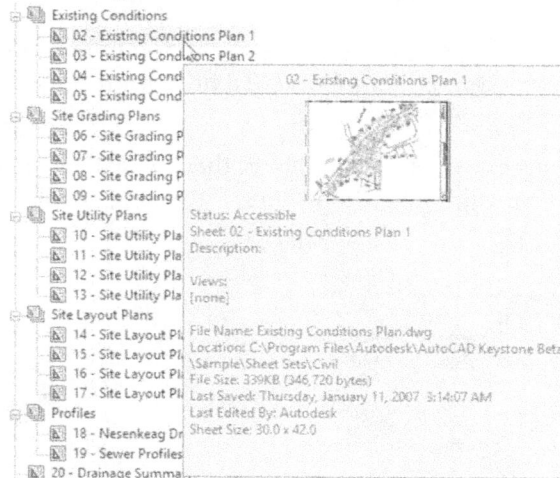

- While you are at Model View tab, if you point to one of the views, you will see the following:

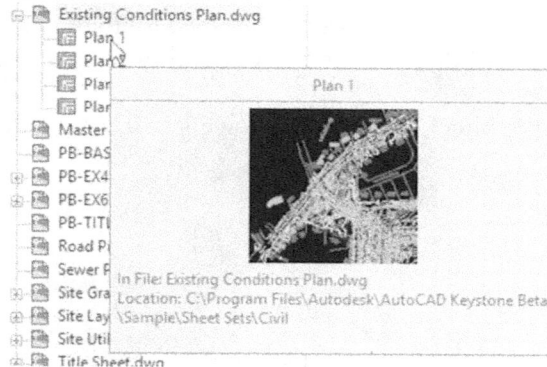

## 23.3   SHEET SET FILE SETUP

AutoCAD will create a DST file, which will hold the association and information that defines the sheet set. The DST file does not contain the files and views, but rather the subsets and links to views (or layouts) of the source files. When you create the sheet set, AutoCAD will ask you about the location of the DST file along with the template files. When the sheet set is created, AutoCAD will create for each sheet a DWG file, containing the sheet and the model views inserted in it. If you go to the folder containing your DST file, you will see the following:

DWG files containing the sheets

DST file

# PRACTICE 23-1    OPENING, MANIPULATING, AND CLOSING SHEET SETS

1. Start AutoCAD 2025

2. Open Sheet Set Manager palette (whether a file is opened or not)

3. Use the Open command to open the following DST file: (depending on the installation of AutoCAD in your machine) Program Files\Autodesk\AutoCAD 2025\Sample\Sheet Sets\Civil\Civil Sample Sheet Set.dst

4. Double-click sheet 00-Cover Sheet to open the sheet

5. Close it without saving

6. Point to sheet 06, which will show more information about it: what is the Sheet size? _____

7. Go to the Model Views tab

8. Click the plus sign to expand the list

9. Click Existing Conditions Plan.dwg; how many views are there? _____

10. Point to several files (or views) to view a preview of each one of them

11. Double-click Erosion Control Plan.dwg; did you open Model Space or layout? _____

12. Close the file without saving

13. Go to the Sheet List tab and close the Civil Sample Sheet Set

## 23.4   SHEET SET USING AN EXAMPLE

This is the first of two methods to create a new sheet set. The merit of this method is the fact that you will have full control over it; hence, it represents the professional choice. Yet this method is lengthy and may take a long time to develop.

In order to create a new sheet set, make sure that it is at least one file (even if it was empty) is opened.

Open the Sheet Set Manager palette, and from the menu at the top, select the New Sheet Set option, and you will see the following dialog box:

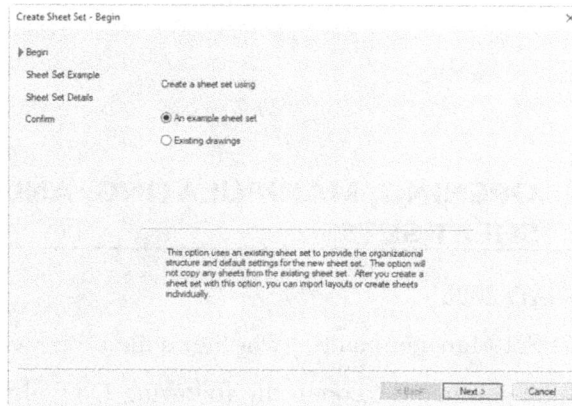

The default option will be **An example sheet set**; click **Next** to go to the second step, and you will see the following dialog box:

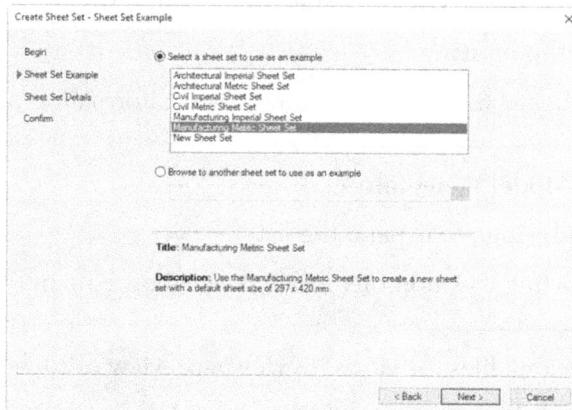

There are six examples: two for architectural (metric and imperial), two for civil (metric and imperial), and manufacturing (metric and imperial). These examples will contain only subsets using templates that included with AutoCAD. Another option is to use a previously user-made sheet set (which can be the company's template). When done, click the Next button to go to the third step, where you will see the following dialog box:

The user should do the following:

- Input the sheet set title
- Input the sheet set description
- Specify the folder to store the DST file
- Specify whether to Create a folder hierarchy based on subsets or not
- Click the Sheet Set Properties button and you will see the following dialog box:

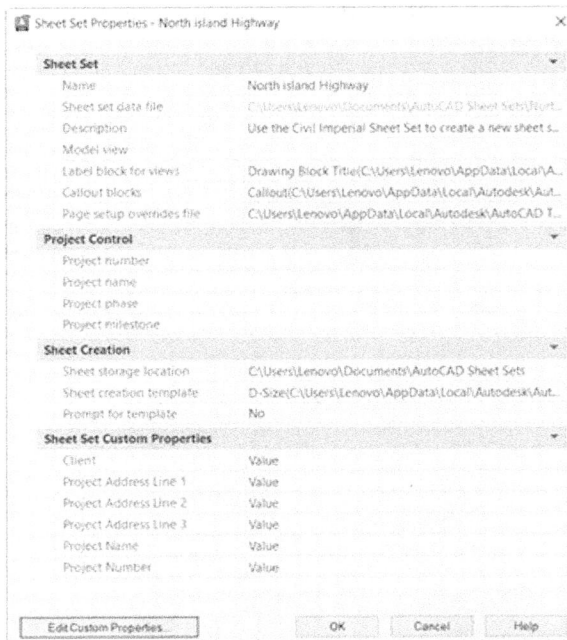

Almost all of the data is already filled using default values. Project Control information are fields related to the sheet set; Sheet Set Custom Properties include information created by AutoCAD. The user can add, delete, or modify them; click the Edit Custom Properties button to see the following dialog box:

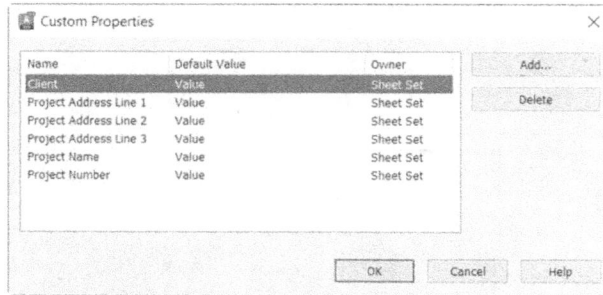

From all of the above there are two different pieces of information you should consider controlling, they are:

- **Sheet storage location**: this piece of information allows you to specify the location of the sheets generated by you. It is preferable to make it the same folder for the DST file
- **Sheet creation template**: this piece of information allows you to specify the location of the DWT file for sheets that will be created when using it

When done, click the OK button to end Sheet Set Properties, then click Next to go to the fourth step.

You will see the following dialog box:

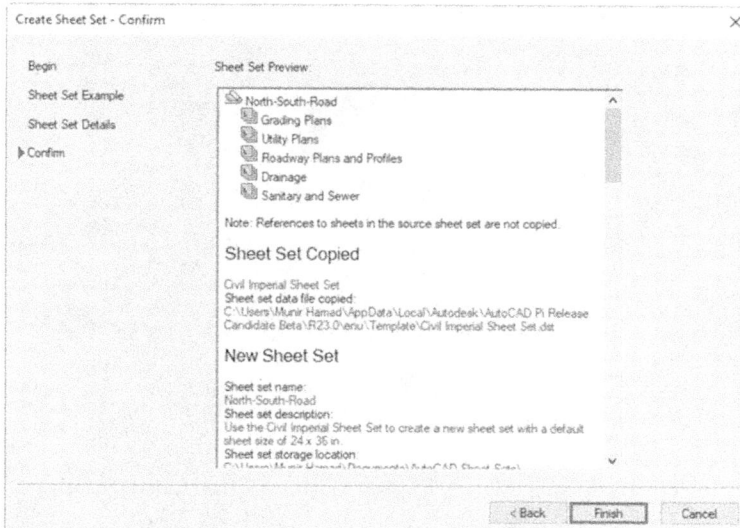

This is the confirmation page, on which you will review your settings; then the user should click Finish to end the creation process. You will return to the Sheet Set Manager palette and see the following:

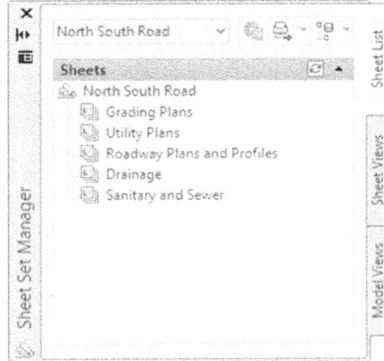

### 23.4.1   Adding a Sheet Into a Subset

To add sheets in subsets, simply right-click the name of the subset, and you will see the following menu:

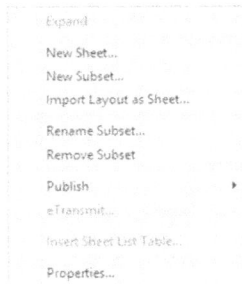

In this menu, the user can:

- To create a new Sheet in the current subset; you will see the following dialog box, fill in the sheet number and title, then click OK, and this will create a new sheet using the template file designated in the creation process:

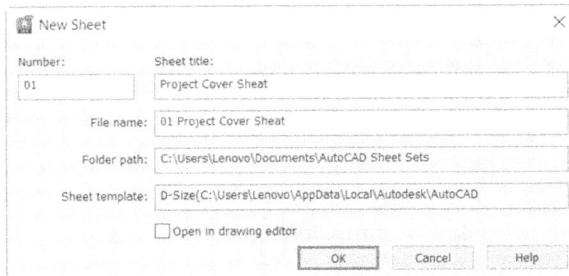

- To create a new subset inside the current subset, you will see the following dialog box, fill in the subset name, sheet location, and template:

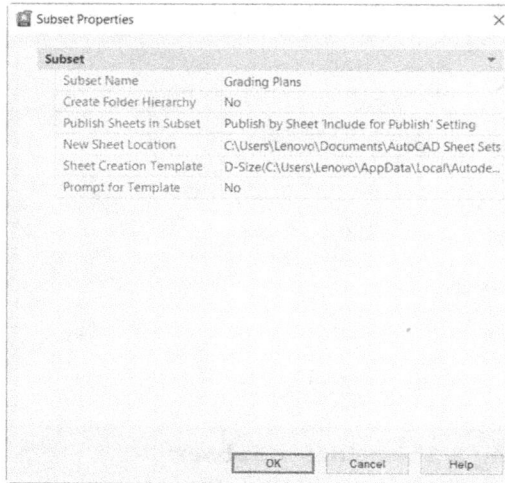

- Import Layout as Sheet
- Rename Subset
- Remove Subset
- Publish
- Properties

### 23.4.2 Sheet Control

If you right-click a sheet, you will see the following menu:

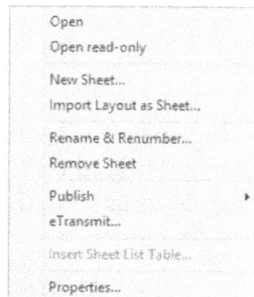

Using this menu the user can:

- Open the sheet (double-clicking the sheet will lead to the same result)
- Open the sheet as read-only
- Create a new sheet in the same subset
- Rename and Renumber a sheet
- Remove Sheet

- Publish the current sheet
- eTransmit
- Properties of the current sheet

## PRACTICE 23-2   CREATING A SHEET SET USING A SHEET SET EXAMPLE

1. Start AutoCAD 2025

2. Make sure a new file is opened

3. Start Sheet Set Manager, and create a new sheet set using an example sheet set

4. Select Civil Imperial Sheet Set

5. Name the new sheet set "North Island Highway"

6. Save the DST file in: Practices\Chapter 23\Sheet Sets\Practice 23-1,23-2

7. Click Sheet Set Properties, and input the following:

   a. Project Number = HW-03

   b. Project Name = Designing North Island Highway

   c. Project Phase = 03

   d. Client = North Island Governorate

8. Finish the creation process

9. Click the name of the sheet set, then right-click and add the following two sheets:

   a. Sheet Number 01 – Title = Project Cover Sheet

   b Sheet Number 02 – Title = Project Master Plan

10. Move the two new sheets to be at the top of the list (using drag and drop technique)

11. Remove the following subsets: Drainage, and Sanitary and Sewer

12. Under Grading Plans subset, create the following sheets:

   a. 03 – Grading Plan 1

   b. 04 – Grading Plan 2

   c. 05 – Grading Plan 3

   d. 06 – Grading Plan 4

**13.** Under Utility Plans subset, create the following sheets:

    **a.** 07 – Utility Plan 1

    **b.** 08 – Utility Plan 2

    **c.** 09 – Utility Plan 3

    **d.** 10 – Utility Plan 4

**14.** Using My Computer, go to the folder where you saved the DST file, and look at the files created by sheet set; how do their naming and numbering compare to the sheets? _____

_____

**15.** Leave the Sheet Set Manager open

## 23.5 ADDING AND SCALING MODEL VIEWS

The next step is to fill the sheets with model views from the source files. Model views should be prepared prior to this step. Our mission will be in four steps:

- Specify folder(s) which contain the files holding the desired model views
- Open the desired sheet
- Drag-and-drop the views in the sheets
- Scale views

Model views will be considered as an external reference in the sheet, hence, any change in the original file will lead to the model views being updated as well.

To accomplish the first step, select the Model Views tab, and you will see the following:

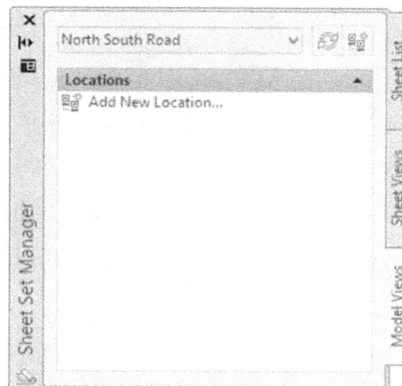

Double-click the Add New Location button to open a normal dialog box to specify the desired folder. You can repeat this step as many times as you prefer to add more folders. Once you are finished, you will receive the following:

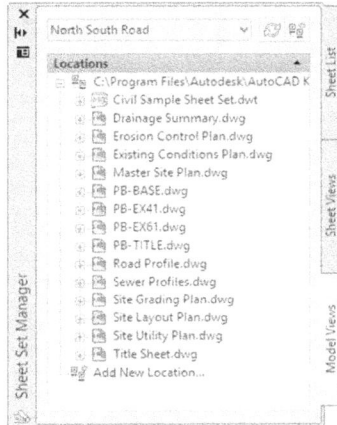

At the left of some of the files there is a plus sign; click it to expand to see the related model views, similar to the following:

Open the desired sheet by double-clicking it, then drag the chosen view inside it; before you decide the exact insertion point, right-click and you will see a list of scales to be used:

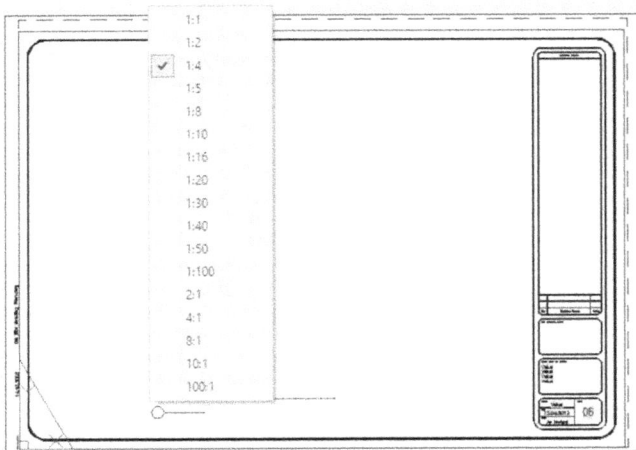

Select the desired scale, and then specify the insertion point. You should receive the following:

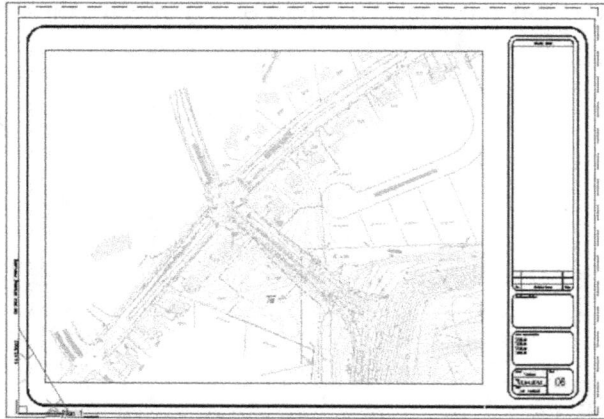

The view you add in the sheet will be considered a sheet view; to prove that, select the Sheet Views tab to see the following:

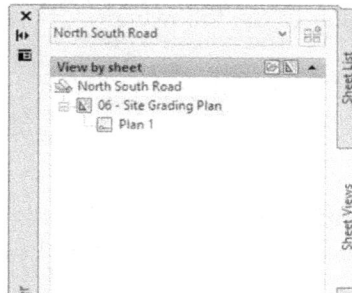

# PRACTICE 23-3    ADDING AND SCALING MODEL VIEWS

1. Start AutoCAD 2025

2. Make sure that the Sheet Set Manager is open and the North Island Highway sheet set is open

3. Double-click sheet number 03, named Grading Plan 1, and notice how a lock is displayed beside it to indicate that this sheet is in use

4. Go to the Model Views tab, double-click Add New Location, and go to \Program Files\Autodesk\AutoCAD 2025\Samples\Sheet Sets\Civil

5. A set of files will be listed; click the plus sign beside the Site Grading Plan.dwg file, and you will see four model views. Drag the Plan view into the current sheet, and scale it 3/16"=1′

**6.** Close and save the file

**7.** Do the same procedure for the following:

    **a.** Plan 2 view inside sheet # 04 (same scale)

    **b.** Plan 3 view inside sheet # 05 (same scale)

    **c.** Plan 4 view inside sheet # 06 (same scale)

    **d.** Plan 1 view in Site Utility Plan.dwg inside sheet # 07 (same scale)

    **e.** Plan 2 view in Site Utility Plan.dwg inside sheet # 08 (same scale)

    **f.** Plan 3 view in Site Utility Plan.dwg inside sheet # 09 (same scale)

    **g.** Plan 4 view in Site Utility Plan.dwg inside sheet # 10 (same scale)

**8.** Close the sheet set

## 23.6   SHEET SET USING EXISTING DRAWINGS

If you are very efficient at creating layouts and do not want to repeat the work, once for creating layouts, and the second to create sheets, you can allow AutoCAD to take all of your layouts and create sheets from them. Also, you can ask AutoCAD to take the folder structure you created to be your subset structure. With this method, whatever is in the layout will be considered part of the sheet, but the viewport will not be an external reference.

While the Sheet Set Manager palette is open, choose the New Sheet Set option, and the following dialog box will appear:

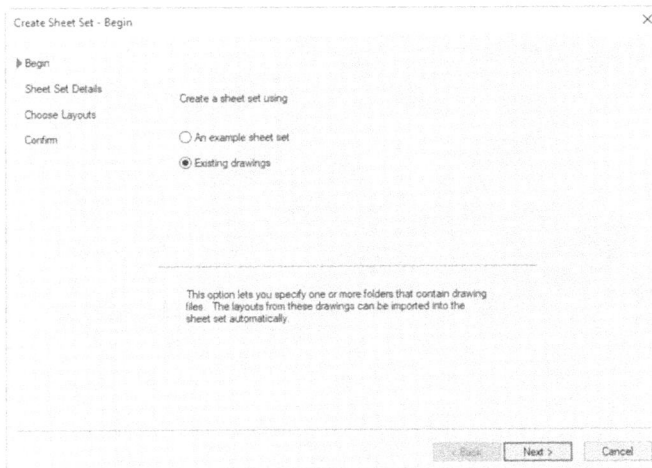

Select the Existing drawings option, click **Next**, and you will see the following dialog box:

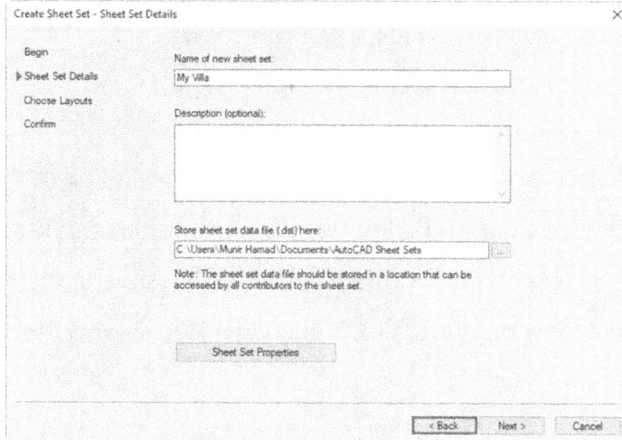

The user should do the following:

- Input the sheet set title
- Input the sheet set description
- Specify the folder to store DST file
- Click the Sheet Set Properties button and you will see the following dialog box

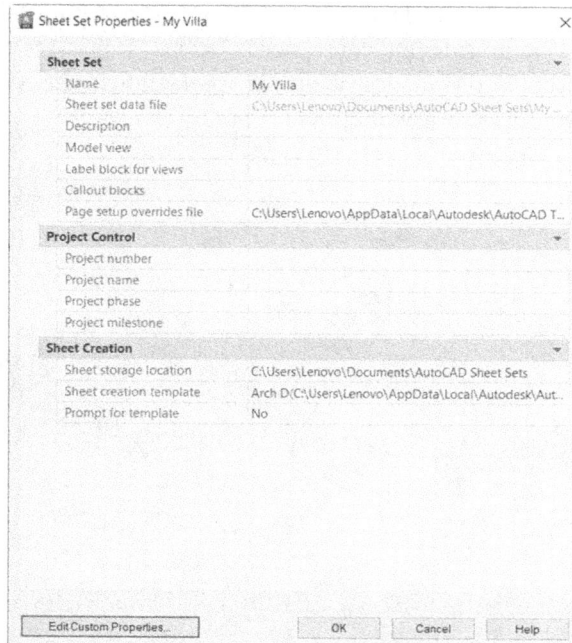

They are the same as the properties discussed in the first method. Click OK, and then click Next, and you will see the following dialog box:

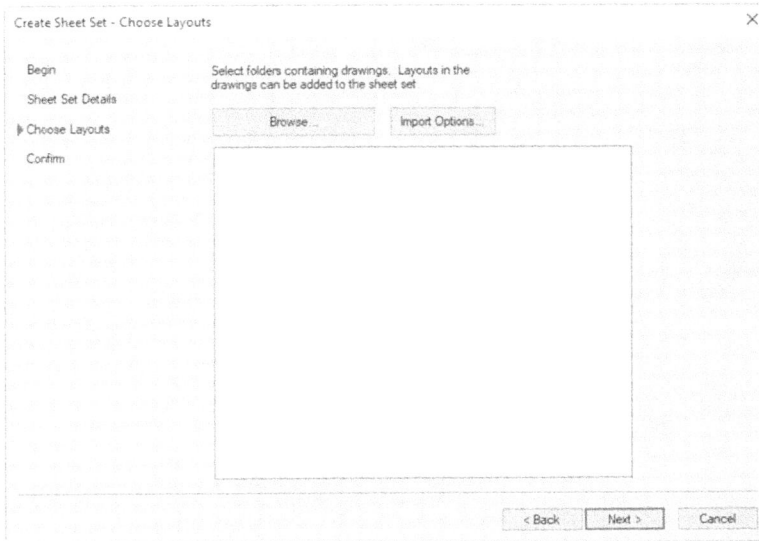

To control the importing of layouts, click the Import Options button to see the following dialog box:

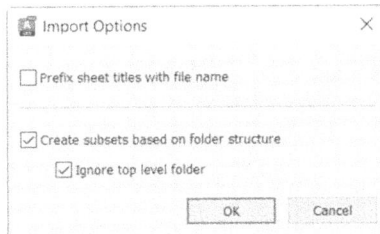

Control all or any of the following:

- Create Prefix sheet titles with file name or not
- Create subsets based on folder structure or not
- Ignore top level folder or not

When done click OK, then click the Browse button to select the folder containing the desired files which contain the desired layout. The user then will able to pick and choose which files/layouts will be included in the importing process. You will see the following:

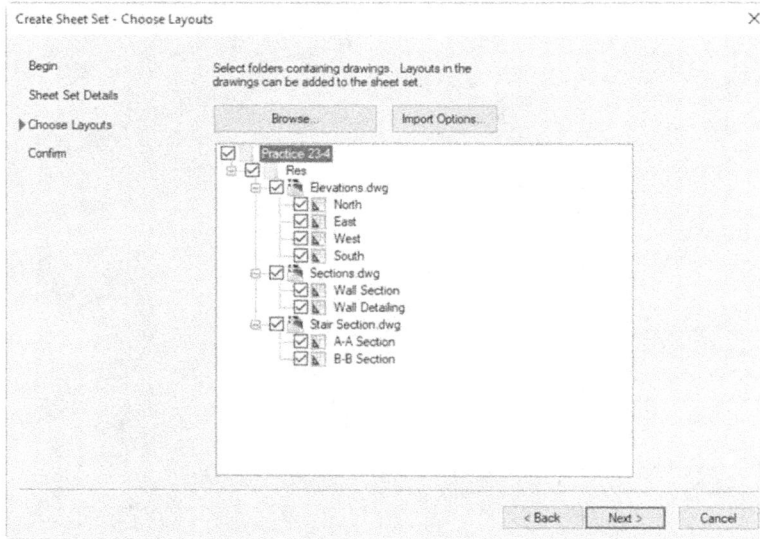

When done, click Next to jump to the final step.

You will see the Confirm page; click the Finish button to finish the creation process:

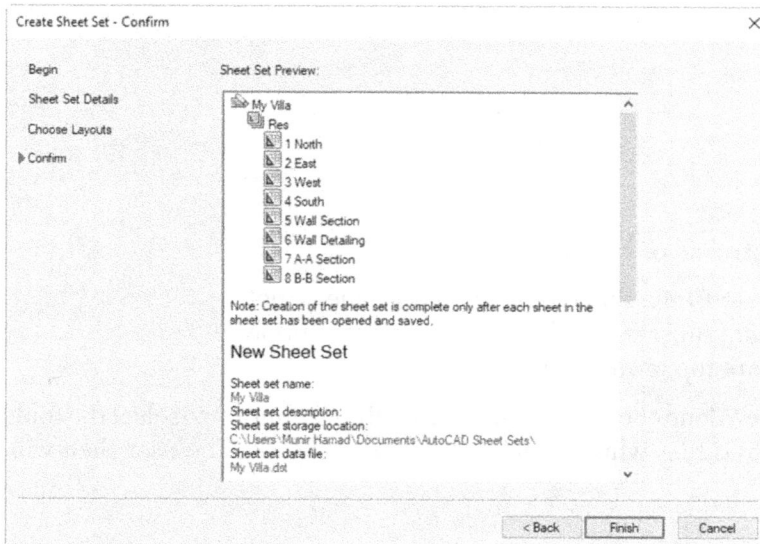

The Sheet Set Manager will look similar to the following:

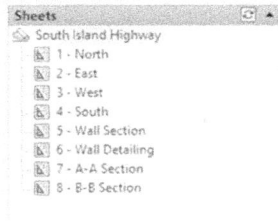

As you can see we brought only sheets, so we need to create subsets to arrange them using the drag-and-drop technique. See the following illustration:

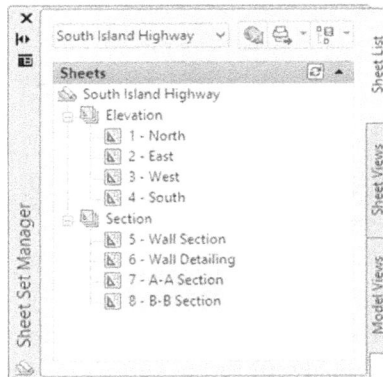

Another helpful tool, which can be used with both the first and the second method, is importing layouts as sheets. Go to the desired subset, then right-click and you will see the following menu; select the Import Layout as Sheet option:

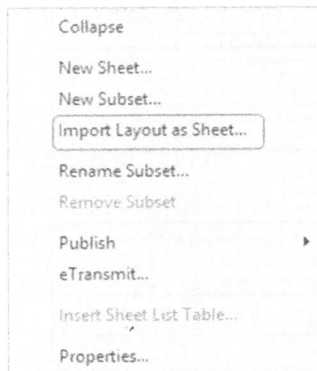

Click the Browse for Drawings button and select the desired file and you will see the following:

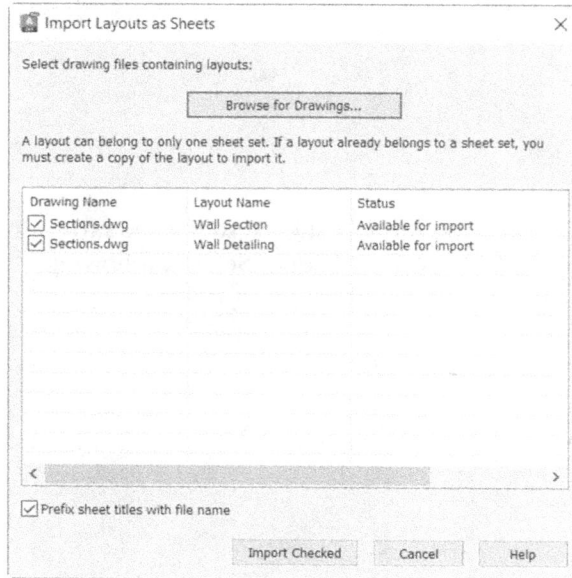

In the example above you selected two files and each contains one layout; you can import either layout or one of them. Click the Import Checked button to end the importing process.

## PRACTICE 23-4   CREATING A SHEET SET USING EXISTING DRAWINGS

**1.** Start AutoCAD 2025

**2.** Make sure that at least one empty file is open

**3.** Create a new sheet set titled "Mira House" using the existing drawing. Save it in Practice\Chapter 5\Sheet Sets\Practice 23-4

**4.** You will find the drawings which contain the desired layouts in Practice\Chapter 5\Sheet Sets\Practice 23-4\Res

**5.** Uncheck the Stair Section.dwg file, taking with you only layouts from Elevations.dwg and Sections.dwg

**6.** Create three subsets: Elevations, Sections, and Stair Sections

7. Using the drag-and-drop technique, move North, East, West, and South to the Elevations subset; move Wall Section and Wall Detailing to Sections

8. Right-click the Stair Sections subset and select Import Layout as a Sheet

9. Browse for the following folder: Practice\Chapter 5\Sheet Sets\Practice 23-4\Res, and select Stair Section.dwg

10. Import both layouts

11. Renumber and Rename the two imported sheets to be 7-A-A Section, and 8-B-B Section

12. Close the sheet set

## 23.7   PUBLISHING SHEET SETS

Using the Sheet Set Manager, if you right-click the name of the sheet set, a menu will appear with the Publish option, and you will see the following sub-menu:

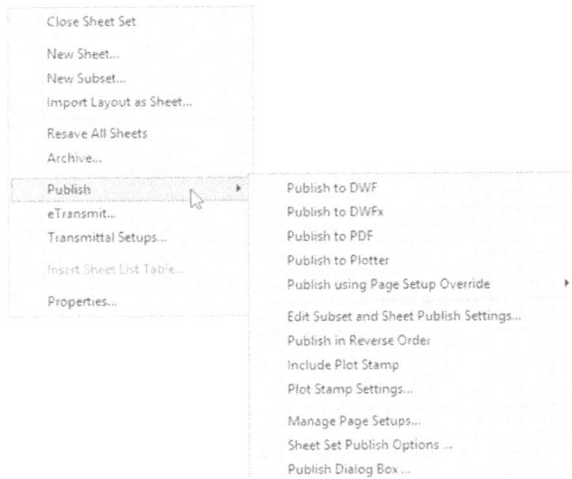

The first four options were discussed previously and they are: Publish to DWF, Publish to DWFx, Publish to PDF, and finally, Publish to Plotter. The fifth is new, which says Publish using Page Setup Override; this will help you to override the default page setup. In order for this option to be available, the user should set the Page setup Override in the Properties dialog box, as in the following illustration:

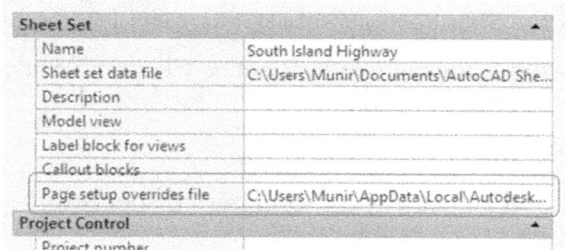

If this option is properly set, you will see the following:

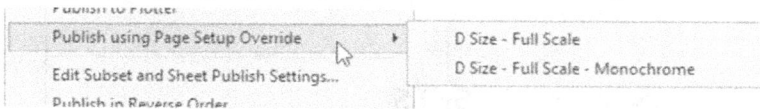

If you select **Edit Subset and Sheet Publish Settings** option, you will see the following dialog box:

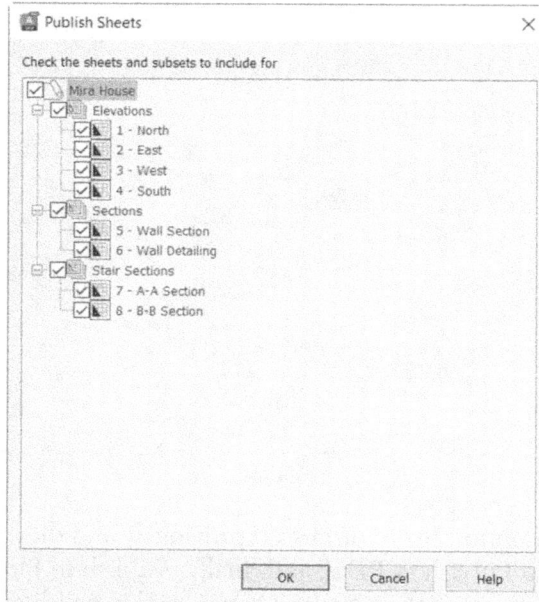

This will allow you to specify subsets/sheets to be included in the Publish command.

The rest of the options are all obvious or have been discussed before.

## 23.8   USING eTRANSMIT AND ARCHIVE COMMANDS

AutoCAD provides two methods to exchange and archive sheet sets, using the eTransmit and Archive commands. Both commands will group all the related files and pack them into a single file. Actually, we do not know why there are two commands to do that because they are identical.

### 23.8.1   Using the eTransmit Command

To create a package of all files of the sheet set, simply right-click the name of the sheet set, and you will see the following menu:

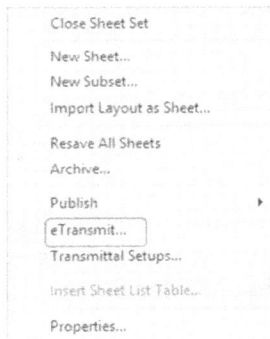

You will see the following dialog box:

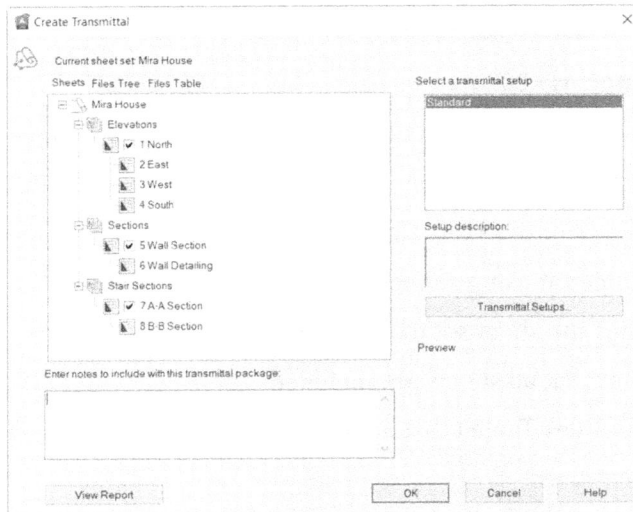

This dialog box is identical to the one discussed in Chapter 22, but with only one change, which is the Sheet tab at the upper left part of the dialog box.

### 23.8.2 Using the Archive Command

To issue this command, right-click the name of the sheet set, and a menu will appear; select the Archive command, and you will see the following dialog box:

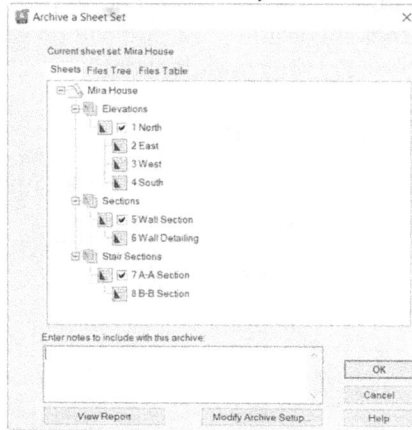

You can see from the above dialog box that it is the same as the eTransmit dialog box. Archive command will not save a setup like we do with eTransmit command, but rather, you have to Modify Archive Setup each time you issue it.

# PRACTICE 23-5   PUBLISHING AND eTRANSMITTING A SHEET SET

1. Start AutoCAD 2025

2. Open the Mira House sheet set that you created in the previous practice

3. Right-click the sheet set name, and select Publish, then Publish to DWF

4. Name the file Mira House.dwf and save it in the same folder as the DST file

5. You can open it using Autodesk Design Review software to view it

6. Right-click the sheet set name again, and select eTransmit

7. Make sure that all sheets are included

8. Go to the Files Table tab, check that the Mira House.dst file is included

9. View the report

10. Produce and eTransmit the package and save it in the same folder as the DST file

11. Close the sheet set

## NOTES

## CHAPTER REVIEW

1. How many methods can create a sheet set?

   **a.** One method

   **b.** Three methods

   **c.** Two methods

   **d.** Four methods

2. _____is a method that helps you to bring in layouts from other drawings, and consider them as sheets

3. Using an Example Sheet Set, you will have viewports ready for you:

   **a.** True

   **a.** False

4. Using an Example Sheet Set:

   **a.** AutoCAD will create sheets

   **b.** AutoCAD will create subsets

   **c.** AutoCAD will create both subsets and sheets

   **d.** AutoCAD will create both subsets and sheets, and then add views

5. Using the Existing drawings method, AutoCAD will consider folders as subsets, and layouts as sheets

   **a.** True

   **b.** False

6. In the Sheet List tab, if you point (without clicking) on any of the sheets, you will see _____ of the sheet

## CHAPTER REVIEW ANSWERS

   **1.** c

   **3.** b

   **5.** a

## CHAPTER 24

# CAD Standards and Advanced Layers

## In This Chapter

- What is CAD Standard?
- How to create a DWS file, configure it, and check it
- What are Layer Filters?
- What is the Layer States Manager?
- What are the advanced functions to control layers?

## 24.1 WHY DO WE NEED CAD STANDARDS?

Since projects are getting larger and more complex, companies that use AutoCAD (or any other software package) are asking for CAD Managers, but who are these people? A CAD Manager has plenty of responsibilities, but among them is to produce standard dimensions, text, and layering systems. AutoCAD is offering a tool to help CAD Managers producing CAD Standard files, linking them to DWG files, then making sure that you did not violate any of the standards.

The process of creating a standard file is lengthy and encompasses the following steps:

- Find an international standard or create your own
- Sit with all the parties whom will be affected by the new standard, and try get their input (stakeholders)

- Put the standard on paper first, getting the approval of all stakeholders
- Create a standard file using AutoCAD
- Put it into a test and take feedback from users
- Make any corrective steps needed
- Put the standard in use and monitor the outcome

Before this tool was introduced in AutoCAD, CAD managers used to check the adherence to the standard file manually using a random check method, but now, CAD Manager has the ability to check several files concurrently.

Another tool is Layer Translator, which will map layers in your current file to compare it to another file, and find the match and mismatch, allowing you to correct any layer names.

## 24.2 HOW TO CREATE A CAD STANDARD FILE

To create a CAD Standard file (*.DWS) try the following steps:

- Open a DWG file containing all the desired layers, dimension styles, text styles, and linetypes. Or, you can create a new file using a template file containing the above
- Go to the Application menu and select the Save As then AutoCAD Drawing Standards option:

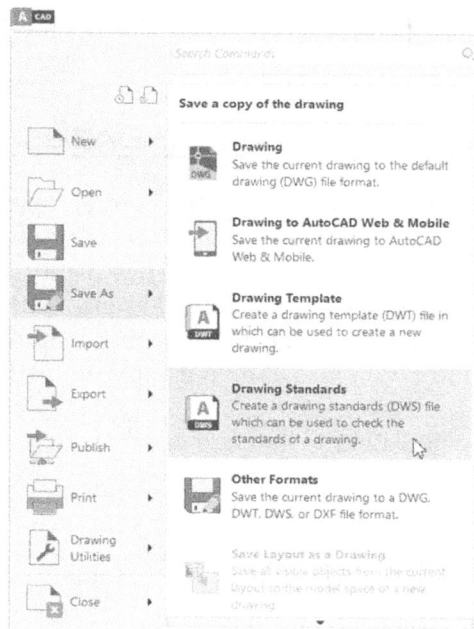

You will the Save As dialog box:

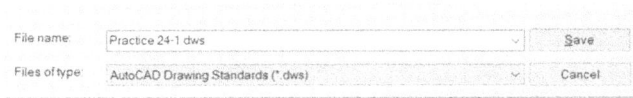

| File name: | Practice 24-1.dws | ∨ | Save |
| Files of type: | AutoCAD Drawing Standards (*.dws) | ∨ | Cancel |

As you can see, AutoCAD will give the file the extension of DWS automatically; input the name of the new standard file then click the Save button. This process will occur only once.

## 24.3 HOW TO LINK DWS TO DWG AND THEN MAKE THE CHECK

After you create the DWS file, you will perform two more steps:

- Configure (link) it to DWG file
- Perform the checking process to find any violation to the standard file

### 24.3.1 Configuring (Linking) DWS to DWG

To create the link, do the following steps:

- Open the desired DWG file to be checked
- Go to Manage Using the **Manage** tab, locate the **CAD Standards** panel, then select the **Configure** button

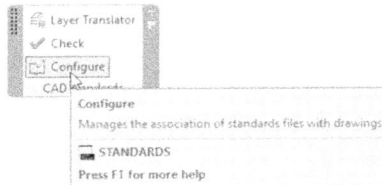

You will see the following dialog box:

Click the plus sign button to link the current DWG with the desired DWS:

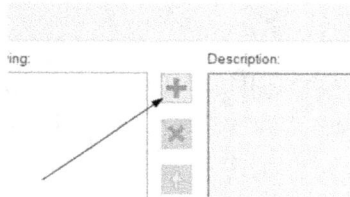

You will see a normal file dialog box; select DWS file, then click OK.

To control what to check for, click the Plug-ins tab, and the following will appear:

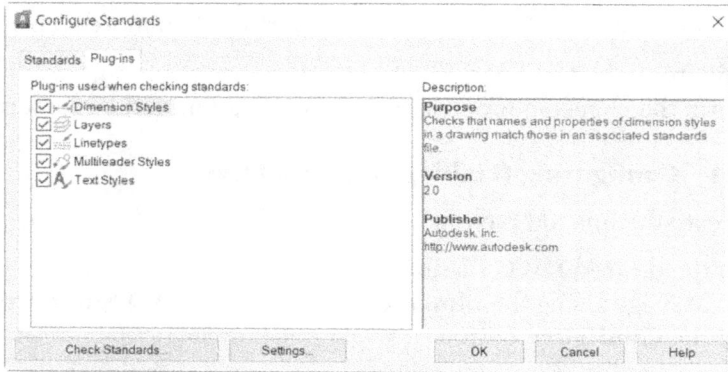

As said before, AutoCAD can check for five things only: Dimension styles, Layers, Linetypes, Multileader styles, and Text styles. To have full control over the process of checking, click the Settings button, and the following dialog box will appear:

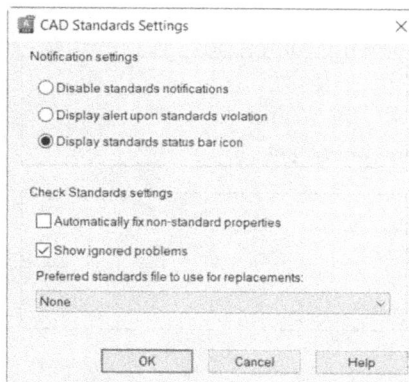

Control the Notification settings by selecting one of the following:

- Disable standards notifications
- Display alert upon standards violation (this is when you are still working with the DWG file)
- Display standards status bar icon (default):

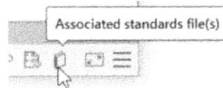

Control Check Standards settings by changing one of the following:

- Automatically non-standard properties, or not?
- Show ignored problems, or not?
- Choose your preferred DWS file

### 24.3.2 Checking a DWG File

The first step should be to open the desired DWG file you want to check for Standard compliance. Then go to **Manage**, locate the **CAD Standards** panel, then select the **Check** button:

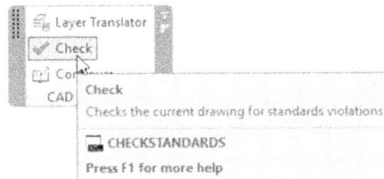

You will see the following dialog box:

You can see that the dialog box is cut into three parts, which are:

- **Problem section**, where AutoCAD will show the problem that violates the standard; in the below example, a name of a layer in your DWG file is not in the standard file:

```
Problem:
Layer 'Dimension'
Name is non-standard
```

- **Replace with section**, where AutoCAD will propose to replace the mistake with something already in the standard file, like in the below example; Dimension layer is to be replaced with A-Dimension layer. In this case, AutoCAD will remove the old layer, and transfer all the objects in it to the new layer:

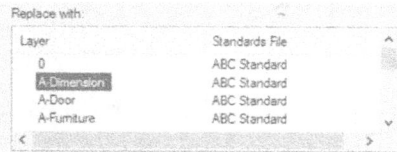

```
Replace with:
Layer            Standards File
0                ABC Standard
A-Dimension      ABC Standard
A-Door           ABC Standard
A-Furniture      ABC Standard
```

- **Preview of changes section**, where AutoCAD will show the changes between the old and the new suggestions. In the below example, beside the name, the color also will change:

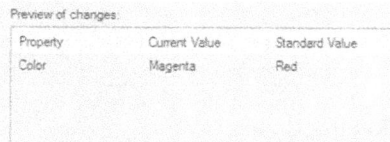

```
Preview of changes:
Property    Current Value    Standard Value
Color       Magenta          Red
```

AutoCAD will offer three ways to deal with any problem:

- **Fix**: To fix the problem using Replace with option
- **Next**: Skip the problem and go to the next problem
- **Ignore**: Ignore this problem and go to the next problem

When this command finishes, it will show the following message:

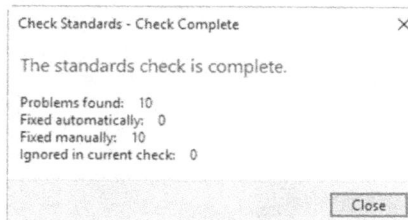

```
Check Standards - Check Complete                    ×

The standards check is complete.

Problems found:   10
Fixed automatically:  0
Fixed manually:   10
Ignored in current check:  0

                                        Close
```

The user can choose to go with another approach, which is to link the DWS file with under-development DWG Do the following steps:

- Go to Configure command, and click on Settings button
- Select the option "Display alert upon standard violation," then click OK
- Whenever the user will make any violation, he/she will see the following message:

Standards - Violation Found ✕

❗ The previous operation created an object that does not comply with standards. What do you want to do?

→ Choose a standard object to replace this one
The Check Standards dialog box will open.

→ Keep the nonstandard object
The object will remain unchanged.

Select whether to replace the violation with an existing setting (in the above case, it was adding a new layer), or keep the nonstandard object.

## PRACTICE 24-1   USING CAD STANDARDS COMMANDS

1. Start AutoCAD 2025

2. Open ABC Standard.dwg, and create from it ABC Standard.dws, then close the file

3. Open **Practice 24-1.dwg**

4. Using the Configure link Practice 24-1.dwg to ABC Standard.dws

5. Go to the Plug-ins tab, and make sure that all the four available options are turned on, then click OK to end the command

6. Before you start checking the compliance, check the color of the layer Dimension, what is the color? _____

7. Start the Check command, and make the following changes

| Problem | Action |
|---|---|
| Dim Style: Outside Walls | Replace it with Outside |
| Dim Style: Inside | Replace it with Inside |
| Layer: Dimension | Replace it with A-Dimension |

| Problem | Action |
|---------|--------|
| Layer: Furniture | Replace it with A-Furniture |
| Layer: Text | Replace it with A-Text |
| Layer: Title Block | Replace it with Title Block |
| Layer: TobeHidden | Replace it with 0 |
| Text Style: Room Titles | Replace it with Room Titles |
| Text Style: TitleBlock_Bold_Mine | Replace it with TitleBlock_Bold |
| Text Style: TitleBlock_Regular | Replace it with TitleBlock_Regular |
| Text Style: Arial_09 | Replace it with Standard from DWS file |

**8.** How many problems did AutoCAD find? _____

**9.** Go to the Layer Properties Manager palette and create a new layer called "Test"; what is the reaction of AutoCAD? _____

**10.** Run Check Standard and replace Test with 0

**11.** Save and close the file

## 24.4   USING THE LAYER TRANSLATOR

Layer Translator is a tool which will help you translate a file's layering system to your layering system using the following steps:

- Open the desired DWG file
- Link it with a file (DWG, DWS, or DWT) holding your layering system
- Use Layer Translator to translate layers (naming, colors, linetype, etc.)
- Save the translation in a file, so you can use it for other files

To issue this command, go to the **Manage** tab, locate the **CAD Standards** panel, then select the **Layer Translator** button:

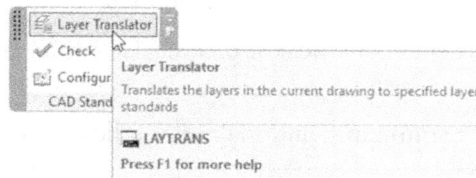

Layer Translator
Translates the layers in the current drawing to specified layer standards

LAYTRANS
Press F1 for more help

You will see the following dialog box:

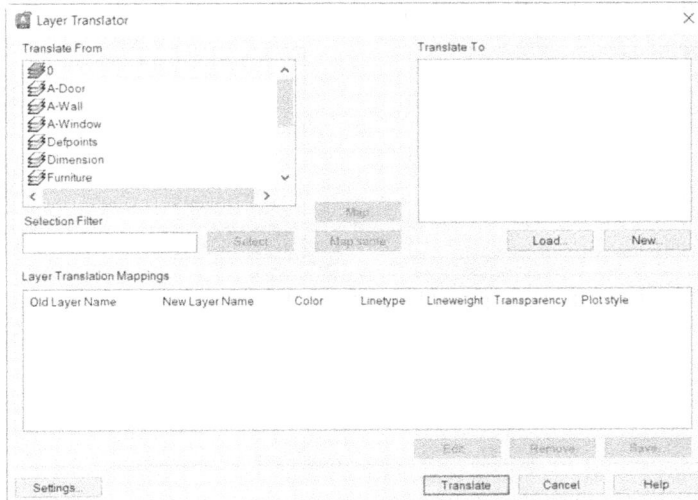

At the top left part, you can see the Translate From list, which contains the list of layers that need to be translated to your layering system. The Translate To list is empty due to the lack of your file. To load your file click the **Load** button; you can have DWG, DWT, and DWS files loaded.

This is what you will see after you load your file:

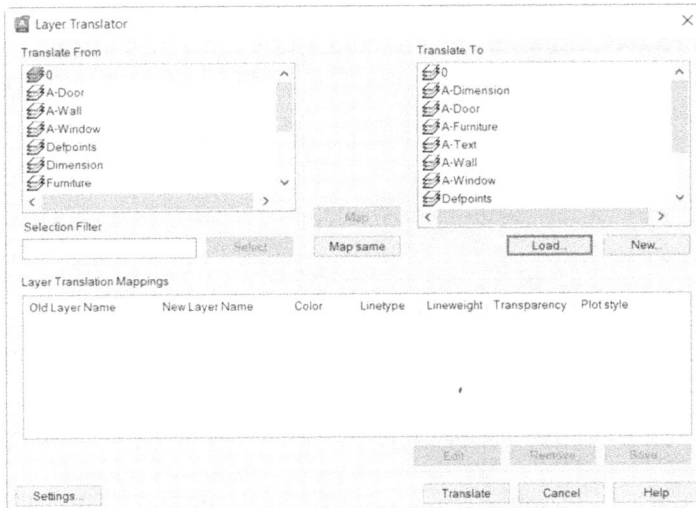

To find the common layers between the two files click the **Map same** button in the middle and you will receive the following:

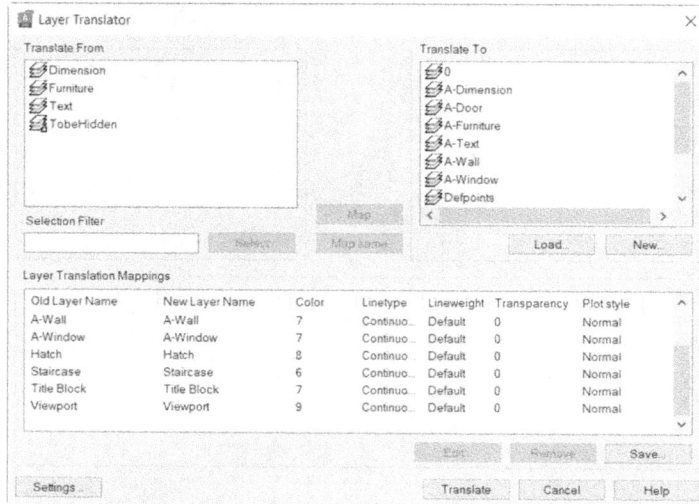

Whatever is left in the **Translate From** part are the layers that do not match with your file. Select one layer from Translate From, and select one layer from Translate To, then click the **Map** button. Keep doing that until you finish all the layers. You can also leave some layers without matching, which means these layers will be left in the file without changing.

Use the **Save** button to save the translation for future use.

To manipulate the process click the Settings button and you will see the following dialog box:

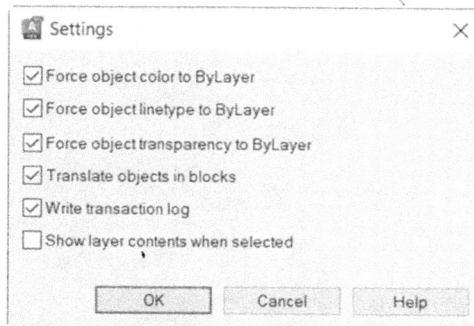

All the settings are self-explanatory. To finalize the command click the Translate button to perform the translation process.

## PRACTICE 24-2    USING THE LAYER TRANSLATOR

**1.** Start AutoCAD 2025

**2.** Open **Practice 24-2-A.dwg**

**3.** Start the Layer Translator command and load ABC Standard.dws

**4.** Using Map Same see if there are any common layers between the two files

**5.** Using the Map button, map the following layers:

     **a.** Dimension to A-Dimension

     **b.** Furniture to A-Furniture

     **c.** Text to A-Text

     **d.** TobeHidden to 0

**6.** Save the translation to the same name of the file with extension DWS

**7.** Close Practice 24-2-A.dwg, and open Practice-2-B.dwg

**8.** Load the DWS file you saved in step(6)

**9.** See how all the layers are already translated

**10.** Save and close the file

## 24.5    DEALING WITH THE LAYER PROPERTIES MANAGER

AutoCAD provided you all of the necessary tools to control the Layer Properties Manager palette. To issue the command, go to the Home tab, locate the **Layers** panel, then select the **Layer Properties** button:

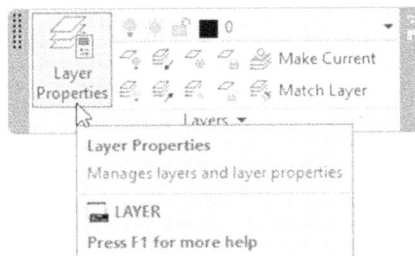

You will see the following palette:

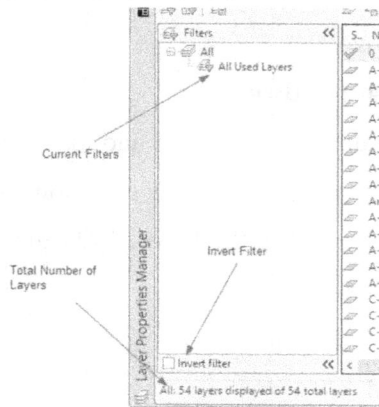

Note the three arrows which identify the three parts of the palette:

- **Total number of layers**: it will show the used and unused layers in the current drawing
- **Current filters**: by default, you will see a single filter which is "All Used Layers" which will show only used layers
- **Invert filter**: will show the inverse of your current filter

While the palette is open, you can do all or any of the following:

- Hide unnecessary columns. Simply right-click one of the columns and a menu will appear; all displayed columns have (✓) at its left, so to hide any of the columns, click the name:

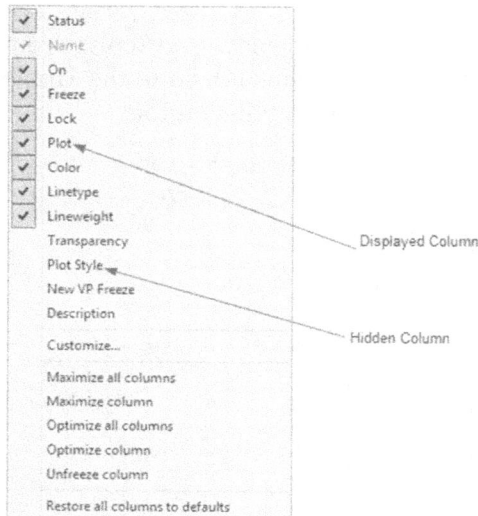

- To sort the layer list based on column either ascending or descending, simply click the title of the column, and arrow will appear; if it was pointing upward, this means ascending, and if it was pointing downward, this means descending:

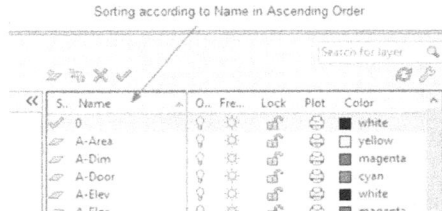

- Relocate the column position comparing to other columns, using the drag-and-drop technique:

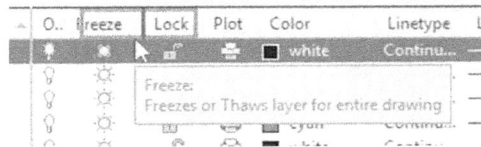

- Change column width or freeze and unfreeze a column
- By right-clicking the heading of a column, the following menu will appear. At the bottom there are six commands: Maximize all columns (show the heading, or the contents whichever larger), Maximize column (the one you right-click), Optimize all columns (show the contents, ignoring the heading), Optimize column, Freeze/Unfreeze column, which means if you horizontally scroll this column will always be shown. Finally, you can restore all columns to defaults:

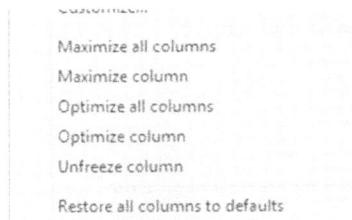

## 24.6 CREATING A PROPERTY FILTER

With the complexity of today's drawings, the user will expect an enormous number of layers in a single drawing. AutoCAD offers you the ability to create a filter to show some of the layers based on a single property or more. Start the Layer Properties Manager, then click the New Property Filter button:

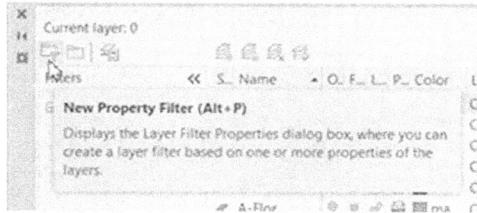

You will see the following dialog box. Do the following steps:

- Type the name of the filter
- Under Filter definition, use the fields below to input one criteria or more. In the below example we used two criteria, the name (A-*) and color (color = Cyan).

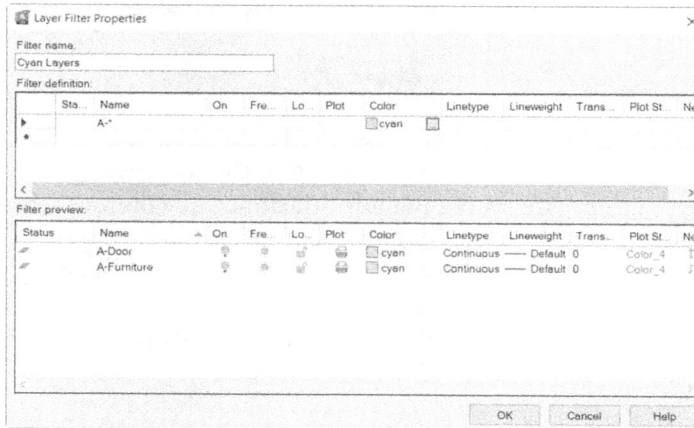

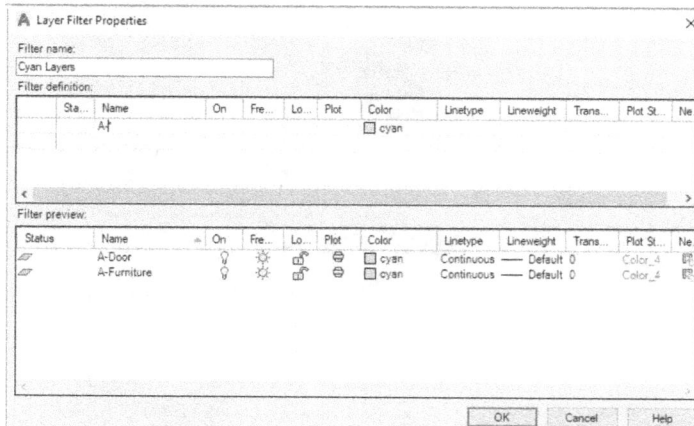

- Automatically AutoCAD will show layers which satisfy the criteria (in our example there are only four layers)
- When you click OK and go back to the Layer Properties Manager, you will see at the left side that the new property filter is added, as shown in the following:

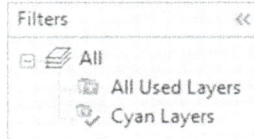

- At the lower left corner, you will see the following:

- At the same time, the layer list in Layers panel will show only the layers of the filter:

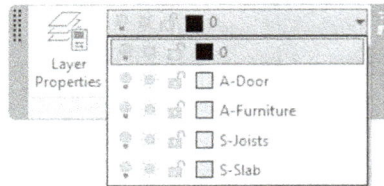

## 24.7 CREATING A GROUP FILTER

A group filter is a filter to group layers which have nothing in common but, for one reason or another, you want to put them together.

While the Layer Properties Manager is open, click the New Group Filter button:

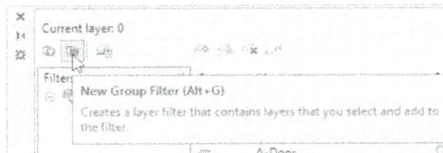

A new filter will be added; type your desired name. Initially this filter is empty. Click All filter, then using the drag-and-drop technique, add the desired layer:

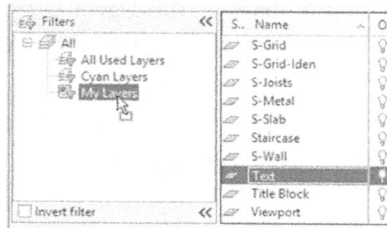

## 24.8 WHAT YOU CAN DO WITH FILTERS

One advantage of filters is to show some of the layers and to hide some but there are benefits to it beyond this single advantage. AutoCAD will allow you to do collective actions on all the layers in the filter in one shot.

When you right-click the filter, a special menu will appear for each type.

### 24.8.1 Property Filter Menu

Right-click the property filter and you will see the following menu:

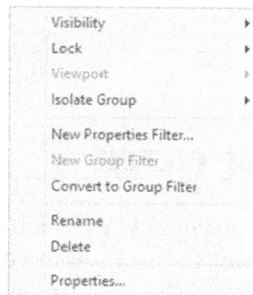

You can do all or any of the following:

- Change the visibility of the layers. You can turn them on or off, freeze them, or thaw them
- Lock or unlock the layers
- If you are in a layout, you can choose to freeze or thaw layers in the current viewport
- If you are in a layout, you can choose to isolate layers in all viewports or the active viewport only
- Create a new property or group filter
- Convert the property filter to group filter
- Rename and delete property filter

- Change the criteria the filter uses by selecting Properties. This option will show you the same dialog box in which you create the property filter

### 24.8.2   Group Filter Menu

Right-click the group filter and you will see the following menu:

Almost all of the options were discussed in the property filter menu, except instead of having a Properties option we have a Select Layers option, as you will see in the following sub-menu:

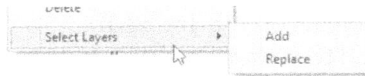

This option will add/replace more layers to the group filter using the normal selecting method. You will see the following prompt:

```
Add layers of selected objects to filter...:
```

Click the desired object so the layer containing this object will be added to group filter.

## PRACTICE 24-3   LAYER ADVANCED FEATURES AND FILTERS

1. Start AutoCAD 2025

2. Open **Practice 24-3.dwg**

3. Open the Layer Properties Manager

4. Layers are sorted by name (ascending), sort them by color descending. What is the name of the first layer in the list? _____

5. Maximize all column widths

6. Show the Description column and hide the Plot Style column

7. Put the Freeze column to the left of the On column

8. Unfreeze the Name column

9. Create a new property filter and name it "Architectural Layers with Cyan color" and set the proper criteria for such a filter

10. Create a new group filter, and call it My Layers. Add Text, Title Block, Frame, and Viewport

11. Right-click the group filter and choose Select Layers/Add option, and select the arrow on the stairs. Check the filter: is there a new layer? Mention the name _____

12. Right-click the property filter and freeze the layers in it

13. Save and close the file

## 24.9 CREATING LAYER STATES

Layer State is used to save and retrieve a set of layers with their current state of color, linetype, lineweight, on/off, freeze/thaw, etc. If the user saved a certain state, retrieving it will lead to retrieving all the states of the layer in one shot, hence, saving a significant amount of time. There are multiple ways to use this feature:

- Make sure the Layer States Manager is displayed, then click the Layer States Manager button:

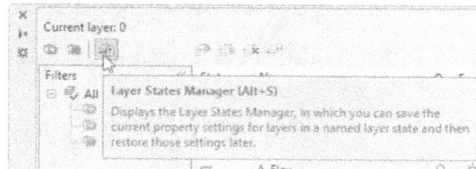

- Or, using the Layers panel, use the first list, and click the New Layer State button:

Either way, the following dialog box will appear:

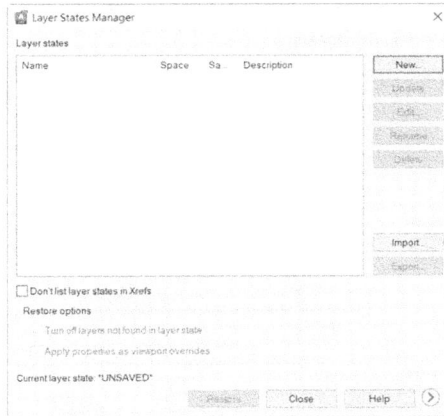

To create a new layer state, click the New button you will see in the following dialog box:

The user should input the name of the new layer state and a good description for its contents. When you click the OK button, the current state of layers will be saved. To see what properties are saved, click the arrow at the lower right corner of the dialog box, and you will see the following:

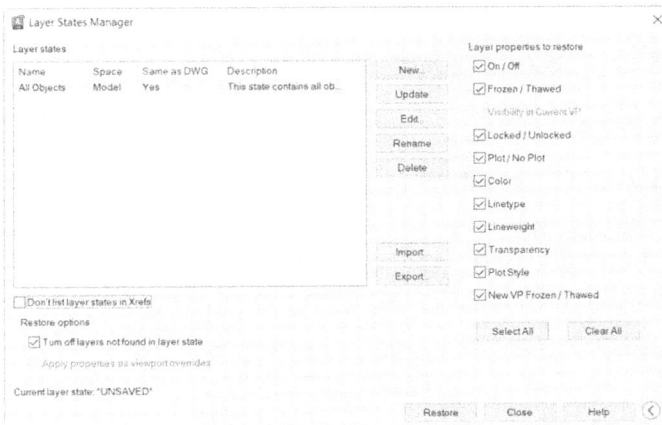

You can see that AutoCAD will save all of the layer properties. If you click the Edit button, you will have the choice to alter the state of the layers saved in layer state, and you will see the following:

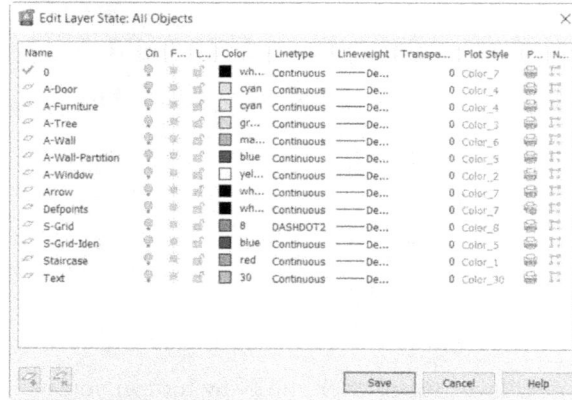

The rest of the keys are self-explanatory.
To retrieve layer states:

- You can use the Restore button at the Layer States Manager:
- You can use the first list in the Layers panel:

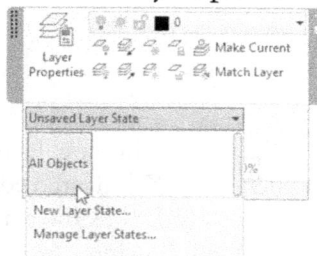

# PRACTICE 24-4   USING LAYER STATES

1. Start AutoCAD 2025

2. Open **Practice 24-4.dwg**

3. Start the Layer States Manager and create a new state from the current state and call it "All Objects," typing at the description the following: "This state contains all layers with their original properties"

4. Freeze layer A-Tree, A-Furniture, and change the A-Window layer's color to blue

5. Save this state under the name "No Trees and Furniture," typing at the description the following: "All Objects except Trees and Furniture

6. Restore the All Objects state

7. Using the Edit button, edit No Tree and Furniture by: freezing layer Arrow, and changing the color of A-Walls-Partition to Cyan

8. Switch between the two states

9. If you have time, create your own state

10. Save and close the file

## 24.10   SETTINGS DIALOG BOX

The Settings dialog box will control three things related to layers: what to do when a new layer will be added, what the settings are for isolated layers, and some dialog settings. To display this dialog box, make sure that the Layer Properties Manager is displayed, then click the Settings button:

You will see the following dialog box:

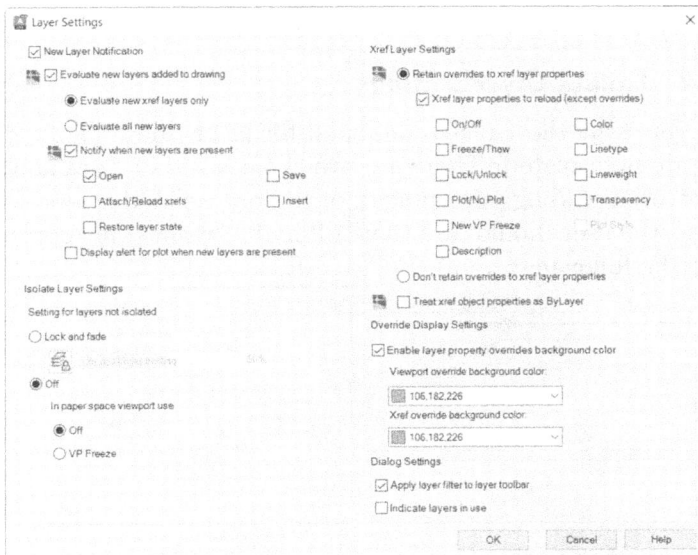

As you can see there are five parts in this dialog box:

- New Layer Notification
- Isolate Layer Settings
- Xref Layer Settings
- Override Display Settings
- Dialog Settings

### 24.10.1   New Layer Notification

In this part, AutoCAD wants to know whether to inform you when a new layer was added. See the following:

Control all or any of the following settings:

- If you want to be notified for new layers
- Should AutoCAD evaluate all layers or only xref layers
- When the notification should happen: at opening the file, saving the file, attaching/reloading the file, inserting the file in other files, or when you restore the layer state
- Should AutoCAD display an alert for plot when new layers are present

### 24.10.2   Isolate Layer Settings

When you issue the Isolate command for layers, how should AutoCAD treat layers that are not isolated? There are two choices: either Lock and Fade (the user should control the fading percentage), or the not isolated layers should be turned off, and if so, the user should turn them off in the layout's viewport or use Viewport freeze. See the following:

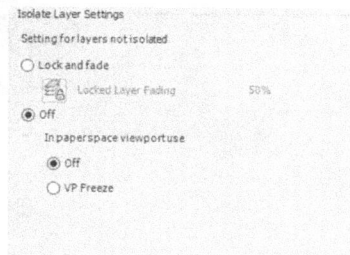

### 24.10.3   Xref Layer Settings

AutoCAD allows you to change the properties of referenced layers (color, linetype, lineweight, etc.) and to change the status of a layer (on/off, freeze/Thaw, etc.) without affecting the original file. Once you save the file, the changes made will be saved. If you want to change this setting, use this part of the dialog box:

Do the following:

- Select whether to Retain overrides to xref layer properties, or not
- If you select to Retain overrides, select which layer properties to reload
- Select whether to Treat Xref object properties as ByLayer, this setting will help you the access to the XREFOVERRIDE system variable; which enables you to force object properties of Xref objects to use ByLayer properties

### 24.10.4   Override Display Settings

If you made layer override, whether in viewport, or xref, would like to see the background of these layers in Layer Properties Manager change to a color of your choice. Use this part of the dialog box:

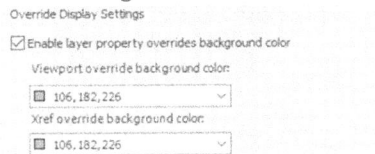

Turn on/off the checkbox Enable layer property overrides background color, then select the desired color

### 24.10.5   Dialog Settings

In this part, the user should tell AutoCAD how to deal with several issues related to layers. They are:

Control the following:

- Apply layer filter to layer toolbar or not?
- Indicate layers in use:

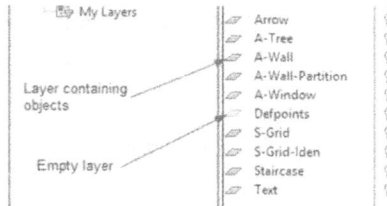

Now what will happen if a new layer was introduced in a drawing? AutoCAD will show you a bubble just like the following, telling you that a new layer was found, and you need to reconcile it:

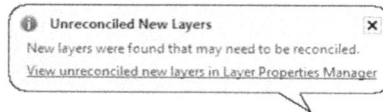

In the Layer Properties Manager, in the Filters part, you will see the following:

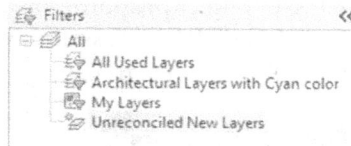

With a new filter named Unreconciled New Layers, which contains all the new layers that appear, you can right-click it and choose the **Reconcile Layer** option to accept these new layers, or Delete Layer, to reject these layers:

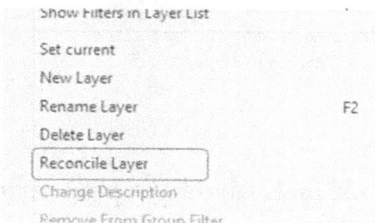

# PRACTICE 24-5    USING SETTINGS DIALOG BOX

**1.** Start AutoCAD 2025

**2.** Open **Practice 24-5.dwg**

**3.** Start Layer Properties Manager

**4.** Start Settings dialog box

**5.** Turn on the New Layer Notification checkbox and make the following settings:

    **a.** Evaluate new layers added to drawing = on

    **b.** Evaluate all new layers = on

    **c.** Notify when new layers are present = on

    **d.** Open & Save = on

    Display alert for plot when new layers are present = on

**6.** Click OK to end the Settings command and close Layer Settings Manager

**7.** Settings for Layer not isolated = Lock and fade and fading percentage = 70%

**8.** From Home tab, locate Layer panel, start the Isolate command, and select one line from the outer wall, then press [Enter]; what happened to other layers? _____. Get closer to one piece of the furniture, and you will notice all objects are locked

**9.** Add a new layer and call it "Test"; save the file. AutoCAD should inform you that new layer was added and you need to reconcile it

**10.** Go to the Layer Properties Manager and you will find a new filter called Unreconciled New Layers; right-click it, and choose to reconcile it

**11.** Save and close the file

## NOTES

## CHAPTER REVIEW

**1.** One of the following statements is wrong:

    **a.** You can use Layer Translator instead of CAD Standards

    **b.** CAD Standard will check for Dimension Style

    **c.** CAD Standard will check for Text Style

    **d.** File extension for Standard file is DWS

**2.** If you right-click _____filter, you will have an option to add layers by selecting

**3.** Property filter will group layers which have the same property(ies):

    **a.** True

    **b.** False

**4.** In Layer Translator, using _____ button will find common layers between two files:

    **a.** Map

    **b.** Map similar

    **c.** Map same

    **d.** Map all

**5.** The settings dialog box contains a method to notify you if there is a new layer is already added:

    **a.** True

    **b.** False

**6.** _____ is to save and retrieve a set of layers with their current state of color, linetype, lineweight, on/off, freeze/thaw, etc.

## CHAPTER REVIEW ANSWERS

    **1.** a

    **3.** a

    **5.** a

# IMPORTING *PDF FILES*, *DESIGN VIEWS*, *AND AUTOCAD WEB/ MOBILE APP*

## In This Chapter

- Importing PDF files to your drawing
- Using Design Views
- AutoCAD Web/Mobile App
- Trace, Markup Import, and Markup Assist

## 25.1    INTRODUCTION

AutoCAD includes two powerful features:

- Importing PDF files to your DWG files as vector files, which will eliminate re-working time and effort. You can also make a partial import of an existing PDF underlay
- Design Views, which will help you share your design with all stakeholders, while keeping your DWG file secure

## 25.2    IMPORTING A PDF FILE

PDF files come from all sorts of software as output. If the origin of your file is AutoCAD, however, you might welcome the chance to convert them to vectors (so you can make the desired changes) and save it as a DWG file. This also applies if you are dealing with PDF underlay (discussed in Chapter 22); then AutoCAD allows you to import the PDF file fully, or partially, to the DWG host file. This does not mean if the PDF file origin was not an AutoCAD file, we cannot import it to a DWG file. We can, but it will be more like an image.

To import a PDF file into your DWG file, use the following steps:

- While you are in the AutoCAD file, go to the **Insert** tab, locate **Import** panel, click the drop-down arrow, and select the **PDF Import** button:

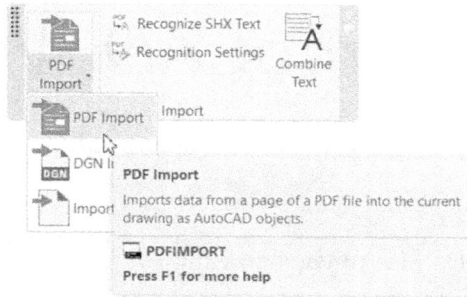

### 25.2.1 Importing a PDF File

To import a PDF file into your current drawing, use the following steps:

- AutoCAD will display the following dialog box:

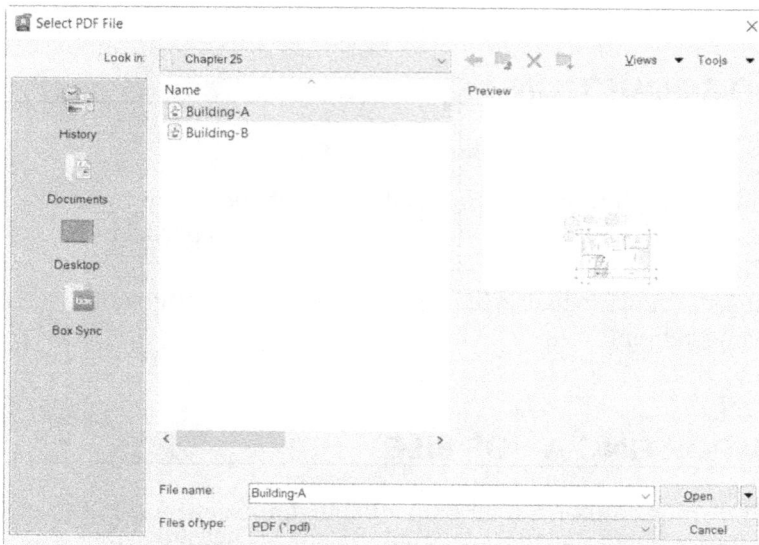

- Select it and click the Open button

- You will see the following dialog box:

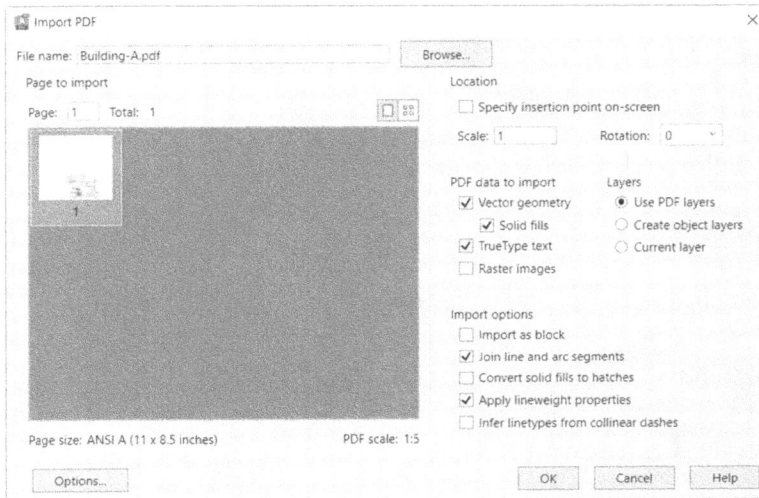

- If your file contains more than one page, specify the pages you are interested in

- Under the **Location** section, select either **Specify insertion point on-screen,** and specify the **Scale** and **Rotation** angle of the import

- Under the **PDF data to import** section, specify what to import: vector geometry (including Solid fills), True Type text (AutoCAD will not recognize SHX fonts), and if there are any raster images

- Under the **Layers** section specify **Use PDF layers** if you want to bring in the layers inside the PDF file (but it will be with prefix PDF example: A-Door, will be PDF_A-Door), or, use **Current object layers** (there will be layers like PDF_Geometry, PDF_Solid Fills, PDF_Text, depending on whether or not these objects are in the PDF), or **Current layer**, which means all objects imported will reside in the current layer

- Under Import options, select whether to **Import as block**, which means all of the import will be considered one block. Whether or not you want to **Join line and arc segments** as polyline. Whether or not to **Convert solid fills to hatches**, and **Apply lineweight properties** (if checked off, all lines will hold the default lineweight), and finally, to **Infer linetypes from collinear** dashes. If this is on, AutoCAD will convert the dashes and dots if they are collinear to a single polyline holding new linetype called, PDF_import

When finished, click OK to complete the import process. Check and identify each object and how AutoCAD converted them.

### 25.2.2 Importing from PDF Underlay

The second option is to import objects from an existing PDF underlay. Use the following steps:

- Make sure you are inside an AutoCAD file containing a PDF underlay. At the command window, type PDFIMPORT and select the PDF underlay
- You will see the following prompt:

```
Specify first corner of area to import or
[Polygonal /All/Settings] <All>:
```

- If you click Settings, you will see the following dialog box:

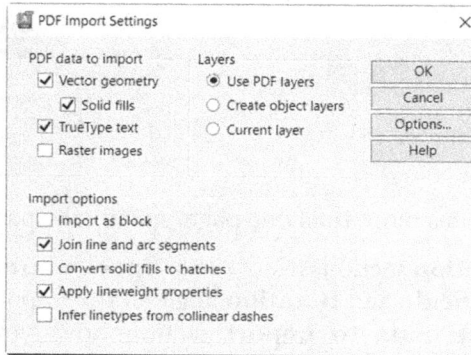

- The above dialog box is identical to what we discussed while importing a PDF file command. Change the settings as you wish and click OK
- Specify two opposite corners to set the area you want from the underlay
- You will see the following dialog box:

```
Keep, Detach or Unload PDF underlay?
[Keep/Detach/ Unload] <Unload>:
```

- AutoCAD is reminding you that it will obtain the selected objects but decide what to do with the PDF underlay; do you wish to Keep it as is, Detach it, or Unload it?
- Input your action and press [Enter] to end the command

### 25.2.3 Dealing with SHX Files

As discussed previously, AutoCAD has two types of fonts, the old one called shape file fonts (*.SHX) and the True Type fonts (*.TTF). By default, when importing PDF to AutoCAD file, AutoCAD will recognize the true type fonts right away, but will consider shape files as geometry. To solve this problem, go to **Insert** tab, locate **Import** panel, click **Recognition Settings** button:

You will see the following dialog box:

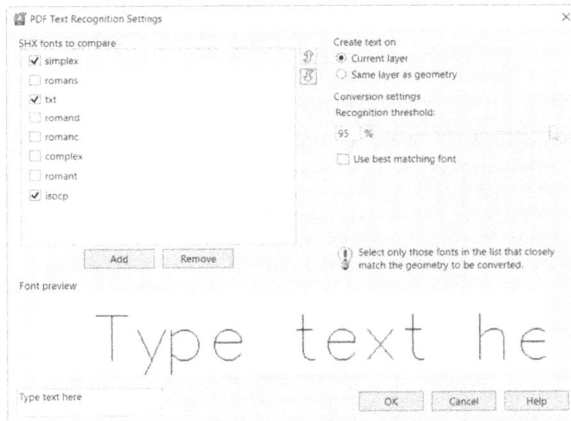

Using the above dialog box, you will help AutoCAD recognize the setup the text in the required way. Do the following steps:

- Under **SHX fonts to compare**, select the shape files you think may exist in your PDF file. This list is not final as you can add or remove from it using **Add** and **Remove** buttons located under the list
- Select in which layer the generated font will reside in: Current layer, or same layer
- If the existing SHX font is not listed in the list, what is the Recognition threshold should AutoCAD use (default is high; 95% similar to one of the listed fonts, you can relax it by setting less percentage). Also, you can select the option of **Use best matching font**
- Click **OK** to end the command

AutoCAD will ask you to select objects (the SHX fonts), if the selected objects contain all or any of the fonts mentioned, you will see the following message:

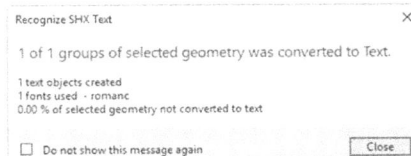

An MTEXT will be created accordingly. AutoCAD will create a new text style holding the name of the font and assign the text to it.

If no matching was found, you will see the following message:

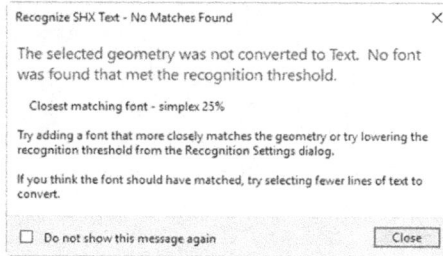

If you want to convert more geometry to fonts in the PDF import, go to **Insert** tab, locate **Import** panel, click **Recognize SHX Text** button:

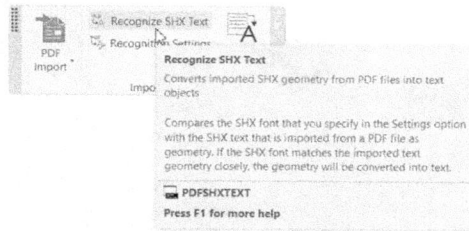

### 25.2.4 Object Detection

When you import a PDF file into your drawing file, most likely, AutoCAD will not identify certain objects as Blocks. To solve this problem, AutoCAD included a technology preview that uses machine learning to scan your drawing for objects that can be converted into blocks.

Do the following steps:

- Go to **Insert tab**, locate **Block** palette, and click **Detect** button:

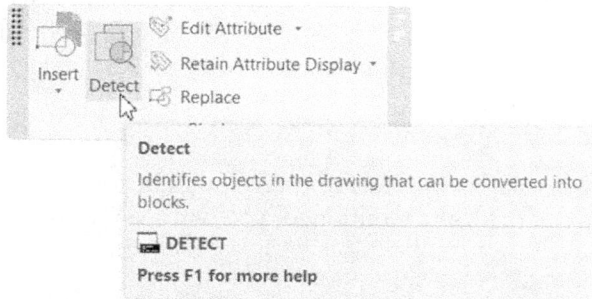

- You will see the following palette:

- This means the detection process started. In short period of time, you will see the result of the detection process:

- Click **Review Objects** to see the results. You will see something similar to the following:

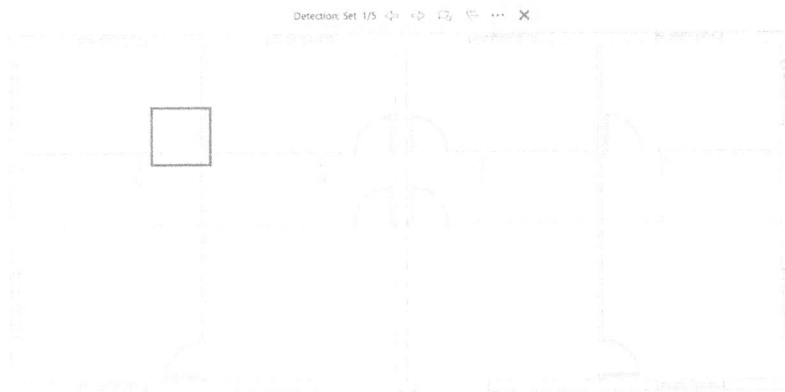

- AutoCAD shows a shape that can be converted to a block, and all the similar shapes in the drawing. If you clicked the blue rectangle, you will see a menu asking you to Convert it to Block, click Convert, you will see the following dialog box:

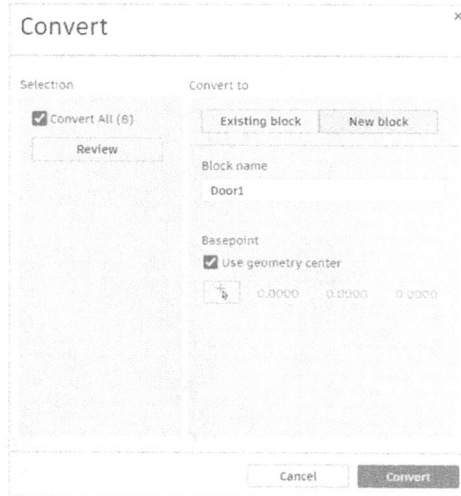

- Select whether it is an Existing block, or New Block. If it was a new block, type in the name of the new block, and click Convert
- Using Detection toolbar, click the left arrow and the right arrow to navigate the different detections, click the three dots to show a menu to allow you to remove suggested instances

## PRACTICE 25-1   IMPORTING PDF FILE

**1.** Start AutoCAD 2025

**2.** Open **Practice 25-1-A.dwg** file

**3.** Start the PDF Import command

**4.** Select Building-A.pdf, and click Open button

**5.** Select the following settings:

    **a.** Vector geometry including Solid fills, and True Type text

    **b.** Use PDF layers

    **c.** Import as block = off

    **d.** Join lines and arc segments = off

    **e.** Convert solid fills to hatches = on

    **f.** Apply lineweight properties = off

    **g.** Infer linetypes from collinear dashes = on

6. Click OK, the file will be inserted @ 0,0

7. View layers; type the name of one layer _____

8. View text styles; what is the name of the text style other than Standard, and Annotative? _____

9. See how AutoCAD converted all lines to polylines, and didn't join them with arcs

10. Start **Recognition Settings** command

11. Make sure that romanc.shx is selected, and select Same layer as geometry

12. Click OK, zoom to the word Hallway, using window selection select the whole word. AutoCAD convert it to MTEXT

13. Save the file and close it

14. Open Practice 25-1-B.dwg

15. You will find a PDF underlay in it

16. Using the keyboard type PDFIMPORT command and select the PDF underlay from its borders. At the command window Select Settings and make the following changes:

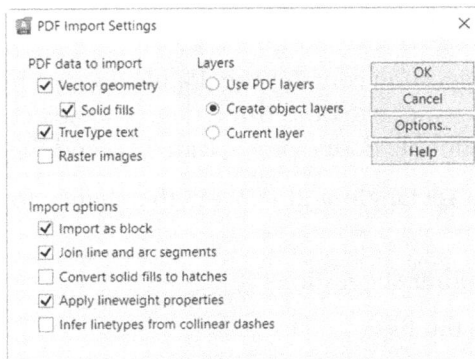

17. When finished, click OK

18. Draw a box containing only the plan with the gridline (title block should <u>not</u> be selected). Then choose to Unload the PDF underlay

19. Check if the import is one block? What is the name of block? _____

20. Explode it

21. Zoom to one of the grid lines, and see how AutoCAD converted each segment to be separate objects

22. Check one of the arcs and see if it is joined to the adjacent polyline

23. Look at layers; how many layers are there and what are their names? _____
_____

24. Save and close the file

## PRACTICE 25-2    OBJECT DETECTION

1. Start AutoCAD 2025

2. Open **Practice 25-2.dwg** file

3. You will see an AutoCAD file with PDF file import exploded

4. Start Detect command

5. What is the result of the detection _____? (5 Sets, and 19 In-stances)

6. Click Review Objects button

7. Click the blue rectangle, and select Convert option, create a NEW block, and call it Door1, click Convert button

8. The second suggestion contains the double door with the single door. In order to correct this issue, click the three dots, and select Remove Instances, and select the two double doors, press [Enter] to end the command

9. Click the blue rectangle and convert the shapes to NEW block and call it Door2

10. Ignore all the other suggestions

11. Save and close the file

## 25.3   CREATING SHARED VIEWS

Sharing your design by exchanging DWG files with other people or firms is very risky. Yet, in today's business environment, you are obliged to send some data to many stakeholders in order for them to view, approve, and reveal the next step. AutoCAD offers us the best solution for this dilemma. You can send stakeholders views of your DWG without sending them the actual one. Using Share Views in AutoCAD will enable you to do the following:

- Not risking your design by sending the original DWG file
- No need for the recipients to have AutoCAD installed in their machines
- No need to start new accounts, no usernames, no passwords
- You need only a browser to view the Design Views

For AutoCAD 2025, if you are a commercial user, or Student/Educator, you will have an account on Autodesk, and you will be signed in by default.

- Go to **Application** menu, locate **Publish** then select **Share View** option, as shown in the following:

- You will see the following dialog box:

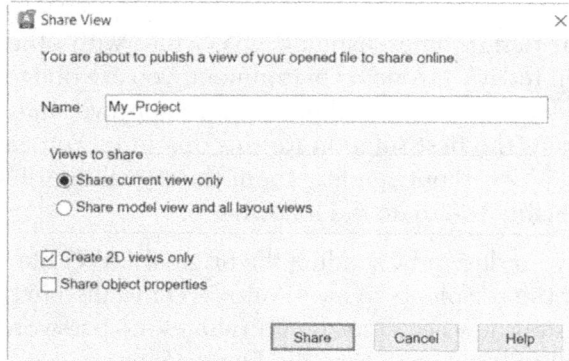

- Type the name of the view you want to share (default value is the name of the current file)
- Select whether you want to share the current view, or you want to share model view, and other layout
- Select if you want to limit view creation to 2D objects only
- Select whether you want to include object properties, or not
- Click Share
- You will see the following message:

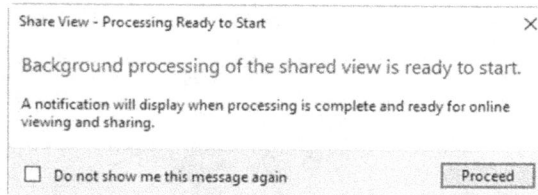

- Click Proceed
- You will see the following message appear as a bubble at the lower right corner of the screen "Share View Upload Complete. You can now share and view the link. View in Browser"

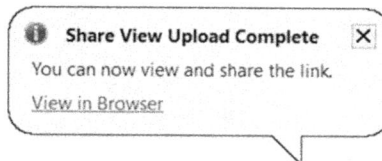

- Clicking on the "View in Browser opens your browser and displays the shared view in the Autodesk Viewer

You will see the following:

Using this screen you can do all the following:

- At the left, choose to view Model, Views, or layouts
- You can zoom in, zoom out, pan, and zoom all using the mouse wheel. You can use Fit, and Home to get the full view
- You can Measure distances, angles, and areas
- Use Markup (Pencil, Arrow, Cloud, and Text)
- At the top left, Control the layers, Properties, and Settings
- At the top right, Add comments, Print, take a screen shot, and Share the view for 30 days

**NOTE**    *You can show Shared Views Palette:*

- Go to **Collaborate** tab, locate **Share** panel, click **Shared Views** button, as shown in the following:

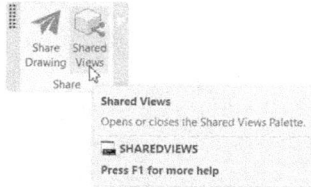

- You will see Shared View Palette:

- At the right of each view, there are three dots, clicking it, will show the following:

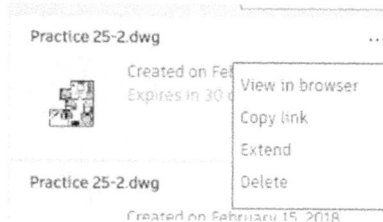

- These options are: to view this shared view in browser, to copy link, so you can send it to stakeholders, Extend to extend the number of days others can view by an extra day, and finally Delete to delete the shared view from cloud.

## PRACTICE 25-3    CREATING DESIGN VIEWS

1. Start AutoCAD 2025

2. Open **Practice 25-3.dwg** file (in order to solve this practice you need to have a username and password for A360)

3. Log in your A360 account from AutoCAD upper right corner

4. Start Share View command

5. Call the Shared View My_Project

6. Select to Share model view and all layout views

7. Turn off Create 2D views only and turn on Share object properties. Click Share, then Proceed

8. When you see Autodesk Viewer, try all available tools

9. Close the browser

10. Start Share View Palette and check the four different functions

11. Save and close

## 25.4    AUTOCAD WEB / MOBILE APP

AutoCAD Web is your way to access and edit your AutoCAD drawings using a browser in your desktop/laptop without the need to install/configure any software. AutoCAD Mobile App is your way to do the same using your mobile/Tablet whether you are running iOS or Android.

All you need to do is access the AutoCAD web or Mobile App and sign in with your Autodesk account.

There is a free-to-use version available for those without an Autodesk subscription or account, but it can only handle opening and viewing DWG files, not editing or reformatting it.

> **NOTE** *What will show here is the AutoCAD Web images, and interface. The Mobile App is similar with slight difference*

In order to utilize this feature, do the following:

- Start your browser
- At the Address bar, type the following: **web.autocad.com**

- You will see the following:

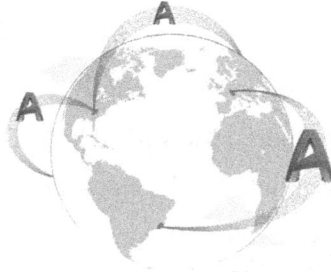

- Click **Sign In** button
- Fill in your Username, and password
- Click Sign In to log into your account
- You will see the following:

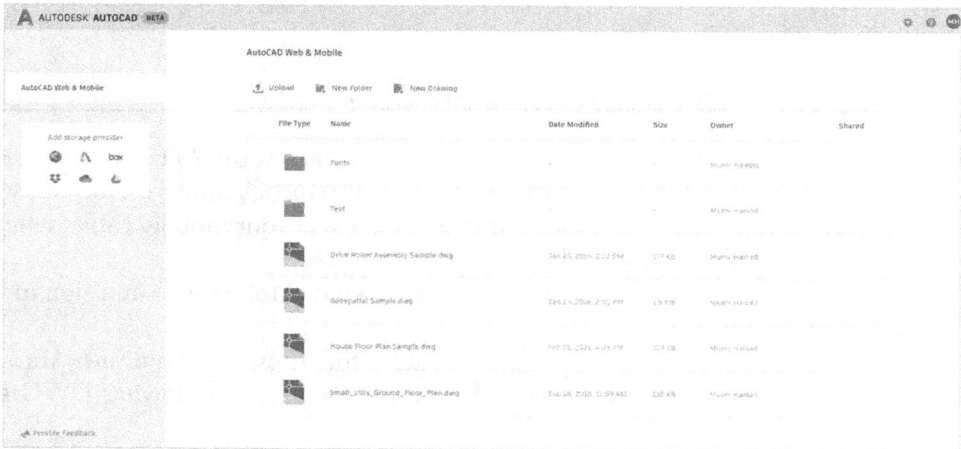

- At the top you can Upload a DWG file from your local hard disk, create a new folder on the web, or start a new drawing from scratch.

- Open one of the existing DWG files, you will see like the following:

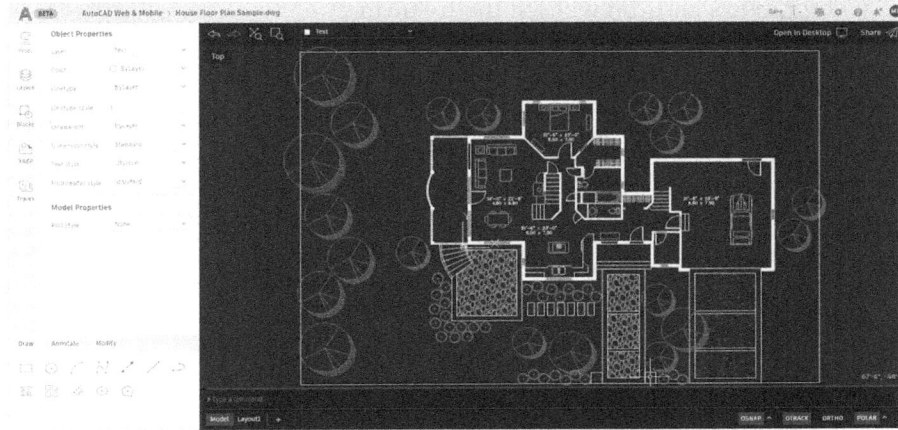

- At the left pane, you have five buttons to select from

## Prop

Prop means Properties. Click **Prop** and you will see the following:

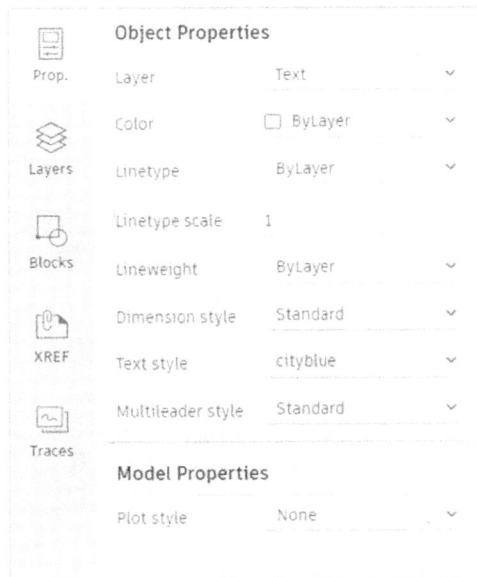

- From the drawing select any object(s) using one-by-one, window, or crossing, to view and change their properties.

**Layers**

Click **Layers** and you will see the following:

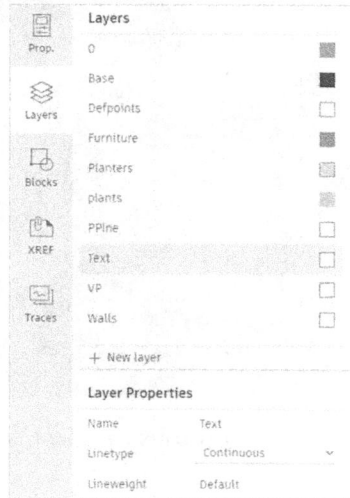

- You can create a new layer, change the current layer's color. Turn it on/off, and lock/unlock. Also, you can change the properties of the layer.

**Blocks**

Click **Blocks** and you will see the following:

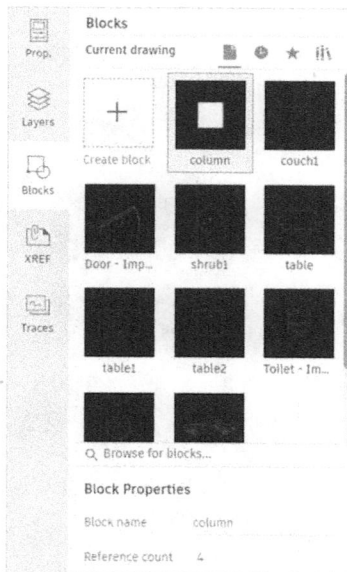

- It will list the Blocks in the current drawing
- The above buttons will show you the Recent, Favorite, and the Block Libraries used. At the bottom, you will see a count of the selected block

### XREF

Click **XREF** and you will see the following:

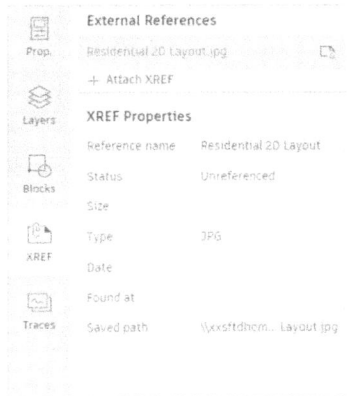

- It will list all xrefs attached to the current file with all its properties

### Traces

Click **Traces** and you will see the following:

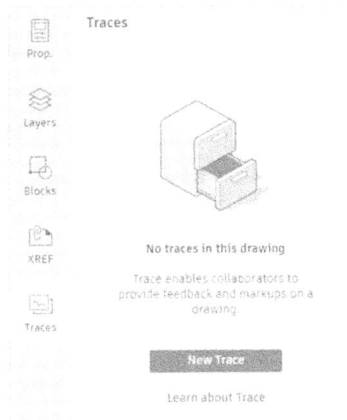

- This part will allow you as contributor to add any object to the drawing as a tracing paper without making any real changes to the drawing
- Click **New Trace** and start adding any remarks you wish to tell the creator of this file what is wrong with it, or, what are the modifications needed

- You will see the following:

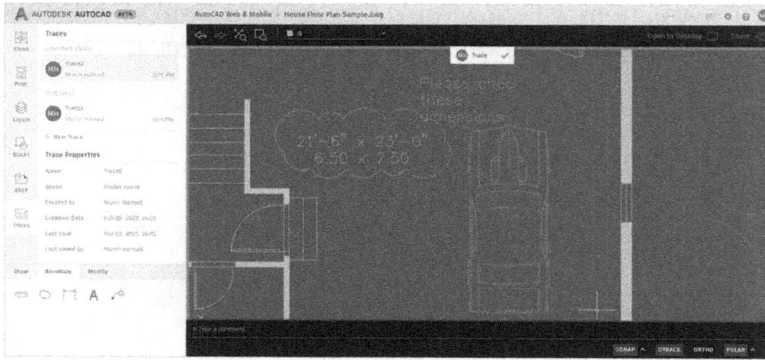

- Once done, click (✓) to end the command
- Of course, these additions will be saved, and the creator of the file will see them
- The lower part of the left pane, will allow you to draw, annotate, and modify as if you are at the normal desktop AutoCAD
- You can use the Zoom In, Zoom Window, Undo and Redo (you can use the mouse wheel to zoom in, and out as well)
- The Status bar contains OSNAP, OTRACK, ORTHO, and POLAR, as you know them from the AutoCAD desktop
- At the top right of the screen, click SAVE to save the changes, or SAVEAS to save it under a different name
- Click The logo at the top left part of the screen to close this drawing

## 25.5   USING AUTOCAD DESKTOP

There are several functions you need to be acquainted with in order for the AutoCAD Web to be a successful story.

### Open from the Web

In order to open a file from AutoCAD Web / Mobile App, do the following steps:

- Start AutoCAD
- Go to Quick Access Toolbar at the top left and click **Open from Web & Mobile** button:

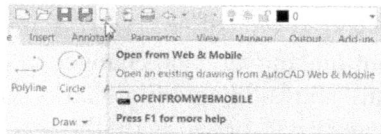

- You will see the following dialog box:

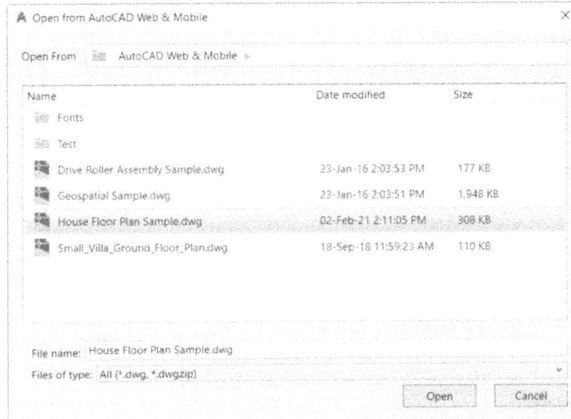

- Select the desired file and click Open
- It will open the file and the XREF tool palette if the file is referenced to any other file

### Follow Traces

While the file is opened from the Web, you can view, if any of your contributors has made any traces to your file:

- Click **Collaborate** tab, locate **Traces** panel, and click **Traces Palette** button:

- You will see the following palette:

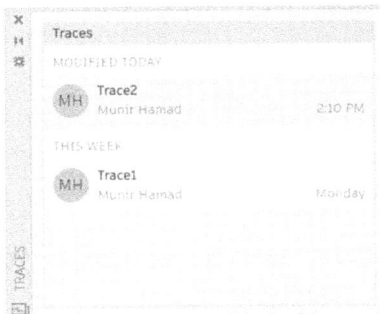

- It will list the Traces on your drawing along with the name of the owners
- Click on any Trace, a thick blue line will surround the screen, and you will see the following:

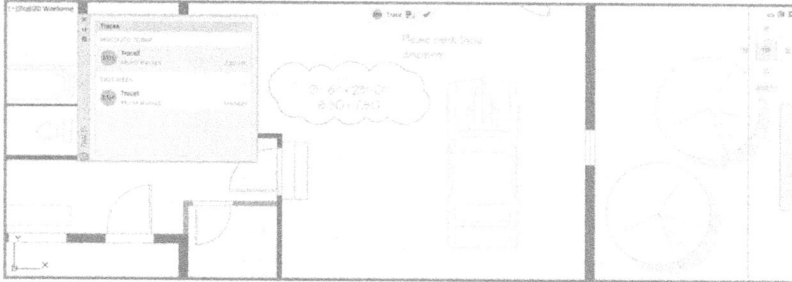

### Save to the Web

In order to save your current file to the Web, do the following steps:

- Open the desired file
- Go to Quick Access Toolbar at the top left and click **Save to Web & Mobile** button:

- You will see the following dialog box:

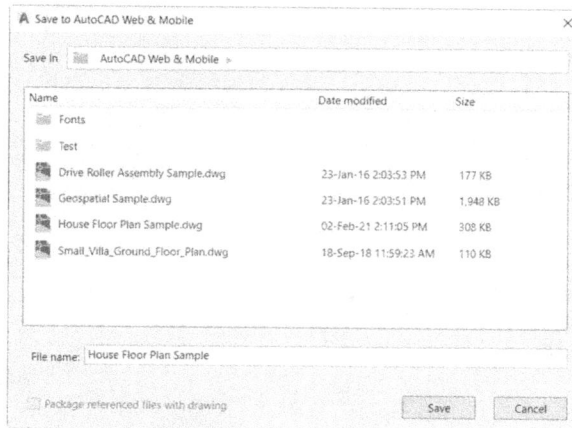

- You can save under the same name, or, you can type a new name
- Click Save to end the command

---

**NOTE**

- *You can create a Trace directly in your files opened by AutoCAD Desktop, using **Collaborate** tab, locate **Traces** palette, click **Trace Palette** button*
- *Use the same method we used in AutoCAD Web*

## 25.6 MARKUP IMPORT AND MARKUP ASSIST

- Using Markup Import you can import PDF, PNG, or JPG files as an overlay in Trace workspace; which contains markups even if it was handwritten.
- Using Markup Assist you can identify text, revision cloud, and Multileader, and convert them to AutoCAD objects

### Markup Import

To work with Markup Import, do the following steps:

- Open your DWG file which contains the original drawing
- Go to **Collaborate** tab, locate **Traces** panel, click **Markup Import** button:

- You will see the following message:

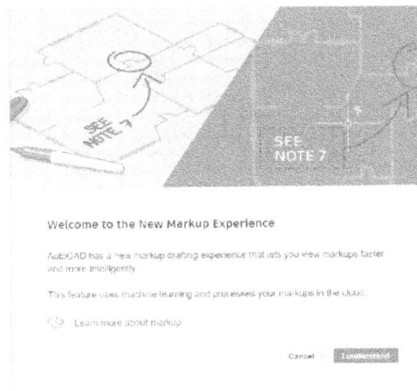

- Read the message and click **I understand** button
- You will see the normal Open File dialog box, select PDF, JPG, PNG file to import as a Markup, click Open
- You will be in Trace workspace with thick blue screen frame, the file will be shown with the following menu to pick a command from:

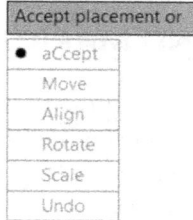

| Accept placement or |
| --- |
| ● aCcept |
| Move |
| Align |
| Rotate |
| Scale |
| Undo |

- You can Accept the import file, Move it, Align it, Rotate it, Scale it, or Undo the process to redo the job again
- After you finish your work, click Accept button to end the import process

**Markup Assist**

- After accepting the Markup Import, you will see at the top of the screen you will see the following menu:

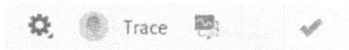

- Hover over the button at the right, it will show you the following:

**TRACEFRONT is on**

Click to switch to TRACEBACK.

With TRACEFRONT on, the active trace is set to edit mode so you can contribute to the trace.

**Press F1 for more help**

- If you click it TRACEBACK will be on. Click it. You will see the following two buttons:

FADEMARKUP          Markup Assist

- FADEMARKUP will fade the markup you have addressed or you want to ignore

- Markup Assist is by default is on, which will allow AutoCAD to assist you to identify the Markups in the imported file, you will see the following:

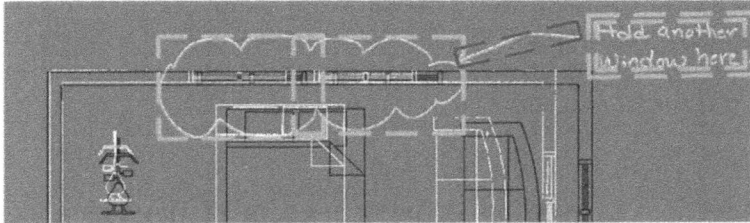

- Hover over the blue dashed line, it will show you a message such as the following: Markup Assist identified a boundary. Click for Options. If you click you will see the following:

- If you click **Insert as Revcloud** option you will add a revision cloud to your file
- If you hover over double dashed blue line, you will see the following:

- Markup Assist identified a text and it will list it to you. If you click, you will see the following:

- You can Insert it as Mleader, Insert it as Mtext, or edit the text in the upper text box, then Update the Existing Text
- If the markup contains words like Move, Markup Assist will suggest you to start the Move command
- If the markup contains words like Remove, Markup Assist will suggest you to start the Erase command

## PRACTICE 25-4   MARKUP IMPORT AND MARKUP ASSIST

1. Start AutoCAD 2025

2. Open **Practice 25-4.dwg** file (do not change the zoom level)

3. Start Markup Import command, and select Gound_Floor_With_Markup. PDF

4. Rotate the Import by 90° using the center of the rectangle

5. Move it to the upper part of your file, using the center of the rectangle. The size of the markup import should be the same size of the original file

6. Click Accept to end the Markup Import process

7. Switch to TRACEBACK

8. Convert the one of the existing revision clouds to revision cloud, and Fade the other

9. Convert the text to Mleader using the same location, and pointing to the revision cloud

10. Click (✓) to end the command

11. Stretch the revision cloud to cover the two windows

12. Save and close

## NOTES

## CHAPTER REVIEW

1. You can import a PDF into my DWG file as an image:

    **a.** True

    **b.** False

2. Import PDF command is in _____ tab

3. One of the following cannot be done in A360:

    **a.** Calculate an area

    **b.** Create a measurement

    **c.** Zoom in, and Zoom out

    **d.** Change layer of an object

4. One of the following statements is wrong:

    **a.** You cannot import PDF files to DWG

    **b.** While importing PDF you can select whether my lines to be polylines or simple lines

    **c.** While importing PDF you can convert solid fills to hatches

    **d.** A & B

5. In the Share View Palette, you can extend the period of viewing more than 30 days:

    **a.** True

    **b.** False

6. While you are using the AutoCAD Web/Mobile App, you can view the names of the Blocks and their counts

    **a.** True

    **b.** False

## CHAPTER REVIEW ANSWERS

**1.** b

**3.** c

**5.** a

## CHAPTER 26

# DRAWING COMPARE

**In This Chapter**

- Drawing Compare Function

## 26.1 INTRODUCTION

Normally drawings will change through the phases of the design cycle; sometimes the change is large and obvious, and sometimes very small that you cannot catch it with the naked eye. AutoCAD provides you with a very handy tool to compare two versions of your drawing (design) and make any necessary changes accordingly.

## 26.2   WHERE TO START

You can start Drawing Compare command using different methods:

- If there are no open files, go to the Application menu, and select **DWG Compare** command:

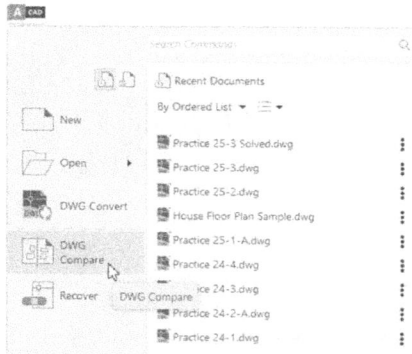

- If there is an open file, go to the Application menu, select **Drawing Utilities**, then choose the **DWG Compare** option:

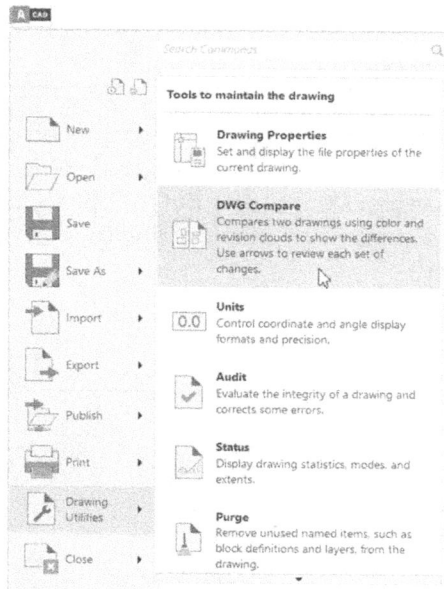

- If there is an open file, go to **Collaborate** tab, locate **Compare** panel, then click the **DWG Compare** button:

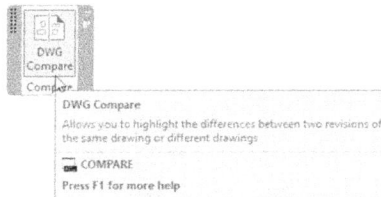

- If you choose the first method, the following dialog box will appear:

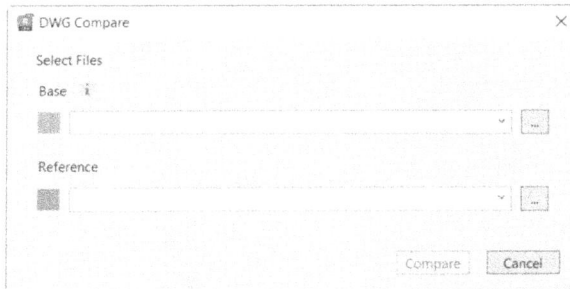

- No files are selected, click the small button with the three dots to select the Base file (which will be green), and then select the Reference file (which will be red), then click the Compare button.
- If you select the second, or third method, the following dialog box will appear:

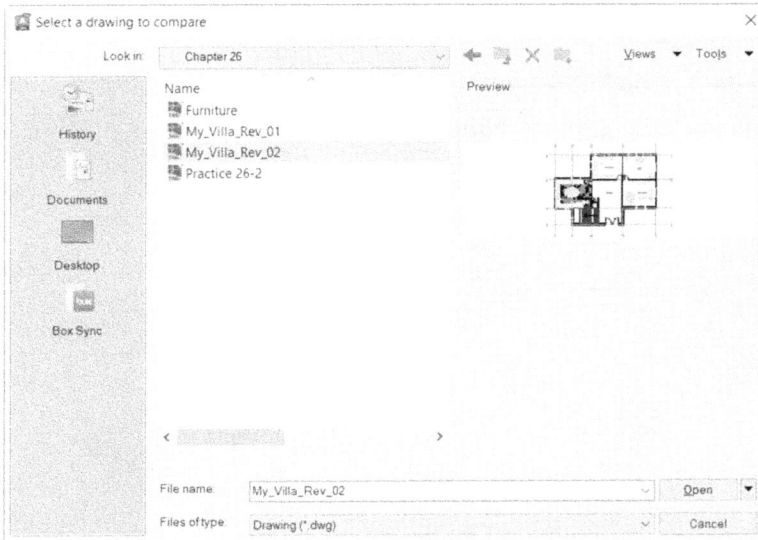

- Select the second file to compare with (Reference file)

## 26.3 COMPARING PROCESS

You will see at the top center of the screen the following docked toolbar:

And the following will happen:
- You will see three colors, Green (Base file), Red (Reference file), and Gray (common to both)
- You will see Revision Cloud marking the differences between the two files

The two drawings will look similar to the following:

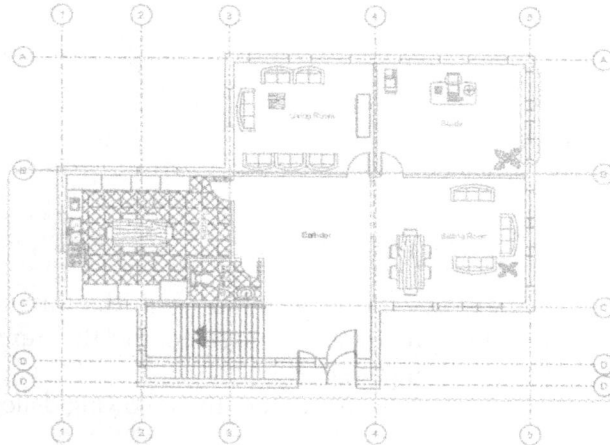

### 26.3.1 Comparison Settings

If you click the Settings button, you will see the following menu:

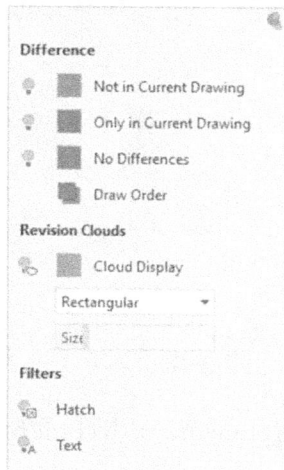

### Under Difference:

- Turn on/off the Reference file (Red), and/or change the color
- Turn on/off the Base file (Green), and/or change the color
- Turn on/off the No Differences, and/or change the color
- Change the Draw Order of the two drawings

### Under Revision Clouds:

- Choose to show/hide the revision clouds
- Change the color of the revision clouds
- Change the shape of the revision clouds from Rectangular to Polygonal
- Change the size of the revision clouds

### Under Filters:

- Click the button to select whether or not hatch objects in the compared drawing are displayed and compared
- Click the button to select whether or not text objects in the compared drawing are displayed and compared

### 26.3.2   The Two Arrows

Next to the Settings button there are two arrows:

Use the two arrows to browse the changes between the two files, AutoCAD will zoom automatically to each part of the drawing which contains changes.

### 26.3.3   Import Objects

Click Import Changes button:

AutoCAD will show the following prompt:

```
Select objects:
```

Select all of the objects of the drawing or some part of the drawing. AutoCAD will import the changes that you select from the Reference drawing, to the Base drawing.

### 26.3.4   Export Snapshot

Click the Export Snapshot button:

You will see the following message:

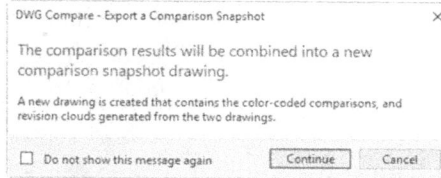

Read Read the message carefully which states that the comparison results will be combined into a new comparison snapshot drawing. Click Continue, and either accept the default name of the new drawing, or type the new name.

When finished, click (✓) to end the comparison command.

## PRACTICE 26-1    DRAWING COMPARE

1. Start AutoCAD 2025

2. Open My_Villa_Rev_01.dwg file (this is your Base file)

3. Start DWG Compare command

4. Select the reference file to be My_Villa_Rev_02.dwg

5. Change the Draw Order to make Rev_02 over Rev_01

6. Turn off Rev_01 and take a good look on the Rev_02 alone

7. Turn on Rev_01

8. Turn off Rev_02 and a look on Rev_01 alone

9. Turn on Rev_02

10. Turn on Hatch and Text to compare them between the base file and the reference file

11. Can you see the differences between the two drawings? Type at least three of these:

    a. _____

    b. _____

    c. _____

12. Change the Revision Cloud Shape to be Polygonal

13. Use the slider to increase the margins to the middle

**14.** Zoom to the sets you created

**15.** Export a Snapshot of the comparison drawing, keeping the default name

**16.** Close the file

## 26.4    XREF COMPARE

You can as well compare XREF DWG file which changed through the attachment life cycle. If the DWG file you attached changed, once you open your file, AutoCAD will inform you that it was changed and will allow you to compare it to the old one.

You will see the following bubble at the lower right corner of the screen:

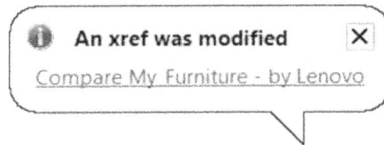

Of course, the name of the computer will appear in the message as shown above. Once you click the link, a thick blue frame around the graphical screen will appear, and the following toolbar will appear:

If you click the first button at the left, you will see the following menu:

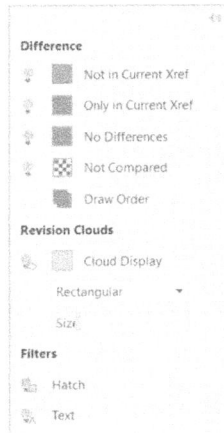

Which is identical to the menu discussed in the Compare command.

Clouds will surround the changes as shown below:

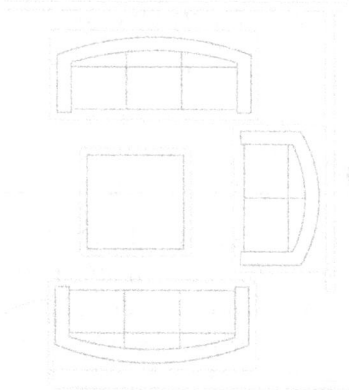

The rest of the buttons are the same as Compare command.

## PRACTICE 26-2   XREF COMPARE

1. Start AutoCAD 2025

2. Open Furniture.dwg, and Save it As My_Furniture.dwg, then close the file

3. Open Practice 26-2.dwg

4. Attach to it My_Furniture.dwg using 0,0 as insertion point

5. Save the file as My_File.dwg

6. Close the file

7. Open My_Furniture.dwg

8. Copy the three sofas and the table at the lower part to the left by 5 units

9. Save the file and close it

10. Open My_File.dwg

11. You will see a balloon at the lower right corner of the screen

12. Click the link to compare the two XREFs

13. Use the arrows to zoom to the new additions

14. Finish the command

15. Save the file and close it

## NOTES

## CHAPTER REVIEW

**1.** You can compare three files in AutoCAD:

    **a.** True

    **b.** False

**2.** You will see three colors: Green for the first file, Red for the second file, and _____ for the common objects.

**3.** The shapes of the Revision Cloud are:

    **a.** Rectangular

    **b.** Circular

    **c.** Polygonal

    **d.** A & C

**4.** You can include or exclude Text, Hatch, and blocks:

    **a.** True

    **b.** False

**5.** You can compare XREF DWG:

    **a.** True

    **b.** False

## CHAPTER REVIEW ANSWERS

**1.** b

**3.** d

**5.** a

# INDEX